Introduction to Flexible Electronics

Introduction to Flexible Electronics

Aftab M. Hussain

CRC Press
Taylor & Francis Group
Boca Raton London New York

CRC Press is an imprint of the
Taylor & Francis Group, an **informa** business

First edition published 2022
by CRC Press
6000 Broken Sound Parkway NW, Suite 300, Boca Raton, FL 33487-2742

and by CRC Press
2 Park Square, Milton Park, Abingdon, Oxon, OX14 4RN

© 2022 Taylor & Francis Group, LLC

First edition published by CRC Press 2022

CRC Press is an imprint of Taylor & Francis Group, LLC

Library of Congress Cataloging-in-Publication Data

Names: Hussain, Aftab M., author.
Title: Introduction to flexible electronics / Aftab M. Hussain.
Description: First edition. | Boca Raton : CRC Press, [2022] | Includes
   bibliographical references and index. | Summary: "The field of flexible
   electronics has grown rapidly in the last two decades, with diverse
   applications including wearable gadgets and medical equipment. This
   textbook comprehensively covers fundamental aspects of flexible
   electronics along with materials and processing techniques. It discusses
   topics including flexural rigidity, flexible PCBs, organic
   semiconductors, nanostructured materials, material reliability,
   electronic reliability, crystalline and polymer materials, semiconductor
   processing and flexible silicon in depth. The text covers advantages,
   disadvantages, and applications of processes such as sol-gel processing
   and ink-jet printing. Pedagogical features including solved problems and
   unsolved exercises are interspersed throughout the text for better
   understanding. The textbook is primarily written for senior
   undergraduate and graduate students in the field of electrical
   engineering, electronics, and communication engineering for a course on
   flexible electronics. The textbook will be accompanied by teaching
   resource including solution manual for the instructors"-- Provided by
   publisher.
Identifiers: LCCN 2021029849 (print) | LCCN 2021029850 (ebook) | ISBN
   9780367439668 (hbk) | ISBN 9781032150437 (pbk) | ISBN 9781003010715
   (ebk)
Subjects: LCSH: Flexible electronics.
Classification: LCC TK7872.F54 H87 2022 (print) | LCC TK7872.F54 (ebook)
   | DDC 621.381--dc23
LC record available at https://lccn.loc.gov/2021029849
LC ebook record available at https://lccn.loc.gov/2021029850

ISBN: 978-0-367-43966-8 (hbk)
ISBN: 978-1-032-15043-7 (pbk)
ISBN: 978-1-003-01071-5 (ebk)

DOI: 10.1201/9781003010715

Typeset in Times
by KnowledgeWorks Global Ltd.

Access the Support Materials: https://www.routledge.com/9780367439668

Contents

PART III   Integration Strategies

PART IV   Applications

PART V   The Road Ahead

Preface

The field of electronics has undergone significant changes in the past few decades. We have come from the era of vacuum tube-based systems to semiconductor devices; from cathode-ray displays to OLED touch screens; from storing a few kilobytes in large floppy disks to storing terabytes of data in the space of a fingernail. All of these developments have led to the proliferation of electronic systems to such an extent that nearly every individual on Earth owns one, if not multiple, "smart" devices. The number of connected systems in the world already outnumbers the number of humans by several times, however, by some estimates, this is only the beginning. The emergence of the Internet of Things (IoT) has been predicted to accelerate the proliferation of electronic systems worldwide. With such a scale having been already achieved, it is inevitable to ask what lies ahead as the "next big thing" in electronics. To many of us, the answer is clear – physical flexibility. The fundamental reason for this belief is the design of systems in nature – all the animal, plant, and bird species have curvilinear body structures, in various shapes and sizes. If electronic systems are to be integrated onto and into the human body, as many science fiction writers believe, then these systems will inevitably need to be free-form, flexible, and in some cases, stretchable. It is with this belief that I take the reader on the journey of discovering the potential of the current state-of-the-art of flexible electronics, and the challenges that still need to be solved.

This textbook has been written as an introduction to the field of flexible electronics, particularly for undergraduate students. All the concepts have been explained in such a way that no prior knowledge of the subject is assumed on the part of the reader, apart from simple high-school physics, calculus, and chemistry. The aim of this book is not to provide a detailed, in-depth literature review of the latest developments in flexible electronics (you have other wonderfully written books for that). Instead, this book aims to build a broad understanding of the basic concepts governing the materials, processes, integration strategies, and applications of flexible electronic systems. A researcher new to this field can first read this book (or a specific chapter herein) for a broad-level introduction, then embark upon reading some of the in-depth literature available on the specific material, device, or application that she is interested in. This book can also be read by somebody not belonging to the field of flexible electronics, but who may be intrigued by the subject (the curious outsider).

The book is divided into five parts. Part I – Introduction – is the introduction of the subject and deals with the basic concepts of physical flexibility and covers the microfabrication processes which are very useful in developing flexible electronics integration strategies. Part II – Materials – covers all the electronic materials, such as silicon, metal oxides, III-V semiconductors, organics, etc., that are typically used in conventional electronic systems. In this part, we discuss the electronic properties of these materials and ways of creating flexible thin films using these materials for electronic applications. Part III – Integration strategies – covers the various processes that have been developed specifically for flexible electronic systems design. These include flexible PCBs, inkjet printing, flexible packaging, 3D printing, and so on. Part IV – Applications – deals with the various individual components, such as processors, memories, displays, batteries, sensors, etc., that need to be fabricated in a flexible form to create a fully flexible electronic system. Finally, Part V – Road ahead – covers an interesting potential addition to the flexible electronic system – stretchability. In this part, we also look at the reliability of flexible electronic systems and the challenges that lie ahead.

This textbook is dedicated to my family – my loving wife and two beautiful kids – for their support for me to work late nights and weekends to complete this book. I sincerely thank my institute – the International Institute of Information Technology Hyderabad – for all the help and support in terms of logistics and access to literature. The development of this book has been quite a journey for me, and I will always fondly remember the days and nights of pouring through information and scientific literature to make sure every sentence written in this book is scientifically accurate.

Author

Aftab M. Hussain is an Assistant Professor at International Institute of Information Technology (IIIT) Hyderabad and the Principal Investigator of the PATRIoT Lab. Aftab earned a bachelor's degree at IIT Roorkee in 2009 and an MS and a PhD at KAUST in 2016. He was awarded a postdoctoral fellowship at the School of Engineering and Applied Sciences (SEAS) at Harvard University in 2017–2018. Aftab has more than a decade of experience working with device fabrication, thin film characterization, and device testing, along with sensor applications in IoT and related areas. His research focus is flexible and stretchable electronics and their applications in various IoT domains such as healthcare, mobility, and smart cities. He has more than 90 scientific publications, holds four granted US patents, along with six others in various stages of filing. He is a member of the Executive Committee (ExeCom) of IEEE CAS/EDS joint chapter in Hyderabad and an Associate Editor (Flexible Electronics) at *Frontiers of Electronics* journal.

Part I

Introduction

1 Introduction

1.1 THE SEMICONDUCTOR REVOLUTION

The last two hundred years have been the most chaotic in human history. We have uncovered laws of the natural world at a pace that their applications have had a hard time keeping up. Before we were done admiring the intricate design of the internal combustion engine, we discovered electricity; and before we were done with the wonders of transatlantic radio communication, we invented airplanes. All of these technologies and their subsequent application in engineering have had a revolutionary impact on human life. One of these technologies is the solid-state semiconductor technology. The solid-state semiconductor technology began with the invention of the transistor in the late 1940s but really took off with the invention of the integrated circuits (ICs) in the early 1960s. Processing systems based on these newly created semiconductor chips were faster and capable of handling more data compared to their vacuum tube counterparts. This led to better designing, simulating, testing, and manufacturing capabilities, leading to even better semiconductor chips. This positive feedback between the product and the manufacturing ecosystem led to an exponential rise in the capabilities of these chips. This came to be known as the Moore's law, named after Gordon Moore who famously predicted in 1965 that the number of transistors per unit area will double approximately every two years.

It can be argued that we are lucky to be alive in the era of semiconductor revolution. We have seen the performance of our desktops, laptops, and smartphones increase every year while becoming ever more affordable. It is easy to mistake this trend as a given. However, as with everything exponential, this is not a sustainable trend. Already, we are seeing a significant slowdown in the increase in CPU performance year-on-year, while the clock speeds have remained plateaued for over a decade. This does not mean that the silicon industry will lose its significance – the world is already too connected for that to happen. However, it does seem that the industry is transitioning and will soon join the other mature (but essential) industries like cement, steel, and mining. For example, we do not expect the tensile strength of steel to double every two years, however, it forms the backbone of any major economy, similarly, we will not expect semiconductor electronics to become significantly faster or cheaper year-on-year, but semiconductor production will be an important industry. The speed at which the semiconductor revolution has taken place has made it difficult for engineers to keep pace with the latest technology. Even before we decide what can be done with a processor of a particular specification, we get one with double the speed/memory at half the cost. Now that the pace of the performance improvement is losing steam, we can reflect on the incredible achievements of the semiconductor revolution and put its results to good use. It is truly remarkable that we can fit billions of switches (transistors) on a piece of silicon about the size of a fingernail. These tiny transistors grouped together as gates make simple decisions that, when propagated through a massive network, can become a significantly complex algorithm.

DOI: 10.1201/9781003010715-2

The areas where semiconductors have made a mark are in the traditional "technology" sector. Fields such as computing, information processing, communications, and data storage were the first areas to have been completely transformed with the semiconductor revolution. Subsequently, all the other fields of human endeavors, such as engineering – with mechanical and civil drawings and simulation being performed on computers; medicine – with the use of digital systems to record patient data, perform tests, and diagnosis; arts – with everything from music to creative designing being done on computers, have come to be dominated by computers. This invasion of our lives by the products of the semiconductor revolution will continue with these devices eventually becoming physically invasive. Many interesting research works are being currently carried out on the introduction of brain-machine interfacing, bioengineering, and biohacking. However, there is a general mechanical incompatibility of electronic systems, which are rigid, rectangular, and brittle, with biological systems, which are soft, curvilinear, and flexible. Even with continuous improvement in performance and reduction in size, the current semiconductor industry still remains two-dimensional because everything happens on the surface of a silicon substrate where all the transistors are patterned into existence using a series of complex process steps. This two-dimensional pattern, its supporting substrate, and the subsequent packaging cause the IC chips to be rigid and brittle.

It is important to understand that the term "flexible electronics" can be used to mean many things ranging from a slightly bent television screen to an electronic device that can be completely crumpled up. These devices and everything in the middle can be rightfully called flexible electronics, hence, the user should investigate and appreciate the amount or level of flexibility in any new device they come across. We will quantify these terms and understand the underlining physics in the following chapter. Further, it is important to differentiate between devices that can be bent to various radii from a starting flat position and devices that are rigidly bent into a particular radius. In some books, the former is referred to as dynamic flexibility, while the latter is referred to as static flexibility. In this book, when we use the term flexible electronics, we refer to the first kind of device – the device that can be reversibly bent to more than one bending radii from a given starting point (dynamic flexibility).

The rise of new applications has caused a demand for electronics that is "free-form", i.e., electronics that can flex, change their shape, and/or is stretchable. This has sparked an entirely new revolution to design processes and material chemistries such that the performance of rigid silicon electronics can be obtained in any size and form. In this book, we will discuss the efforts being made in the fields of electronics design, process design, material science, and chemistry to understand the challenges and opportunities in free-form electronics. A major focus will be on the materials that can be used to make these free-form electronic systems, followed by the manufacturing processes specific to free-form electronics. We will also discuss the strides made in research and the applications already demonstrated to be flexible.

1.2   HISTORY OF FLEXIBLE ELECTRONICS

The field of flexible electronics can be thought of as a natural extension of the semiconductor revolution. However, if we include the world of interconnects and passive components, there have been excursions into the world of flexible electronics right from the beginning of the twentieth century. For instance, in 1904, Albert Parker Hanson was granted a patent by the United States patent office for an assembly of interconnected metal strips on layers of paraffin paper [1]. The image in Figure 1.1 is from the original patent application. While this is a simple assembly, it demonstrates that the concepts of electronics and flexible electronics have always gone together and that flexible electronics is not an after-thought of the semiconductor revolution. According to the inventor, the purpose of the original patent was to make the product compact and low-cost. Indeed, these considerations remain the same as we build flexible counterparts of some circuits even today. Over the years, many design and

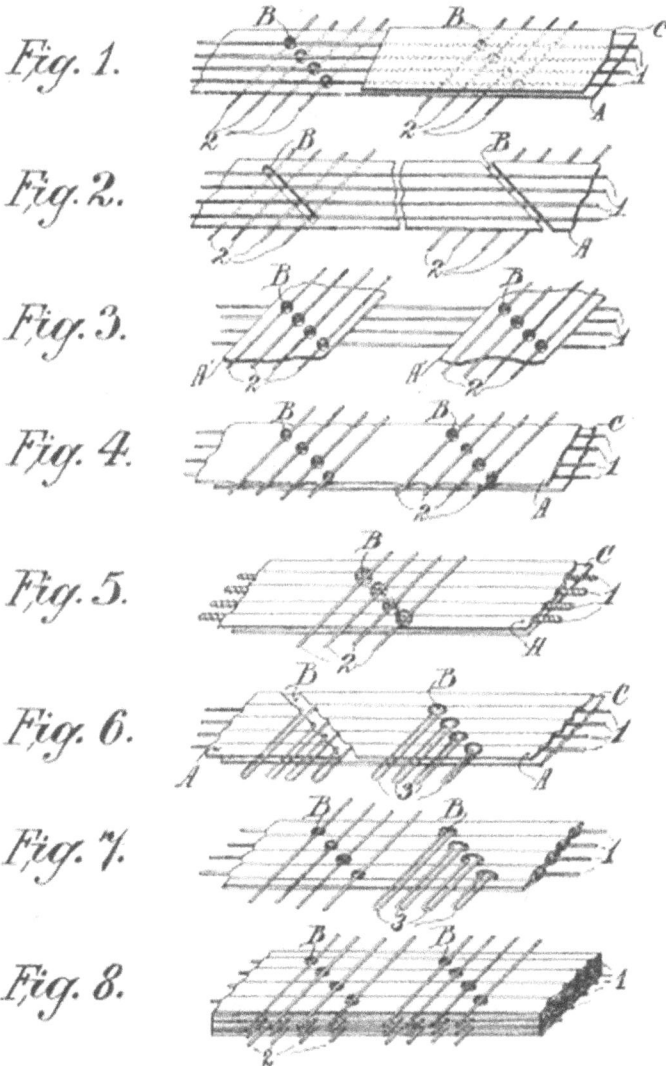

FIGURE 1.1 A figure from the patent filed by Albert Hanson in 1904 with the USPTO, titled "Electric cable". Patent granted in 1905, US782391A [1].

material innovations were made to the concept of having single or multi-layer metal interconnection. These were generally led by the printed circuit board industry, even though their main focus remained metal interconnections on rigid boards.

Studies on stand-alone flexible electronic circuits can be traced back to the late 1970s. This was motivated by the movement from rigid single-crystalline solar cells to amorphous thin-film silicon solar cells. This simple move suddenly made it possible for these devices to be fabricated on plastic substrates because of the low temperature required for deposition and subsequent processing of amorphous materials. Thus, both the substrate and the active thin film were potentially bendable up to a certain bending radius, giving rise to flexible solar cells. In fact, some of the first studies on the effect of bending and reliable operation of such flexible solar cells can be found in the 1980s. In the 1990s, several organic thin films were shown to have semiconducting properties. Because these thin films could be deposited using low-temperature processes such as evaporation or using solution-based processing, any substrate material (compatible with the chemistry of the thin film) could be used to fabricate these devices. Again, because of this freedom, both the active layer and the substrate could be flexible, giving rise to flexible organic electronics. In the 2000s, several "transfer" techniques were developed in which a thin film of a crystalline inorganic material was transferred from a base substrate to another host substrate. Several processes for removal of the thin film from the base substrate and for adhesion to the host substrate were introduced. These and many other themes are still common in flexible and stretchable electronics research. Apart from this, several applications, where flexible and stretchable electronics can be relevant, have also been reported and studied over time. In the 2010s, techniques such as paper electronics have joined age-old application areas such as the development of artificial skin. Novel fabrication techniques such as 3D printing and inkjet printing have enabled a host of applications to be demonstrated in cost-effective ways. Further, many interesting applications have been proposed for new age flexible electronics such as electrocorticography (ECoG), heart monitoring, dental health monitoring, thermotherapy, and so on (Figure 1.2).

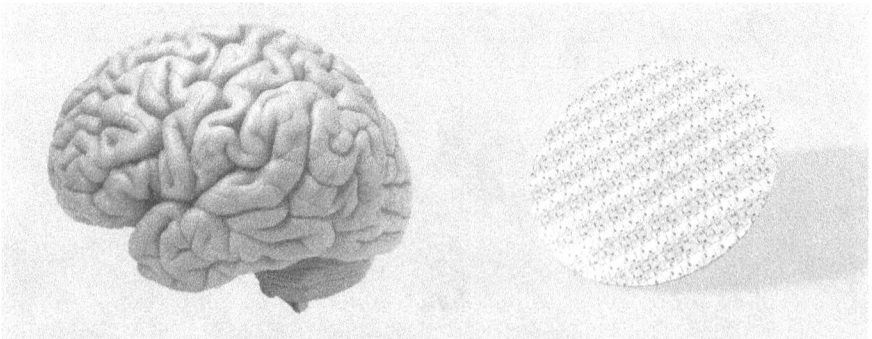

FIGURE 1.2   The difference between man-made and natural computing systems is stark. One of the goals of flexible electronics research is to bridge this gap.

1.3 NEED FOR FLEXIBLE ELECTRONICS

The traditional electronics industry is largely based on microprocessors and memories that employ silicon as the base material. There are several reasons why silicon is popular with the electronics industry. First, it is available in abundance in the Earth's crust at a relatively low cost. Second, it can be obtained at very high levels of purification and crystallized into ingots using various highly scalable processes. Third, the electronic properties of silicon are well defined and consistent over its entire lattice and can be locally modified by adding suitable impurities at controlled levels. Fourth, it is simple to form high-quality contacts between the external electronic circuit and silicon-based circuits for seamless transfer of information between individual components. With such a large array of advantages, silicon dominates as the material of choice for all electronic components manufactured today. However, electronic circuits based on silicon are rigid and brittle. While the engineering that goes into making today's electronic gadgets is indeed remarkable, it pales in comparison to the engineering seen in nature. Some of the microchips fabricated these days have several billion transistors squeezed into an area of a few square centimeters. These processors can perform tens of billions of operations per second and can store vast amounts of information. However, even this is unimpressive compared to the engineering that goes into our very own brains. We have billions of neurons connected to each other in a vast network with trillions of individual connections. This "circuit" consumes less power than an incandescent bulb and can make complex decisions, generate new ideas, provide inspiration in millions of other "circuits" like it, and even study itself. Further, the brain is able to grow, heal, and regenerate if needed, which, unfortunately, our iPhones cannot do.

An ultimate goal of many science fiction scenarios is the fusion of the human brain with the internet. However, given the vast differences between silicon-based electronics and the biological world, this mythical unification remains in the science fiction realm. While silicon-based transistors are tinier and faster compared to an individual neuron, a network of transistors fails to compare with a network of neurons because the latter can support many more connections, can form complex structures in three dimensions (compared to the 2D circuits of transistors), and can process and interact with multiple signals such as chemical, electrical, and mechanical. In order to achieve interconnection of the electronics world with the biological, it is essential to fabricate circuits that have the same mechanical form factors as those of biological systems. It is important to use material systems that can interact electronically but can also respond to mechanical or chemical stimuli. The vast difference between the silicon wafer and the brain needs to be eliminated by making electronics with materials similar to the ones found in biological systems. This is one of the reasons why flexible electronics is gathering momentum.

1.4 EXAMPLE APPLICATIONS

An obvious application of flexible electronics is its use in the wearable industry. There are already many smartwatches and wristbands for health monitoring applications. These devices include temperature sensors, blood oxygen level sensors, heart-rate monitors, and even blood pressure sensors, along with processing, display, and wireless communication capabilities. However, even with a small bending radius and apparent conformal design, these devices cannot be called flexible, because the internal components used in fabricating these devices are still rigid. These designs take advantage of the fact that individual chips for processing, communication, and display control are tiny and, thus, if mounted on a flexible substrate, give the illusion of the circuit being completely flexible. Indeed, these gadgets are predecessors to truly conformal electronic circuits wherein all the individual components can also be flexible, providing a much better user experience and comfort.

As stated in the previous section, the ultimate goal of flexible electronics research is to tap into the vast potential of connecting the human brain to the internet. The surface of the brain can be a good starting point to put electrodes for sensing brain activity. At present, this is done using metallic sensor nodes connected to a large processing unit using wires. This delicate procedure requires the patients to be in an operation theatre under expert supervision and hence cannot be done easily in everyday life. One of the goals of flexible electronics research is to fabricate an electronic "sleeve" for the brain that can be placed on the cortical surface and information about the activities of the brain can be obtained wirelessly in real-time. These devices, once implanted in the brain, can remain there for several years passing on brain activity information for processing while obtaining information from the outside for actuation. Such a device will need to have tremendous flexibility and a large area along with processing, storing, and wireless communication capabilities.

Other examples include the creation of foldable phones and tablets. Such devices have the potential to revolutionize the way smartphones and even laptops are used, and how we interact with technology and each other. There are some examples of foldable phones already in the market, but these devices are only foldable along a particular axis. Truly foldable phones will be completely flexible so that they can be folded along any axis and can be packed into a tight space as per our convenience (similar to a textile accessory). These and many other potential applications have spurred research interest in flexible electronics over the past few years. Several novel materials, processes, and manufacturing techniques have been developed to cater to these potential requirements. In the subsequent chapters, we will discuss these developments and analyze the gaps that are still remaining to realize the true potential of flexible electronic circuits.

2 Physics of Flexible Electronics

2.1 WHY ARE THINGS FLEXIBLE?

A common everyday observation is that making materials thinner increases their flexibility. For example, a large aluminum block cannot be easily flexed; however, the same material, when made into a thin foil can be very easily flexed. Another typical observation is that different objects with the same thickness can have different flexibility depending on the material they are made of. These simple observations tell us that there are two different properties of an object that determine its flexibility – the material and the thickness in the direction normal to the axis of flexing. It is also important to note that both these parameters are independent of each other. We can change the material and its thickness independently subject to the constraints of fabrication. Thus, in terms of obtaining flexible thin films for specific applications, such as electronics, we can either reduce the thickness of known electronic materials or change the material used to fabricate electronics to a material that is more flexible at the given thickness. The former approach involves exploring processes for use of silicon thin films to fabricate electronics on a flexible substrate, while the latter involves exploring novel materials for the fabrication of electronics with similar or enhanced properties compared to conventional silicon electronics. For either approach, it is important to understand the physics of flexing in regular lattice materials (crystals) as well as polymer materials in order to design electronics with specific flexing capabilities.

In the case of materials with regular crystal lattice (such as silicon and most metals), we will refer to the ball-spring model. It is a simplified model portraying the behavior of atoms in a regular lattice and can be useful for an intuitive and basic understanding of the behavior of crystalline materials. In the case of polymers or amorphous organic materials, the elastic behavior is deeply influenced by the chemistry of the structure. Parameters such as molecular weight, cross-linking percentage, and hydrogen bonding can determine the elastic properties of the material. Thus, for these materials, it is prudent to study each material (or at least each class of material) separately. Notwithstanding, we will investigate commonly known mechanisms of elasticity and hyper-elasticity for polymers in this chapter. However, before moving forward, it is important to quantify the relationship between the parameters of a device and its flexibility in order to design flexible electronics pertaining to some specifications. However, before we quantify this relationship, we need to quantify flexibility itself. The perceived flexibility, or the lack thereof, of an object around a particular axis of bending, is called its flexural rigidity. We define the flexural rigidity of an object as the bending moment required to obtain a unit curvature in a particular bending axis. This is the resistance offered by a material to bending, thus, the lesser the flexural rigidity, the higher is the perceived flexibility of the object.

DOI: 10.1201/9781003010715-3

2.2   BENDING RADIUS AND CURVATURE

To understand the definition of flexural rigidity and obtain a mathematical relationship in terms of the material parameters, it is important to understand the concept
of curvature. For any structure or surface, the curvature provides a measure of how
much the structure bends (or deviates from a straight line or plane). Mathematically,
the curvature of any function is defined using a circle, which has constant curvature
(or bending) at all points. Hence, for a general two-dimensional curve $g(x,y)$,
the curvature at any point is defined as the curvature of a circle approximating the
curve at that point. Consider the function $y = f(x)$, as shown in Figure 2.1. The curvature of this function at any point P is given by:

$$\kappa = \frac{|y''|}{\left(1+y'^2\right)^{\frac{3}{2}}}$$

If the function can be represented in the parametric form $x(t)$ and $y(t)$, the curvature at any point is given by:

$$\kappa = \frac{|x'y'' - y'x''|}{\left(x'^2 + y'^2\right)^{3/2}}$$

One of the most important parameters in the field of flexible electronics is the
minimum bending radius a particular device can achieve. The bending radius or
radius of curvature is defined as the inverse of the curvature of a structure. Thus,

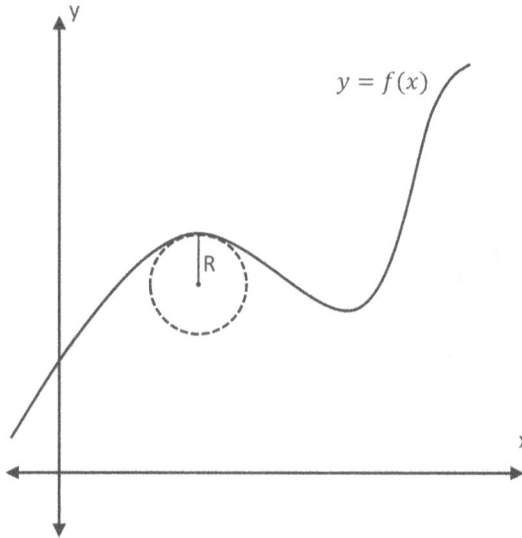

FIGURE 2.1   The curvature of the function $y = f(x)$ at a point P is defined as the curvature
(or the inverse radius) of the circle approximating the function at that point.

bending radius R is given by $1/\kappa$. It is important to note that the radius of curvature decreases for increased curvature, i.e., when bending is more. Thus, a lower minimum bending radius is desirable for flexible electronic devices because it shows that the device can function even when it is subjected to a large curvature. The curvature of a straight line is zero and its bending radius is infinite. In case of a circle, the curvature at a point (and hence the radius of curvature) can be calculated using the parametric equations $x(t) = r\sin(t)$ and $y(t) = r\cos(t)$. From the equation, the curvature of the circle is obtained as $1/r$, independent of the value of t. Thus, the curvature of a circle is always constant, and its bending radius is the same as its radius.

The general equation for curvature can be simplified further if the curvature of the structure is small compared to its dimensions (i.e., if the radius of curvature is large with respect to its dimensions). Consider a function $f(x)$ going through a small curvature at point P (Figure 2.2). The radius of curvature is the radius of the arc approximating the function at point P. The small angle dA is the angle of the arc formed by the small arc of length l between points P and Q. Because the curvature of the function is small, the length of the arc can be approximated as dx and the angle formed by the tangent at point P is given by, $A = df/dx$. Thus, the radius of curvature is obtained as:

$$\frac{1}{R} = \frac{dA}{l} = \frac{dA}{dx} = \frac{d}{dx}\left(\frac{df}{dx}\right) = \frac{d^2 f}{dx^2}$$

This equation is the simplified version of the curvature equation, for small curvatures. We can also arrive upon this equation with the argument that if the curvature is small, the term $\left(\frac{dy}{dx}\right)^2$ in the denominator will be small compared to unity and thus can be neglected. Hence, the bending radius for small curvatures is given by:

$$\frac{1}{R} = \frac{d^2 y}{dx^2}$$

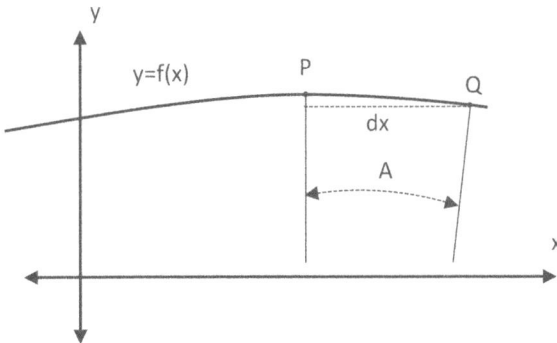

FIGURE 2.2 For small curvatures, the curvature at point P can be approximated as the second derivative of the function at that point.

2.3 PLATE FLEXURE

In general, the field of flexible electronics deals with thin films or planar substrates subjected to a torque that causes flexing. Thus, it is important for us to understand the theory of plate flexure. Consider a simple uniform plate of thickness h and length L, pinned at both ends, being flexed due to an applied force F at the center (Figure 2.3). The plate is considered to be infinitely wide along the axis going into the page. Because the plate is pinned at both ends and cannot move in the direction of the applied force, we have vertical forces of $F/2$ at both ends of the plate. Because of the thin-film nature of the plate, we assume $h \ll L$. This is a reasonable assumption for most of the practical flexible electronics applications. Because of the application of the force moment, there is a deflection in the plate. The amount of deflection varies as we go from the center of the plate to the ends and is represented by a function $w(x)$, with its maximum value at the center of the plate. It is assumed that this deflection is small enough for the material of the plate to be in the linear elastic region of its stress-strain curve.

To determine the deflection of the plate, it is important to derive $w(x)$ as a function of the applied force, the dimensions, and the material properties of the plate. This can be done by requiring the forces and torques on the plate to be in equilibrium so that there is no further movement (or rotation) of any section of the plate. We consider a small section of the plate from the point x to $x + dx$. This infinitesimal section has a downward load per unit area that can be represented as $f(x)$. This area denotes the plane formed in the x-z direction. Because all the forces are considered to be per unit length along the z-axis, that description is omitted for the following forces and moments. Thus, the total downward force on this section is $f(x)dx$. A shear force acts on the cross-section of the plate because of the shear elastic forces inside the plate. The magnitudes of the forces are Q at x and $Q + dQ$ at $x + dx$. Thus, the total downward force acting on the plate due to this shear is dQ. The balance of these forces gives the relation:

$$f\ dx + dQ = 0$$

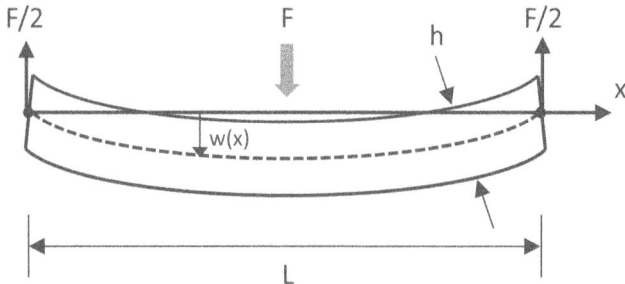

FIGURE 2.3 A vertical force F creates a downward bend in a plate fixed at both ends. The downward displacement w can be modeled as a function of x for a given force distribution on the plate.

Or,

$$\frac{dQ}{dx} = -f$$

The section is also under a bending moment M at position x and $M + dM$ at position $x + dx$ due to the elastic restoration forces inside the plate, which can be related to the material properties of the plate. The total moment because of these forces is dM. This moment is balanced by the moment created due to the vertical shear forces on the cross-section of the plate. Thus,

$$dM = Qdx$$

Or,

$$\frac{dM}{dx} = Q$$

Thus,

$$\frac{d^2M}{dx^2} = -f$$

This is a simplified expression for the bending of a plate due to applied moment or vertical load force. It assumes there are no horizontal forces applied to the plate. In the more general case, if a horizontal force N is applied to the plate, the expression becomes:

$$\frac{d^2M}{dx^2} = -f + N\frac{d^2w}{dx^2}$$

This expression contains information about vertical and horizontal forces and relates the bending (w) to the bending moment (M). However, we require the bending of the plate given the applied forces, material properties, and dimensions of the plate. Thus, it is important to determine the bending moment in terms of the bending and material properties of the plate.

For a plate bent downward as shown in Figure 2.3, the top half of the plate is under compression and the bottom half is under expansion. There is a plane inside the plate with no linear expansion or compression called the neutral plane. The longitudinal stress σ on the plate varies with the distance from the neutral plane, as shown in Figure 2.4. The net result of these stresses is to exert the bending moment M on the cross-section of the plate. The force per unit length (along the z-axis) on a small vertical section of this plate is given by σdy. This force produces a torque around the neutral plane on a point located at a distance of y from the neutral plane. Thus, the net moment is given by:

$$M = \int_{-h/2}^{h/2} \sigma\, y\, dy$$

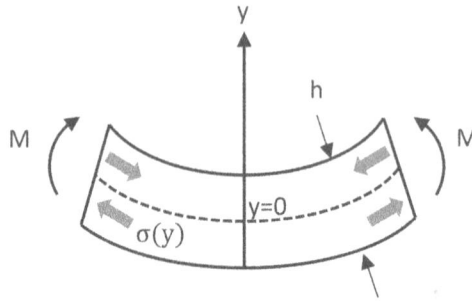

FIGURE 2.4 The longitudinal stress σ on the plate varies with the distance from the neutral plane, for a given bending moment M applied to a section of the plate.

Now, we need to obtain the value of σ in terms of the known material properties and the bending of the plate. The contraction of the upper part and expansion of the lower part of the plate creates a strain ε in the longitudinal direction. Further, because we assume that the plate has a very large width, the strain in the z-direction is considered to be zero. It is known that a principle stress component σ produces a strain of σ / E in the direction of stress and $v\sigma / E$ in the two directions normal to it, where E is the Young's modulus of the material and v is the Poisson's ratio. Thus,

$$\varepsilon = \frac{\left(1-v^2\right)}{E}\sigma$$

The bending moment can be written as:

$$M = \frac{E}{\left(1-v^2\right)} \int_{-h/2}^{h/2} \varepsilon\, y\, dy$$

Now, for a small curvature, the longitudinal strain can be approximated using the radius of curvature. Consider a small section dx that makes a small angle $d\theta$ with the center of curvature. The length dx at the neutral plane can be written as:

$$dx = Rd\theta$$

At a point y-distance away from the neutral plane, the length of the small section is $(R+y)d\theta$. Thus, the strain in the small section at a distance y from the neutral plane is given by:

$$\varepsilon = \frac{(R+y)d\theta - dx}{dx} = \frac{(R+y)d\theta - Rd\theta}{Rd\theta} = \frac{y}{R}$$

Thus, the bending moment is given by:

$$M = \frac{E}{R\left(1-v^2\right)} \int_{-h/2}^{h/2} y^2\, dy$$

Solving the integration, we get:

$$M = \frac{Eh^3}{12 \, R \, (1 - v^2)}$$

By definition, the flexural rigidity of a plate is the bending moment required to obtain a unit curvature. Thus,

$$D \equiv MR = \frac{Eh^3}{12 \, (1 - v^2)}$$

This is the basic equation for the flexural rigidity of a simple uniform plate under bending. Clearly, the thickness of the plate plays a major role in determining its flexural rigidity, along with material properties like Young's modulus and Poisson's ratio. Further, because the radius of curvature is assumed to be large compared to the thickness of the plate, the radius of curvature can be approximated as:

$$\frac{1}{R} = -\frac{d^2 w}{dx^2}$$

Thus,

$$M = -D \frac{d^2 w}{dx^2}$$

From Figure 2.4, it is clear that the distribution of strain varies with the distance from the neutral plane. For a uniform thin film, the neutral plane is at the center of the thickness, while the maximum strain occurs at the surface of the film. Now, we know that the maximum strain experienced by a thin film of thickness t, bent at a bending radius of R, can be approximated as:

$$\varepsilon_{max} = \frac{t}{2R}$$

This is intuitive because lower bending radius will result in a higher strain on the thin film, while reduction in thickness leads to lower strain. This is a useful first-order approximation for obtaining values of possible bending radii for materials with various thicknesses. It should be noted that this equation only holds for the situation when $t \ll R$. The maximum strain that a material can be subjected to before fracture is called fracture strain, and is a constant for a given material. If the maximum strain exceeds the fracture strain of the material, the material surface will undergo cracking, leading to a possible catastrophic failure of the system. Thus, for successful and repeated flexing, it is important that ε_{max} does not exceed the fracture strain. Hence,

$$\varepsilon_{max} = \frac{t}{2R} < \varepsilon_f$$

This provides a limit to the bending radius for a substrate of a given material for a given thickness. Conversely, we can use this equation to determine the thickness of a material if a specific bending radius is required.

2.3.1 FLEXURE UNDER FIXED BENDING MOMENT

The simplest case of flexing takes place when the flexible electronic device is applied a certain bending moment from one end with respect to another. This can be approximately done by pinching the ends of a device and bending it (action is shown in Figure 2.5a). In this case, the distributed load on the plate is zero (thus, $f = 0$). Further, because there is no applied force on the plate, the shear force, $Q(x)$, should also be zero. Because $Q(x)$ is a derivative of the moment of force experienced by the plate, the moment $M(x)$ is a constant. This constant bending moment is the one that is applied at the end of the device, $M(x) = M_a$. Thus, the displacement of the plate can be obtained by double integrating the moment equation. The boundary conditions, in this case, can be obtained by considering one of the ends of the plate to be fixed. Thus, for $x = 0$, we have $w = 0$. Further, because this device is clamped at one end, the slope at this end cannot change, thus, $dw / dx = 0$ for $x = 0$. Using these boundary conditions, the bending of the device under the applied bending moment can be obtained as:

$$w = -\frac{M_a x^2}{2D}$$

Thus, the plate bends in the shape of a parabola. The maximum displacement is obtained at the end where the bending moment is applied. The extent of bending is proportional to the magnitude of the applied moment, the square of the length of the plate, and is inversely proportional to the flexural rigidity of the plate.

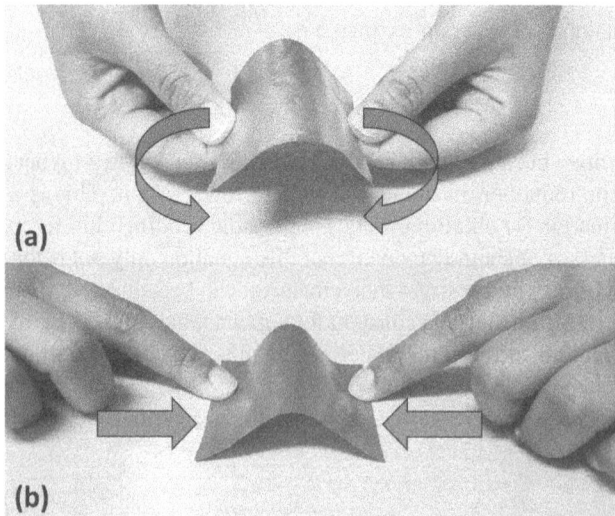

FIGURE 2.5   (a) The force moment applied at the ends of a flexible plate creates a displacement along the plate. (b) A horizontal force applied to the ends of a flexible plate causes buckling if the force exceeds a critical point.

2.3.2 BUCKLING

Another very common phenomenon seen in flexible electronic devices is the buckling of the device due to applied horizontal forces. In practice, it is seen that a flexible device does not bend immediately when a certain force is applied. The out-of-plane buckling only takes place if the applied force exceeds a threshold. This critical force can be calculated using the mathematical model developed for the bending of plates. Figure 2.5b shows the buckling of a flexible electronic substrate under a horizontal force. There is no distributed load on the device, thus, $f = 0$. Hence, we have:

$$D\frac{d^4w}{dx^4} + N\frac{dw^2}{dx^2} = 0$$

We can double integrate this equation with respect to x, to obtain:

$$D\frac{dw^2}{dx^2} + Nw = C_1x + C_2$$

where C_1 and C_2 are constants. However, because there are no vertical movements of the plate edges, we have $w = 0$ at $x = 0$ and $x = L$. Thus, constants C_1 and C_2 have to be zero. The differential equation thus takes the form:

$$D\frac{dw^2}{dx^2} + Nw = 0$$

This equation has a general solution of the form:

$$w = A\,\sin\sqrt{\frac{N}{D}}x + B\cos\sqrt{\frac{N}{D}}x$$

For w to be zero at $x = 0$, we need to have $B = 0$. However, because w is also zero at $x = L$, we can either have $A = 0$ or the sine function equal to zero at $x = L$. With $A = 0$, the solution is that w is zero for all x, i.e., there is no buckling. However, for A to be non-zero, we have the condition that:

$$\sin\sqrt{\frac{N}{D}}L = 0$$

Thus, the variable inside the sine function should be an integral multiple of π.

$$\sqrt{\frac{N}{D}}L = k\pi$$

We can solve this for the applied horizontal force N, in terms of the flexural rigidity of the plate and its length to obtain:

$$N = \frac{k^2\pi^2 D}{L^2}$$

The minimum force required to have a non-zero solution for $w(x)$ is obtained for $k = 1$. This is the critical horizontal force required for the onset of buckling in a flexible device.

$$N_c = \frac{\pi^2 D}{L^2}$$

When this force is applied, the plate bends into a half-sine function. The phenomenon of buckling is unique because no displacement is seen in the flexible plate until the critical force is applied, thereafter, the plate suddenly bends into its final position. The amplitude of this bend, A, can be calculated considering the strain potential energy stored in the device for the work done by the applied force.

2.4 THE BALL-SPRING MODEL

The concepts of elasticity and flexural rigidity stem from the bonding of individual atoms to form the lattice of a material. To understand these concepts, it is important to understand the state of equilibrium the atoms are in. Chemical bonds are said to be formed when two atoms have an electrostatic attraction toward each other. The reasons for this attraction could be many; however, whatever may be the reason, this attraction leads to a force that brings the atoms together. However, if the atoms get too close to each other, electrostatic repulsion starts taking place between the two nuclei, leading to a repulsive force. Thus, in all cases of chemical bonding, there is a stable distance at which atoms feel no attractive or repulsive force. This distance is generally referred to as the bond length. This equilibrium position is said to be stable because if the atoms are pushed closer to each other, they are repelled back to their original position, and if they are pulled apart, they are attracted back to their original position. Figure 2.6 shows the potential energy drop in the system because of the movement of hydrogen atoms toward each other. It is assumed that the system has zero potential energy when the distance between the two atoms is infinite. One can differentiate the energy curve with respect to position to find out the magnitude of force at a given point. The force is clearly zero when the atoms are in their minimum energy position. Any perturbation in either direction leads the atoms back to this equilibrium position. However, the atoms can be permanently separated if they are provided with sufficient energy to go above the zero-potential level at infinity. This energy is the strength of the chemical bond between the two atoms.

There is a simple and intuitive way to imagine this scenario. Consider two tiny balls attached to a spring. When the spring is unstrained, there is no force on either ball. This is the equilibrium position. However, if the balls are moved away from each other, the spring gets elongated and exerts a force driving the balls back to the equilibrium position. Conversely, if the balls are pushed toward each other, the spring exerts a force driving them away. In both cases, the magnitude of the force is proportional to the strain and the direction of the force is such that the equilibrium position is restored. This simple model describing the behavior of atoms in a chemical bond is called the ball-spring model. As seen in Figure 2.6, the energy profile for the ball-spring model is similar to the actual energy profile for atomic bonding, for small strains. In the case of large strains, the model can differ from the actual energy profile.

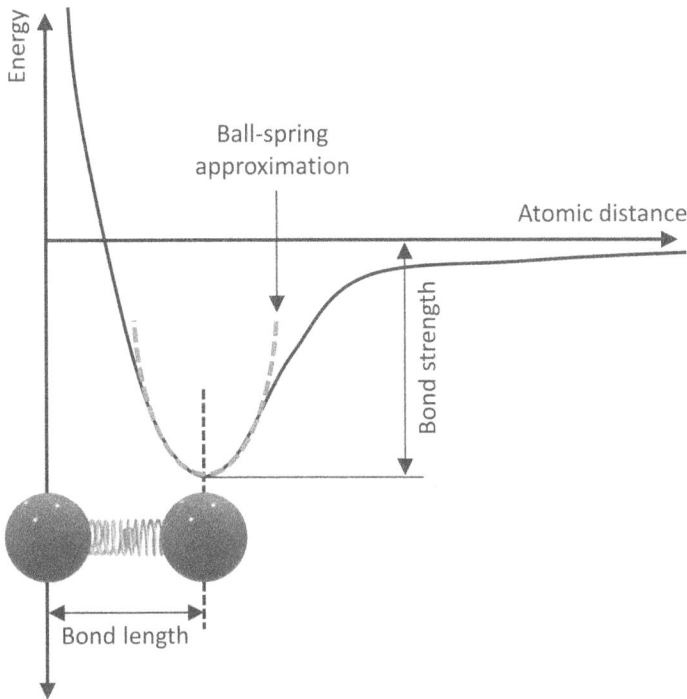

FIGURE 2.6   The energy versus distance curve for the ball-spring model closely follows the observed energy profile for atomic boding. The minimum energy position defines the equilibrium for the bond, providing information about the bond length and the bond strength.

The concept of two atoms bonded together by a spring can be extended to an entire lattice, as shown in Figure 2.7a. All the atoms in a lattice are bonded to their neighboring atoms through a chemical bond. This bond is represented by a spring. When the lattice atoms are strained, the strain is distributed through the lattice, as shown in Figure 2.7b. Further, as in the case of a spring, the displacement of the lattice is proportional to the force applied. Thus, the underlying principles of stress and strain and that of Hooke's law of elasticity can be described effectively using the ball-spring model. Another interesting consequence of the stable equilibrium is that the atoms oscillate around the equilibrium if a small perturbation is applied. This oscillation is in free space and there is very little damping to it. This is also the underlying cause of thermal energy, or what we call "temperature". The oscillation can propagate from one set of atoms to the next, which is one of the underlying causes of thermal conduction. The oscillation also produces electromagnetic waves because charges are constantly accelerating. This is the underlying cause of black body radiation. All the associated theories related to thermodynamics and black body radiation can be traced down to the oscillation of atoms around the equilibrium position because of thermal energy. In the case of the ball-spring model, this oscillation can be easily understood intuitively because any mass attached to a spring is under stable equilibrium and a small perturbation indeed creates oscillations around the equilibrium position.

(a)

(b)

Force

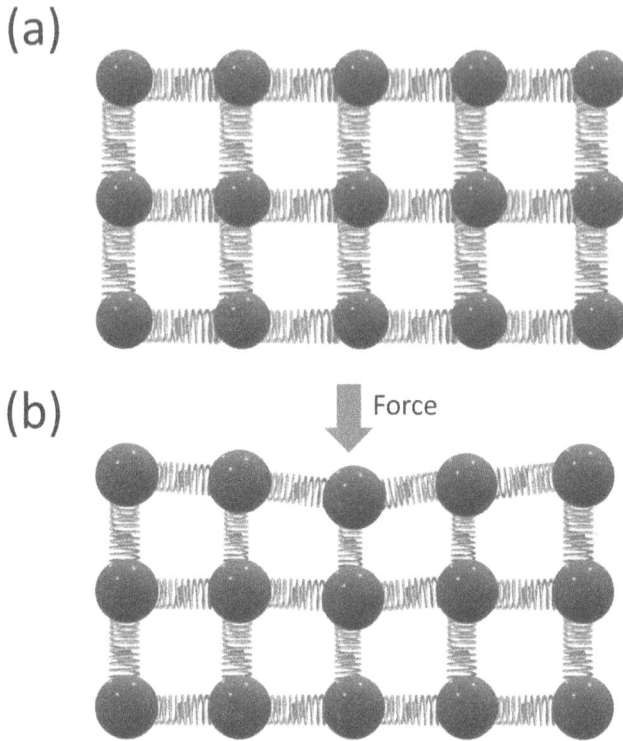

FIGURE 2.7   (a) An entire lattice can be modeled using the concept of neighboring atoms bonded together by a spring. All the atoms in a lattice are bonded to their neighboring atoms through a chemical bond. (b) When an external force is applied, the lattice atoms are strained, and the strain is distributed throughout the lattice springs.

However, in general experience, the oscillations are damped due to friction or drag, causing them to eventually stop. At the atomic scale, the oscillations typically stop because of the outflow of energy due to the radiation of energy packets from the oscillating charged particles (black body radiation). The ball-spring model also makes it easy to imagine the flexibility of a material in terms of the compression and expansion of the lattice springs, as shown in Figure 2.8. When a plate is flexed, the strain in the layers parallel to the neutral plane is proportional to the normal distance from the neutral plane (as discussed in the previous section). Most metals, non-metals, ceramics, and non-polymeric solids can be roughly modeled using the ball-spring model.

2.5   FLEXIBILITY IN POLYMERS

We generally associate a large amount of flexibility and stretchability with polymer and organic materials. This intuition comes from the various plastic, rubber, and organic substances we encounter every day. However, it is important to understand that the underlying physics of flexibility and stretchability is very different in polymers as compared to other materials. Polymers are generally made of long chains of

Expansion

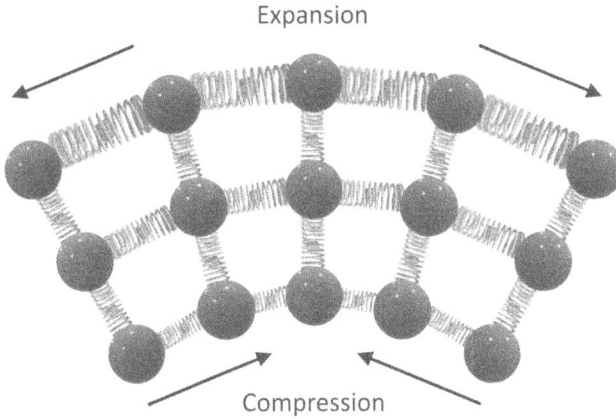

Compression

FIGURE 2.8 The distribution of compressive and tensile strain when a material lattice is flexed can be easily visualized using the ball-spring model. There is a neutral plane that does not undergo any strain.

"monomers" attached to each other through chemical bonds. We can still apply the ball-spring model to these strands of polymers and can obtain the mechanical behavior of each individual strand based on that. However, upon analysis, you will find that the carbon-carbon bonds required to make the polymer strands are very strong, making the associated spring very stiff. However, polymers as bulk materials behave differently – they are soft and elastic even for large strains. This apparent difference comes from the fact that the individual polymer strands are only weakly attached to each other through van der Waal's forces or hydrogen bonding. These can also be modeled as springs, but with very low stiffness. When an external force is applied, the individual polymer chains simply slide against each other against the restoring force of these weak bonds. Thus, it is easy to obtain large strains in polymer materials even with limited applied stress.

However, there is one trick with which we can "toughen" a polymer material. This can be done by introducing cross-links into the polymer material. Cross-links are chemical bonds that bind one polymer chain to another. These are as strong as the bonds within the polymer chain itself. If a large number of cross-links are established, the bulk polymer effectively becomes one giant molecule. Thus, if an external force is now applied, the resulting strain is opposed by these strong chemical cross-links, which means significantly less strain is obtained for a given applied stress, in the case of a well cross-linked polymer. The most well-known example of this process is the vulcanization of rubber. In this process, natural rubber (polyisoprene family) is hardened by treating it with sulfur. The sulfur atoms form bonds with carbon atoms in two different polymer chains, effectively forming a bridge between the two chains. The process can be controlled to determine the amount of cross-linking and the number of sulfur atoms per cross-link. These parameters strongly influence the mechanical properties of the final rubber. Other methods to vulcanize (also called cure) polymers include exposure to heat or UV radiation and exposure to specific chemicals.

EXERCISES

2.1. Consider a glass substrate that needs to be flexed up to the bending radius of 5 cm. We require the substrate to be a 2 cm × 2 cm square piece. Given the fracture stress of glass is 70 MPa and Young's modulus is 70 GPa, what should be the thickness of the substrate?

2.2. We have a chain of n atoms of radius r bonded together by a chemical bond. The bond is approximated as a spring of constant k (N/m). What is the Young's modulus of this material chain along the axis of bonding?

2.3. Consider a square plate of size 5 cm × 5 cm and thickness of 0.1 mm. The Young's modulus of the material is 1 GPa and the Poisson's ratio is 0.4. Calculate the flexural rigidity in the direction normal to the thickness. What is the minimum horizontal force required to buckle this plate?

2.4. For the plate in question 2.3, obtain the moment of force required to flex the plate to the bending radius of 20 cm.

2.5. Obtain equation for the curvature for the curve $y = ax^n$ at the origin, where a and n are constants.

3 Semiconductor Basics

3.1 THE SILICON ATOM

We saw in the last chapter how two hydrogen atoms come together and form a bond with a particular interatomic distance and with a specific bond energy. However, it is even more interesting to consider how an individual hydrogen atom is formed. The hydrogen atom consists of one proton as its nucleus and one electron "orbiting" the nucleus. The word orbiting can be misleading here because the electron doesn't really move inside the nucleus. While the picture of an electron revolving around a nucleus is very intuitive (because we have grown used to the planetary model of the solar system), there are fundamental problems to consider with this model. Even at the time when the planetary model of the atom was proposed (early twentieth century), it was well known that a charged particle under acceleration will produce an electromagnetic wave (or a photon). This radiation should gradually reduce the potential energy of the electron, thus causing it to go closer to the nucleus and then annihilate. This model predicted that all the atoms in the Universe should collapse in a few picoseconds.

The current understanding of electron behavior in an atom involves the wave-particle duality, the uncertainty principle, and the Schrodinger equation. The electron is in a negative potential well when it comes close to a nucleus. This means that the energy of an electron is lower than when it was a free particle. From a thermodynamic standpoint, this lowering of energy is desirable for the electron, and it causes the electron to stay close to the nucleus until a sufficient amount of energy is provided for it to escape. This situation is similar to the gravitational potential well we are all in because of our proximity to the Earth, requiring a specific escape velocity (or energy) to move out of the gravitational influence of the Earth. However, the difference in the case of an electron is that it is not free to possess any amount of energy. There are certain energy levels that the electron can occupy. These energy levels are given by the solution of the Schrodinger's equation. The negative potential because of the presence of the nucleus, or the negative "potential well", and the rest mass of the particle determine the particular energy levels an electron is allowed to be in. In the case of a particle in a 1-D infinite rectangular well, the allowed energy levels are given by:

$$E = \frac{\hbar^2 n^2 \pi^2}{2 m_0 L^2}$$

where, $\hbar$ is the Plank's constant divided by 2π ($h / 2\pi$), m_0 is the mass of the particle, L is the length of the well, and n is a positive integer. It should be noted that this is true for all particles in a potential well – including macroscopic particles, say marbles. However, if we imagine a marble placed in a gravitational potential well such as a bucket, we generally don't associate discrete energy levels with it. This is because

DOI: 10.1201/9781003010715-4

the width of the well is so large, and the mass of the marbles is so high (compared to the width of the nuclear potential well and the rest mass of electrons) that the difference in energy from one energy level to the next is imperceptibly small. However, in the case of the subatomic system, the allowed energy levels for an electron trapped in the electrostatic pull of a nucleus are easily distinguishable. For example, for an electron trapped in the potential well of a single proton (such as in the case of a hydrogen atom), the lowest allowed energy level is -13.6 eV. This energy level resembles Bohr's atomic model and provides a quantum mechanical explanation for the experimental success of the Bohr's model. This is the amount of energy needed to release the electron from the bondage of the proton and make it a free particle.

Once the solution to the wavefunction of the electron is known, we can obtain all the information about it, for example, its energy level, the probability of finding the electron at a particular distance from the nucleus, and so on. The distance with the maximum probability of finding the electron can be considered the "orbit" of the electron. However, it is worth remembering that the electron does not actually revolve around the nucleus. It is present at all points, at all times, with a certain probability. In the case of atoms with multiple protons and neutrons, the negative potential well becomes deeper and steeper. An electron trapped in it may require a lot more energy to be released from the electrostatic field of the nucleus. As n is increased in the equation, we get all the energy levels that can be occupied by an electron. In some cases, the probability distribution of the electron around the nucleus takes certain orientations (such as p- and d- orbitals). If only one electron exists inside the potential well, then only one of these many energy levels is filled. Further, if we increase the number of electrons in the potential well, we have to consider their effect on each other and postulate their behavior accordingly. To this effect, the Pauli's exclusion principle states that any quantum state can only be occupied by a single electron (which are classified as "Fermions"). The quantum state includes the energy level of the particle along with its orbital orientation and spin information. The spin of an electron is a hypothetical construct (again, electrons are not moving or revolving or spinning inside an atom) that can have one of two possible values. Thus, once we have an electron in a particular energy level and orbital orientation, there can only be one more electron in that orbital, necessarily with the opposite spin value. Because electrons always look to minimize their energy, they would have been happy at the lowest allowable energy level. However, Pauli's exclusion principle makes the electrons occupy higher energy levels (two at a time). This forms the basis of the construction of an atom (Figure 3.1). The electrons that have the highest energy (or the least negative energy), have the least influence of the nucleus on them, are most susceptible to be lured away from the nucleus. These "valence" electrons are the ones that are the most reactive in a chemical reaction and participate in forming chemical bonds.

With this, we can construct the electronic structure of atoms of any element given the number of protons and neutrons in the nucleus and the number of electrons surrounding it. Because silicon is the most widely used material for all commercial consumer electronics, we will discuss, in this chapter, the electronic structure of the silicon atom and its lattice. Therein lie the reasons for the use of silicon as the workhorse of the semiconductor industry. Silicon belongs to group IV(a) of the Periodic

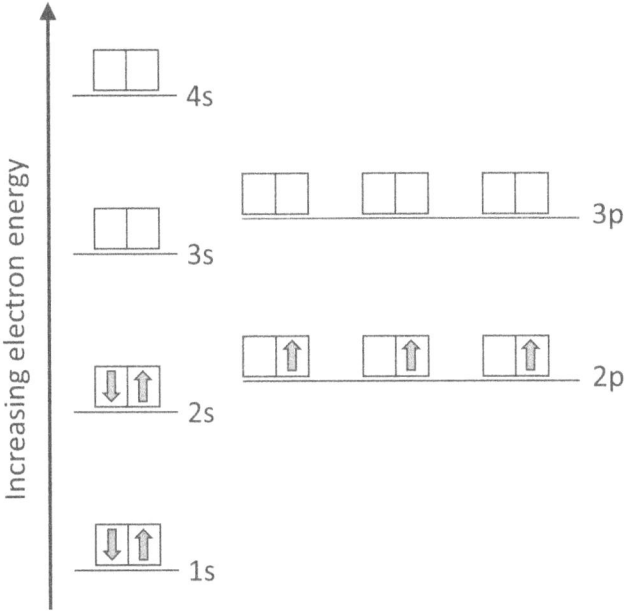

FIGURE 3.1 Thermodynamic principles dictate that electrons should occupy the lowest possible energy state; however, Pauli's exclusion principle causes the electrons to occupy higher energy levels. The electrons with the highest energy are least under the nuclear influence, causing them to participate in chemical bonding.

Table. It has 14 neutrons and 14 protons in its nucleus, and 14 electrons surrounding the nucleus. The electrons are distributed in different orbitals as: $1s^2\,2s^2\,2p^6\,3s^2\,3p^2$. All these orbitals are at different energy levels, 1s being the lowest and 3p being the highest. The electrons in the 3s and 3p levels have the least negative energy and are the most reactive in a chemical reaction. These electrons also contribute to electronic conduction through the silicon lattice. It is important to understand the fact that because electronic conduction involves the physical movement of charge carriers through space, it is necessary to have a medium for them to move. Thus, a single atom floating in space can never be classified as a good conductor of electricity (even if it is a gold atom), because there is nowhere for electrons to travel. Hence, electronic conduction, or lack thereof, is a property of the material lattice structure, and not of the individual element or atom.

3.2 THE SILICON LATTICE

We saw in the last chapter while discussing the ball-spring model, how two hydrogen atoms come together and form a bond with a specific interatomic distance and with a specific bonding energy. Silicon forms a solid lattice such that every silicon atom is bonded to four neighboring atoms. The outermost 3s and 3p orbitals of silicon hybridize into four sp^3 orbitals. These symmetric orbitals arrange themselves such that their lobes are furthest away from each other. Thus, the four bonds are formed such that silicon atoms are at the corners of a regular tetrahedron, with a silicon atom at the center. It is important to note the fact that the atom does not "decide" to hybridize its orbitals, it is again the solution to the Schrodinger equation that changes because of the presence of other atoms (because of more potential wells around). The new solution also changes the number of allowed energy states for the electrons in the system. In fact, the allowed energy levels that electrons can take increase significantly as more and more atoms come together – until the discrete levels form a pseudo-band of energy in which the electrons can take a particular value. The number of available states in this band depends on the number of atoms in the cluster (which is easily in the millions even for a nanoparticle). Indeed, these states only represent the available or allowed energy states. The way these are "filled up" or occupied by electrons depends on other considerations. This is similar to a classroom full of chairs where students can sit (allowed energy states), however, whether a student is occupying a particular chair at a particular time depends on many factors.

The distribution of electrons in allowed energy states is generally determined by the total energy in the system and the distribution of this energy among electrons. Let us consider a simple example. Figure 3.2 shows a small number of allowed energy states in a hypothetical lattice. These energy levels, called E1, E2, and E3, have energies 1, 2, and 3 units. Thus, if an electron is occupying one of the states of E3, it has 3 units of energy. Now, suppose these states are to be filled by five electrons so that the total energy is 11 units. While filling, we cannot change the number of electrons (conservation of mass), or the total energy they have (conservation of energy). There are three distributions which satisfy these constraints: (E1, E2, E3) = (2, 0, 3), (1, 2, 2), and (0, 4, 1). Among these distributions, the number of ways of

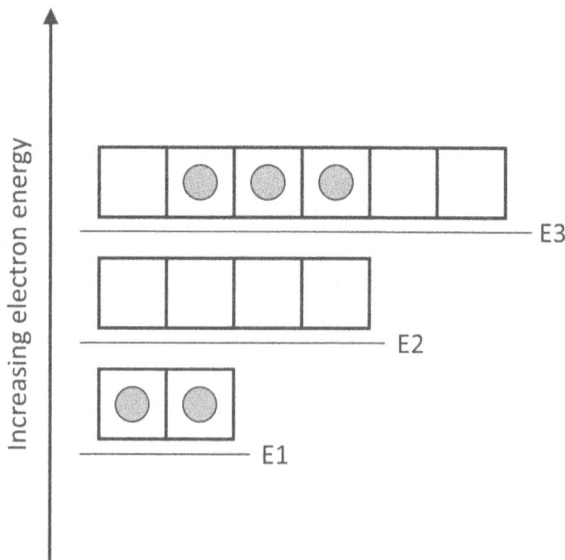

FIGURE 3.2 One of the ways of filling five electrons into three energy levels such that their total energy is 11 units.

rearranging electrons is based on the total number of available states and the electrons occupying a particular state. Thus, for configuration (2,0,3), the first energy level with two states is occupied by two electrons. Thus, there is only one way of rearranging electrons. However, the third energy level has six states occupied by three electrons. The number of ways of rearranging them is $6!/3!(6-3)! = 20$. In this way, we can obtain the total ways in which electrons can be rearranged for all the three allowed distributions. They are 20, 180, and 6 for distributions (2, 0, 3), (1, 2, 2), and (0, 4, 1) respectively. Thus, the total ways in which electrons can be arranged are 206 and if completely left to chance, we can conclude that electrons will most likely occupy the (1,2,2) distribution because of the number of ways in which they can arrange themselves in this configuration is the highest. Thus, given a completely random distribution of electrons such that both the number of electrons and the total energy of the system are conserved, there is a distribution that has a much higher probability of existence. This distribution is the Fermi-Dirac distribution given as:

$$F(E) = \frac{1}{1 + e^{\frac{E-E_F}{kT}}}$$

where k is the Boltzman constant, T is the temperature and E_F is the Fermi energy level, which gives the energy level with an electron occupation probability of 0.5. The Fermi-Dirac distribution tails off exponentially on either side of the Fermi energy, thus, for energy levels sufficiently below the Fermi energy, the electron occupation probability is 1, while for those sufficiently above, the electron occupation probability is 0.

In the case of silicon, the Fermi energy falls in a bandgap, i.e., there are no allowed states that have the occupation probability of 50%. Thus, when the allowed energy levels start on either side of Fermi energy, we have an almost completely filled band below the Fermi level, called the valence band, and an almost completely empty band above the Fermi level called the conduction band. The electrons in the conduction band are free to move in the lattice, but in the case of pure silicon (intrinsic silicon), there are not too many electrons there (one electron in conduction band for every 5 trillion silicon atoms in the lattice). Thus, at room temperature, pure silicon does not conduct electricity well. However, there are ways in which silicon can be made to conduct through specific pathways. This technique leads to the fabrication of most of the electronic systems around us.

3.3   DOPING SILICON

At room temperature, pure silicon, or "intrinsic" silicon is a poor conductor of electricity because of the lack of electrons in the conduction band. However, silicon can be made into a good conductor of electricity by adding some "impurities" into the silicon lattice. These impurity atoms contribute a carrier for electronic conduction through the silicon lattice. If this is done selectively, it can create pathways inside a silicon chip that are much more conductive than the rest of the material. The process of deliberately introducing impurities into silicon lattice to improve its conductivity is known as doping. Doping in silicon is done using elements from the group

Excess electron   Excess hole

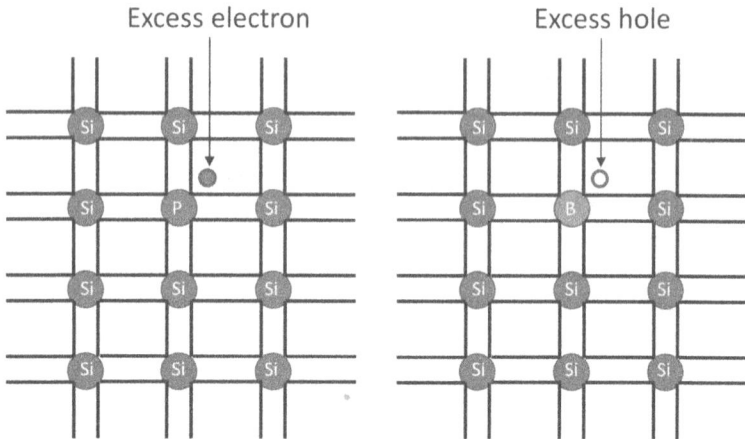

FIGURE 3.3   A simplified 2D representation of the silicon lattice with a donor (P) and acceptor impurity added, and the corresponding free charge carrier generated.

V(a) and III(a) of the Periodic Table. Elements from group V(a) are known as donor dopants (for example, Phosphorus, Arsenic), while those from III(a) are known as acceptor dopants (for example, Boron). When these elements are introduced in the lattice, they replace the silicon atom and occupy that lattice position. This is generally depicted using a simplified 2D lattice with silicon atoms forming bonds with the four neighboring silicon atoms (Figure 3.3). One should remember that this is merely a representation, the actual silicon lattice has silicon forming bonds with four silicon atoms at the four corners of a regular tetrahedron.

Donor dopants have a total of five electrons in their outermost band (valence electrons). When donor dopants are introduced into the lattice, they form four covalent bonds with the four neighboring silicon atoms, while the fifth electron is loosely bonded to the nucleus of the donor at an electronic energy state somewhere in the middle of the bandgap of silicon, as shown in Figure 3.4. The difference between the

Free electron

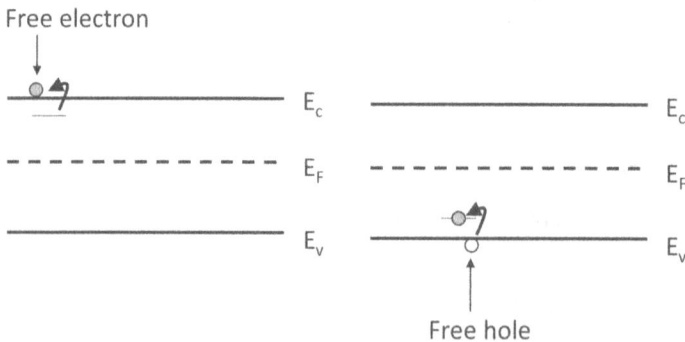

Free hole

FIGURE 3.4   The energy levels of the acceptor and donor dopant atoms are close to the valence and conduction bands respectively, allowing the charge carriers to be generated with the thermal excitation energy available at room temperature.

energy level of the excess electron and that of the conduction band is easily overcome by the thermal energy present in the lattice at room temperature. Thus, the electron enters the conduction band and can contribute to electron conduction in the lattice. Because the number of electrons in the conduction band depends on the number of dopant atoms, the conductivity of the silicon lattice can be precisely controlled by controlling the number of dopant atoms inserted into the lattice. For example, a phosphorus dopant concentration of 10^{14}/cm^3 leads to a resistivity of 44.5 Ohm-cm, while a concentration of 10^{20}/cm^3 leads to a resistivity of 0.85 mOhm-cm.

In the case of acceptor dopants, there are only three electrons in the outermost electronic band. Thus, these atoms form three covalent bonds with the neighboring silicon atoms. These atoms provide a low energy electronic state close to the valence band energy of the silicon lattice, as shown in Figure 3.4. This state can be occupied by an electron from the lattice at room temperature, creating a "hole" in the valence band. An electron from the valence band of a surrounding atom can "jump" into this hole, effectively making it appear as if the hole is traveling through the lattice. Thus, once created, a hole can appear to travel in different directions in the lattice because of subsequent electron jumps in the opposite direction. Because a hole is clearly a "lack of electron", it is said to carry a positive charge and can contribute to electrical conductivity inside the lattice, similar to an electron in the conduction band.

While the mechanism of conduction is different in the case of donor and acceptor dopants, the end result is that the silicon lattice becomes much more conductive. Dopant atoms are inserted into the silicon lattice through thermal diffusion (dopant diffusion), or by bombarding the lattice with charged ions of a particular element (ion implantation). In the case of dopant diffusion, the extent and depth of doping depend on the temperature, dopant element, and diffusion time. The doping process is generally carried out in two steps: deposition and drive-in diffusion. The deposition step is carried out in a quartz vacuum chamber where a dopant source is introduced to silicon wafers at high temperatures. The dopant source can be a gaseous, liquid, or solid depending on the dopant species and the process requirements. A gaseous source can be directly introduced in the chamber, while a solid source is generally placed in the form of wafers adjacent to the silicon wafers to be doped. In the case of a liquid source, a carrier gas is used to transport the vapors of the liquid to the doping chamber. Once the deposition step is complete, the dopant sources are removed, and the silicon wafers are subjected to a high thermal budget to "drive-in" the dopant atoms to the requisite depth and concentration. The process conditions used for these steps have to be modeled precisely to obtain a particular concentration, depth, and profile of doping.

Another technique used to dope silicon wafers is the ion implantation process. In this case, a stream of high-energy ions of a specific dopant species is bombarded on the silicon substrate. The ions are accelerated toward the silicon substrate using an electric field. The depth of doping depends on the ion energy, while the extent of doping depends on the ion current. These independent controls over depth and extent have made ion implantation a popular choice for the CMOS fabrication industry. Usually, the top surface of silicon lattice suffers some damage due to the incoming high-energy particles. The extent of damage depends on the mass of the dopant species and the total concentration of dopants in the substrate. This surface damage is generally repaired by annealing the wafer, i.e., subjecting the wafer to a high-temperature cycle. The energy associated with the thermal vibrations at high temperatures provides silicon and dopant atoms the activation energy needed to reform the broken lattice bonds.

3.4 SEMICONDUCTOR FABRICATION TECHNIQUES

3.4.1 THE SILICON WAFER

Silicon chips that are at the heart of all electronic systems, start their journey as a wafer of pure silicon material. These wafers are manufactured using state-of-the-art processes in ultra-clean factories, followed by thousands of carefully crafted steps to "print" circuits on them to make microprocessors, memories, transceivers, and so on. The ultra-pure silicon used in the fabrication of these components starts its journey as quartzite (an ore of silicon). Quartzite is silicon dioxide with trace amounts of iron and other contaminant oxides. High purity quartzite is reduced to silicon using coke at high temperatures. Generally, an excess of quartzite is maintained to limit the formation of silicon carbide. However, the resulting silicon is only around 98% pure and thus has to be purified before being made into silicon chips. The purification process is based on the distillation of trichlorosilane ($SiHCl_3$) and is known as the Siemen's process. Powdered silicon is reacted with hydrochloric acid (HCl) to form $SiHCl_3$, which has a low boiling point of 31.8°C. The $SiHCl_3$ vapor is distilled and is reduced to pure silicon using hydrogen gas. Several iterations of the Siemen's process lead to ultra-pure silicon, known as electronic-grade silicon. The purity of this silicon is approximately 99.999998% and is one of the purest substances known to man. However, the silicon thus obtained is in polycrystalline form, i.e., there are only small areas of ordered lattice structure. These areas are called grains and are separated by grain boundaries. For the electron distribution to be as theoretically predicted in the previous section, we require a crystalline silicon structure with all the atoms arranged in a periodic lattice arrangement. A crystalline silicon ingot is obtained from the electronic grade silicon using the Czochralski (CZ) process. This involves melting the silicon in a quartz crucible and recrystallizing it according to a seed crystal structure. The seed crystal is dipped into the silicon melt and pulled out of the melt, giving a crystalline ingot. The cylindrical ingot is then cut into individual silicon wafers. In case a silicon wafer with a particular dopant concentration is required, an appropriate amount of dopant element can be added to the melt.

The silicon wafers thus obtained have very high purity and crystalline integrity. However, because the silicon melt is inside a quartz crucible for a long time, some impurities from the crucible can diffuse into the melt, and subsequently into the silicon ingot. These impurities are generally atoms of oxygen or carbon, which produce an energy state inside the silicon bandgap. This mid-gap state can cause free electrons to fall and be trapped inside, causing loss of electrical conduction. Further, the presence of impurity atoms can also lead to dislocation defects in the lattice, or cause a vacancy defect (absence of a silicon atom from a lattice site). These defects give rise to more mid-gap states leading to further loss of carriers. Thus, in some cases, the pure polycrystalline silicon obtained from the Siemen's process is crystallized through a process called the float zone (FZ) process. In this process, an RF-based heating coil is passed over a vertically oriented polysilicon cylinder. A seed crystal is attached at one end of the cylinder. As the silicon melts and recrystallizes, the silicon atoms orient themselves according to the seed crystal, thus forming a single-crystal lattice. The RF heating coil is slowly moved away from the seed end toward the other end of the cylinder, melting and recrystallizing silicon along the path. At the end of the process, the complete polysilicon cylinder is recrystallized according to the seed crystal orientation. The advantage of the FZ process over the CZ process is that the silicon melt never comes into contact with any surface, thus greatly reducing the chances of contamination. However, because the melt under the RF heating coil has to support its own weight while being attached to the cylinder because of surface tension effects, there is a limitation on the diameter of the cylinder that can be crystallized using the FZ process.

3.4.2 DEPOSITION

The deposition of a thin film on a substrate is one of the processes that form the basis of the semiconductor manufacturing industry. In general, deposition entails covering a particular substrate with a uniform thin film of another material. Broadly speaking, thin-film deposition techniques can be divided into physical vapor deposition (PVD) and chemical vapor deposition (CVD) processes. In the case of PVD, the process of adsorption of the thin-film material on the substrate material happens while the thin-film material is in its final form. In the case of CVD, the precursors of the thin-film material are brought into the deposition chamber and are converted into their final form through chemical processes.

Physical vapor deposition techniques entail the use of the thin-film material itself as the source material. The general idea of PVD is the ejection of material from a target (same as the thin film to be deposited), transportation of the ejected material in a vacuum, and deposition of the material on a substrate in a thin-film form. Atoms can be ejected from the target by various means, such as resistive heating (thermal evaporation), electron beam heating (electron beam evaporation), ion bombardment (sputtering), or laser beam bombardment (laser ablation). In thermal evaporation, the target material is melted and evaporated under very high vacuum conditions, generally in the 10^{-9} torr range. The ultra-high vacuum ensures a low melting point for the target, reduced contamination of the thin film, and collision-free transport of evaporated material molecules. The linear inertial transport of material ensures that the deposition is extremely directional. Electron beam (or E-beam) evaporation is similar to thermal evaporation, except that the thin-film material is heated by bombarding it with a high-energy electron beam. In the case of sputtering, the target material is bombarded with argon ions to physically eject chunks of material from the target and deposit them on the substrate. While the target material does get heated in sputtering, it is not melted and evaporated, thus providing a means to deposit very high melting point materials such as tungsten (W). The stream of argon ions (Ar^+) is created by striking a plasma inside an argon-filled chamber using an RF power source and providing a DC electric field to accelerate the ions toward the target. The flow rate of argon, RF power, DC field strength, and the target material contribute to determining the deposition rate in sputtering. Laser ablation is similar to sputtering, except that it involves bombarding a target with a laser beam to eject material.

In the case of CVD, usually, gas-phase reactants are used to create the thin-film material in the deposition chamber through a chemical reaction. The reaction occurs due to thermal activation (temperature of the chamber) or because of plasma excitation. The reaction is designed such that it produces one solid and one or more gas-phase reaction products. The solid deposits on the substrate as a thin film, while the gas phase reaction products are ejected from the chamber through the vacuum pump. An example of CVD is the deposition of silicon thin film using silane (SiH_4) as a gaseous reactant. Silane decomposes into silicon and hydrogen at high temperatures (> 600°C). The silicon deposits on the substrate as a thin film, while hydrogen gas can be pumped out of the chamber. In case the temperature for thermal CVD is too high for a substrate or pre-existing thin films to handle, the gaseous reactants are decomposed using a plasma. This process is known as plasma-enhanced CVD or PECVD. An example of PECVD is the deposition of silicon dioxide using SiH_4 and N_2O as gaseous reactants. An important variation of CVD deposition is the atomic layer deposition (ALD) process. In this process, the gaseous reactants are introduced in the chamber one at a time, separated by an inert gas purge. The introduction of a gas in a chamber causes it to adsorb on all the surfaces of the chamber (including the substrate surface to be deposited). This is followed by purging the chamber with an inert gas such as nitrogen. Thus, the chamber is free of the reactant gas except for all the surfaces where the reactant was adsorbed. Now, another reactant gas is introduced in the chamber. Because the first reactant is only present on the surfaces, the chemical reaction only occurs there, and the reaction product is adsorbed by the surfaces in the form of a thin film. The chamber is purged once more, and the cycle is repeated. Thus, a single atomic layer is deposited in one cycle, giving precise control over the deposition rate. Further, because all the surfaces are uniformly covered, there is a high degree of conformality in ALD thin films.

3.4.3 LITHOGRAPHY

Lithography is a process used to pattern thin films deposited or grown on a substrate or to create a masking film on the substrate for subsequent processing. Lithography comes from the words *lithos* meaning "stone" and *graphy* meaning "to write". Thus, lithography literally means to write in stone. Originally, the process was used to transfer patterns onto a paper using an engraved mask coated with ink. The process is still used, with some modifications, to print currency notes using an engraved silver block. The process essentially entails the use of a highly accurate, complicated, and painstakingly made mask to define a pattern on substrates at an industrial scale. In the semiconductor manufacturing industry, the process is carried out with the help of light exposure.

Modern-day lithography equipment relies on the exposure of a specific thin film (called resist) to a specific medium (usually ultraviolet light) which causes chemical changes in the resist making it either more or less soluble in a particular solvent compared to the rest of the film. The exposure medium can be ultraviolet light, electron beam, proton beam, or X-ray light. Of these, ultraviolet light is the most commonly used medium for lithography, the process then known as photolithography.

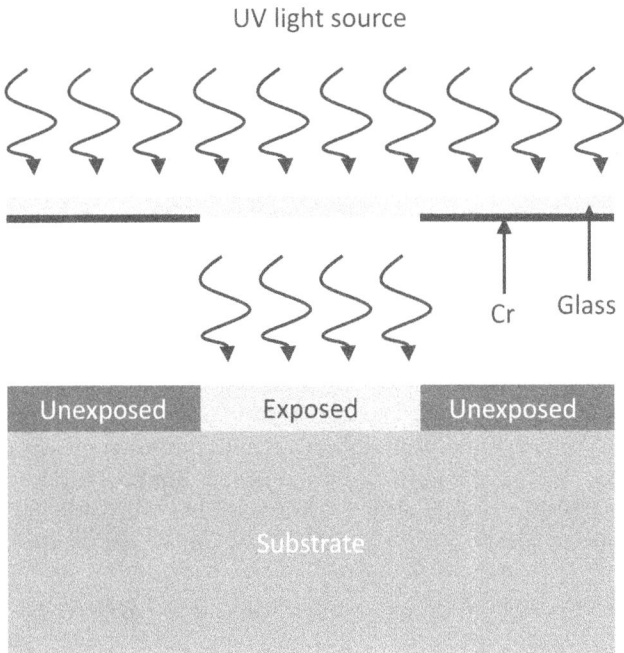

FIGURE 3.5 The process of photolithography involves the exposure of selected areas of a photosensitive thin film (photoresist) to UV light in order to alter its solubility compared to the unexposed parts of the film.

In principle, a photosensitive thin film (or photoresist) is deposited on the substrate and selectively exposed to ultraviolet radiation (as shown in Figure 3.5). The selective exposure is done using a mask that has a pattern with clear and dark areas. The substrate area covered by the clear gets exposed to the radiation, while the dark areas on the mask protect the substrate from radiation. The radiation causes chemical changes in the structure of the resist, which changes its solubility in a particular solvent, called the developer. A developer bath is used to remove the soluble resist while keeping the rest of the resist thin film in place. Thus, the desired pattern is transferred into the thin film which can now be used as a masking layer for further processing. Generally, photoresists are classified as positive or negative photoresists. Positive resists become soluble in the developer after exposure to UV radiation, while unexposed positive resist is insoluble. On the other hand, negative resists become insoluble on exposure to UV radiation.

An important consideration in lithography is the minimum feature that can be reliably obtained using the process. Because photolithography relies on UV radiation, the minimum feature is limited by the wavelength of the light used due to diffraction effects. To overcome this problem, the industry has progressively moved toward higher energy UV radiation as the exposure medium. However, as the wavelength is reduced, the complexity of the optics required to control and focus the beam increases, thus increasing the cost of the lithography equipment. Exposure media such as E-beam and X-rays have been experimented with because of their much lower wavelengths, and consequently much better resolution capabilities. However, E-beam lithography suffers from the problem of low throughput because it can expose a single point on the substrate at a time, as opposed to the flood exposure feature of UV radiation. In the case of X-rays, the problems include obtaining an X-ray source with a precise and stable wavelength, and fabrication of optics for manipulating X-ray light.

3.4.4 ETCHING

Etching is the selective removal of a thin film from certain locations on the substrate. The locations from where the film is to be removed are determined using a mask film. The mask can be photoresist patterned using lithography techniques or another thin film that was in turn patterned using lithography. Etching can be carried out using liquid etchants (called "wet" etching) or using plasma or gas phase etchants (called "dry" etching). In general, a given thin film can be etched using several process techniques and etchants. The selection of a particular process and etchant depends on many factors such as the substrate in question, the masking thin film, throughput required, cost constraints, and so on. To select a specific etchant, the process is characterized using parameters such as selectivity, etch rate, and isotropy.

Selectivity of a particular etchant with respect to a thin film is the ratio of the etch rates for the thin film compared to the target thin film. For example, let us assume we want to etch an amorphous silicon thin film deposited on a glass substrate. If we chose to mask this thin film using a photoresist, we need to find an etchant that has very high selectivity, i.e., it does not attack the photoresist (or in other words, has a very low etch rate for photoresist), while etching silicon with a reasonable rate. In this case, we can use XeF_2 gas to etch the silicon thin film. However, if we decide to use KOH as an etchant for silicon, then we cannot use photoresist thin film as a mask because KOH is not selective to it, i.e., has a high etch rate for both silicon and photoresist. In this case, we may have to put another thin film as a masking layer (known as a hard mask) before going for the etch. Thus, it is important to know the etchants available and their selectivities before planning a lithography and etching process.

The isotropy of an etch defines how directional an etch is with respect to the thin-film lattice (as shown in Figure 3.6). In case the etch progresses at the same rate in the vertical as well as the horizontal direction, the etch is said to be isotropic. Wet chemical etches and gas-based etches are examples of isotropic etches in most cases. A famous and notable exception is the etching of single crystal silicon using KOH bath, which progresses extremely slowly along the {111} plane of the silicon lattice. In case there is no horizontal etching (known as undercut), the etch is said to be perfectly anisotropic. Plasma-based etches are examples of anisotropic etches – their directionality emerging from the acceleration of ions in the plasma toward the substrate using a DC field inside the chamber. The degree of anisotropy of an etching process is defined as:

$$A = 1 - \frac{d_h}{d_v}$$

where d_h is the horizontal etch distance, while d_v is the vertical etch distance.

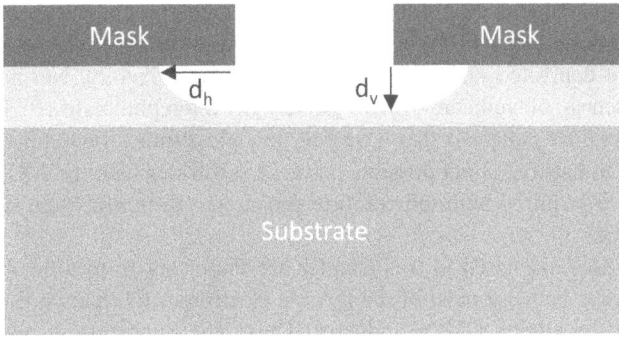

FIGURE 3.6   The isotropy of an etch defines how directional an etch is. In case the etch progresses at the same rate in the vertical as well as the horizontal direction, the etch is said to be completely isotropic.

3.5   SILICON TRANSISTORS

The processes of doping, deposition, lithography, and etching are carried out multiple times to obtain electronic devices called transistors, that can be interconnected to form complex digital circuits. Transistors are devices that can control the amount of current flow through a channel based on the applied voltage difference between its terminals. This is achieved by manipulating the number of charge carriers in the channel based on the applied electric field at the gate. This "field effect" at the metal-oxide-semiconductor structure is the reason why silicon transistors are commonly referred to as MOSFETs (metal oxide semiconductor field effect transistors). They generally consist of three terminals – source, drain, and gate, with a fourth contact to the substrate of the semiconductor. Silicon MOSFETs are typically fabricated using single crystal silicon, using the doping process to obtain the source, drain, and channel; while using the deposition, etching, and lithography processes to obtain the gate stack, which consists of an insulator (gate dielectric) between the semiconductor channel and the gate metal contact.

3.5.1   THIN-FILM TRANSISTORS

An important variation of the process for obtaining MOSFETs is the use of amorphous or polycrystalline materials as semiconductors to fabricate transistors. Such transistors are known as thin-film transistors (TFTs), because they can be directly fabricated using semiconductor thin films deposited using PVD or CVD processes, as opposed to the single crystal wafers required for MOSFET fabrication. This opens the possibility of creating transistors on any surface where a semiconductor thin film can be deposited, and subsequent processing can be done. However, the TFTs, formed using amorphous and polycrystalline semiconductors, do not perform as well as the MOSFETs because of the scattering of charge carriers at the grain boundaries. Further, the deposited semiconductors typically have more defect density compared to their single crystal versions, leading to trap states in the band structure. Nevertheless, TFTs form an integral part of the electronics industry, particularly in the digital display application. Indeed, the first demonstrations of flexible electronic systems were done using TFTs as their basis because the complete transistor structure could be deposited and fabricated on a flexible substrate. Normally, flexible substrates such as polyethene (PE) or polyethylene terephthalate (PET) are based on organic polymer materials that have very low glass transition temperature. Thus, formation of transistors is not possible on these substrates with the MOSFET structure because it requires high-temperature processing to produce the single crystal semiconductor.

The processes discussed in this chapter are important from a flexible electronics point-of-view because most of the flexible electronics fabrication uses processes developed in the silicon CMOS industry. Thus, it is important to understand the basic terminology associated with these processes before we can discuss the materials, integration strategies, and applications of flexible electronics. In the following chapters, we will consider various material systems that can be used to form transistors and other electronic structures on flexible substrates.

EXERCISES

3.1. Consider a particle of mass 1 gram in an infinite 1D potential well of 1 cm length. Obtain an expression for the allowed energy levels for the particle, in terms of the quantum number, n. Assume that the particle has 1 Joule energy. Which energy level will it occupy? Calculate the difference in energy between this and the next higher energy allowed level.

3.2. The bandgap of silicon is 1.1 eV and the Fermi level is midway between the valence and conduction band. What is the probability of occupation of the valence and conduction band edge at room temperature (300 K)?

3.3. Describe the process steps involved in obtaining silicon crystal silicon ingots from silicon ore quartzite.

3.4. What are the advantages and disadvantages of E-beam lithography over conventional UV photolithography?

3.5. Assume that a wet etch has a time-dependent vertical etch rate of $r_v(t) = \frac{k}{t+\tau}$ nm/s, where τ and k are constants with units of time and distance, respectively. What is the time taken to etch through a thin film of thickness $2k$? If the anisotropy of the etch has to follow the equation $A = e^{-t/\tau}$, what should be the horizontal etch rate as a function of time?

Part II

Materials

4 Flexible Silicon

4.1 THE STORY OF SILICON

The rise of Complementary Metal Oxide Semiconductor (CMOS) technology has led to a digital revolution. It all started with the invention of the transistor by John Bardeen, Walter Brattain, and William Shockley in 1947 at Bell Labs. The three inventors were awarded the Nobel prize in Physics in 1956, "for their researches on semiconductors and their discovery of the transistor effect". This was followed by the demonstration of the first integrated circuit by Jack Kilby in 1958. While these advancements provided the platform for the silicon age, it would have been impossible to foresee at that time, the paradigm shift they will bring in the world. What followed has been a succession of technological breakthroughs in fabrication leading to some of the most advanced manufacturing practices known to humans. By 1965, it was clear (at least to Gordon Moore) that the rate of advancement in the semiconductor manufacturing industry is going to be exponential. He famously predicted that the number of transistors in an integrated circuit will double every two years. This law, which later came to be known as the Moore's law, has held true for close to four decades. Indeed, an exponential advancement over the course of the last four decades has led the semiconductor fabrication industry to be inarguably the most advanced manufacturing infrastructure developed by mankind.

The success of the CMOS industry can be gauged from the fact that every other industry in the world now relies on electronics for everything including sourcing, operations, distribution, marketing, and sales. The world is now moving toward a future with "smart" devices at the heart of every activity. Even traditional industries that have been unchanged for thousands of years such as agriculture, animal husbandry, textile, and mining have been completely transformed because of the tools and technology made possible by the silicon CMOS industry. The very fact that electronics can influence and transform an industry, leads to exponential advancement in the semiconductor industry itself. Better electronics leads to better performing and faster processors, which leads to more computational power in the hands of designers and more precision in the hands of the manufacturers. This leads to better quality of materials, better process techniques, and better connectivity for the dissemination of ideas. This leads to even better electronics, thus forming a positive feedback loop leading to an exponential explosion in computational capacity. The key mechanism for the development of faster transistors has been dimensional scaling, i.e., reduction in the size of the transistor in every subsequent generation of integrated circuits. This leads to more transistors being incorporated in a given area, which increases functionality. Dimensional scaling has reached such unimaginable proportions that a microprocessor of area 2 cm x 2 cm can easily have a few billion transistors. Serendipitously, dimensional scaling also increases the speed of individual transistors without affecting power density. Hence, with every subsequent generation of electronics, we had integrated circuits that are more compact, and can perform more

DOI: 10.1201/9781003010715-6

functions at a faster rate. Further, the success of the semiconductor manufacturing industry and the use of electronics in every conceivable activity has led to a massive scale of operation which has significantly reduced the cost per transistor. The scale of the industry is such that in 2014 the number of transistors produced in the World reached 8 trillion every *second*!

Dimensional scaling of transistors drove the advancement of the semiconductor industry from the late 1960s to the 1990s. However, early in the twenty-first century, it was realized that we cannot rely on dimensional scaling alone for the advancement of electronics. We reached a saturation point in the dimensional scaling capabilities of the industry because of the cost associated with fabricating such tiny structures, the reliability of the fabrication process, and most importantly because of the limitations of physics itself. The issue is that the channel of the transistor has become so small (of the order of a few nanometers), because of decades of dimensional scaling, that electrons now have a finite probability of tunneling through the channel and through the gate dielectric. This causes unwanted leakage current and leads to unsustainably large power dissipation densities. This led to innovations in the structure of the transistor itself (introduction of FinFETs) and a change in the materials used to fabricate electronics (high-κ dielectrics and metal gates). While these innovations extended the exponential growth in the performance of electronic devices, we cannot expect the trend to continue forever.

The exponential increase in the performance of electronics has led to an interesting consumer behavior pattern as well. We, as consumers, *assume* that the next generation of electronics will be much better and faster than the previous. We have become accustomed to the fact that every year, a new generation of smartphones will be launched which will be faster, slimmer, lighter, smarter, and cheaper. Indeed, that should not necessarily be true. Concepts like dimensional scaling, FinFET architecture, and high-κ/metal gate have kept the performance of silicon chips close to an exponential growth trajectory. However, going forward, it might become unrealistic to expect the microprocessors to keep becoming faster and smaller. This seems odd because we have grown up in the era of exponential advancement of the silicon industry. This is not to say that the silicon industry will come to an end. In fact, the silicon manufacturing industry is now reaching a maturity that can be associated with, say the cement or steel industry. We know a significant amount about steel – its various types, effects of annealing on the microstructure, effects of various contaminants, and so on – however, nobody expects the tensile strength of steel to double every two years. The industry still quietly produces millions of tons of steel, we still use it in all walks of life, and it still employs millions of people. This seems to be the future of the semiconductor manufacturing industry as well.

With the advancement in performance saturating, the electronic industry may look to enhance other aspects of user experience for the next generation of gadgets. This includes physical flexibility and stretchability – giving electronic devices the freedom of shape so that they can be rolled up or folded after use. This will lead to wearable, implantable, or large-area electronics with wide-ranging applications such as displays, energy harvesters, sensor networks, and so on. However, these "free form" electronic devices will still have to perform at the same, if not more, level as that of the previous generation of devices. Further, we will expect these flexible and stretchable devices to last as long, if not more, as the conventional devices. Thus, the challenge is to produce flexible and stretchable electronics while maintaining the performance and reliability of conventional electronic devices. The development of a generation of such free-form electronics can lead to an information revolution because each individual, thing, animal, and plant on the planet could potentially be covered with sensors. An example is the development of BlueFinn by mmh Labs, which uses flexible sensor arrays mounted onto marine animals to track their movement. The development of free-form electronics is important for a world where electronic systems are deeply integrated with the biosphere. This is because intelligent systems made by humans (such as smartphones and computers) are very different from natural intelligent systems (humans and animals). While man-made electronic systems are rigid, rectilinear, clean, and dry; the brains that make these devices are soft, curvilinear, wet, and squishy. All naturally occurring systems – including all humans, animals, birds, and plants – have curvilinear form, flexibility in movement, and stretchability in tissue. This makes the integration of rectilinear and rigid electronic systems with the natural world challenging. Thus, the next breakthrough in electronic systems should be the development of flexible and stretchable devices that can conform to natural systems.

4.2 SILICON SUBSTRATES

Although the discovery of the transistor action and the first demonstration of integrated circuits took place using germanium as the semiconductor, silicon is the most widely used semiconductor for the state-of-the-art CMOS fabrication industry. In Chapter 3, we saw the development of pristine silicon wafers from the naturally occurring ore quartzite. The ore is reduced to obtain silicon, which is then purified through distillation and then crystallized through the Czochralski or float zone processes. The silicon wafers thus fabricated have high purity and incredible crystal integrity. However, there are many more unique qualities that have made silicon the workhorse of the semiconductor fabrication industry. In this section, we will understand the qualities and characteristics of the silicon lattice in detail and explore the reasons why silicon is so widely used for electronics manufacturing.

4.2.1 CRYSTALLOGRAPHY

Silicon forms a diamond lattice structure with each silicon atom at the center of a regular tetrahedron bonding with four silicon atoms in the corners. Crystal structures are generally characterized by a unit cell that can be repeated over 3D space to form the complete crystal structure. In the case of silicon, the unit cell is cubic with a side length of 0.543 nm. The structure of the unit cell is shown in Figure 4.1. The structure of the silicon crystal is also known as the diamond crystal structure because carbon also forms a similar lattice in diamonds. The number of atoms along a particular plane or direction in a crystal can differ leading to directionality in the physical properties of a crystal. For example, the tensile strength of a single-crystal material is dependent on the direction in which the tension is applied. Hence, it is important to define specific directions and planes in a crystal structure to compare relative properties. This is done through Miller indices, which are a set of numbers denoting

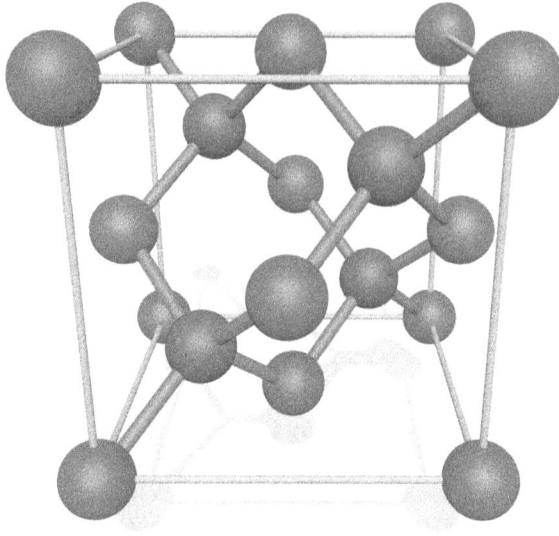

FIGURE 4.1   The silicon lattice unit cell is face-centered cubic (FCC), with lattice constant of 0.543 nm and four silicon atoms internal to the unit cell.

a specific plane, set of planes, or a direction in a crystal lattice. In the case of silicon, because the unit cell is a cube, the primitive directions are chosen as the three orthogonal edges of the cube intersecting in a point and the primitive vectors are of equal length. To obtain the Miller indices of a given plane, the three intercepts of the plane, in terms of the lattice vector, along these directions are noted (Figure 4.2).

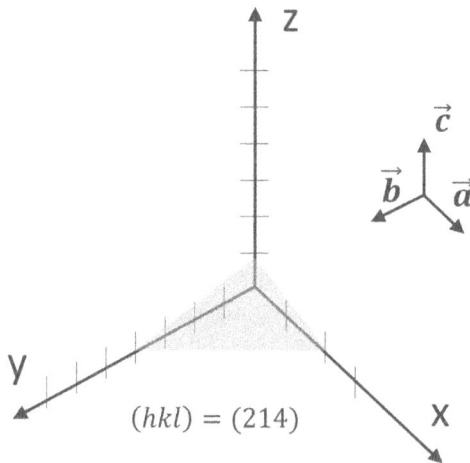

$(hkl) = (214)$

FIGURE 4.2   The process for obtaining the Miller indices for any plane includes finding the intercepts with lattice vectors (a, b, c), inverting them and normalizing. In the given example, the intercepts are x = 2, y = 4, z = 1. Inverting them gives 1/2, 1/4, 1 respectively. They are normalized by multiplying by 4 to obtain 2, 4, and 1.

The intercepts are inverted and cleared of fractions by multiplying with an appropriate integer. The resulting numbers are generally written together in brackets, such as (hkl). There are several notations related to the Miller indices describing different aspects of a crystal. The indices in curly brackets, such as $\{hkl\}$, represent the family of planes parallel to the plane (hkl) and equivalent to it by crystal symmetry. The indices in square brackets, such as $[hkl]$, represent the direction normal to the place (hkl), while indices in angle brackets, such as $< hkl >$, represent the set of all directions equivalent to $[hkl]$ through crystal symmetry.

4.2.2 Charge Transport

In the case of silicon, the most commonly used wafers for CMOS fabrication are the (100) wafers (the silicon lattice is oriented such that the [100] direction is normal to the surface of the wafer). Different crystal orientations have different atomic distributions, which leads to a difference in key properties such as surface defect density, electron mobility, reactivity and so on. The orientation of the crystal with respect to the wafer surface is of primary importance in CMOS fabrication because the Metal Oxide Semiconductor (MOS) structure depends on the properties of the semiconductor/oxide interface for its characteristics. One of these properties is the effective mass of a carrier which indicates the ease with which a carrier can accelerate in the presence of an electric field (carrier mobility). The difference in effective mass arises due to the difference in band structures along the different directions in the crystal, which in turn, arise because the distribution of the atoms on the surface varies with the crystal orientation. This changes the distribution of the potential wells as experienced by the charge carriers, thus, the local solution to the Schrodinger equation, and hence, the band structure also varies. For silicon, the electron mobility along the (100) direction is known to be more than that along the (110) or (111) direction. However, it is interesting to note that hole mobility is better along the (110) direction. Thus, to determine which direction should be used for electronics fabrication, we have to consider another important parameter – the defect density at the semiconductor/oxide interface.

4.2.3 Semiconductor/Oxide Interface

The MOS structure consists of a semiconductor channel and metal gate separated by a dielectric oxide. It resembles a capacitance structure with the difference that one of the electrodes of the capacitor is a semiconductor. When an electric field is applied across the gate dielectric (through charges on the metal gate), the conduction of the semiconductor channel changes allowing the transistor to be "switched on". Because the current conduction takes place along the semiconductor surface very close to the oxide layer, the defect density at the semiconductor/oxide interface places a major role in determining the quality of the transistor. The presence of defects at the interface causes charge carriers to get trapped, reducing the available charge carrier density. This leads to a lower current for the same applied field, in other words, higher resistance. Further, roughness at the interface surface can lead to carriers being scattered causing further reduction in current. Thus, a smooth defect-free semiconductor oxide interface is important for efficient charge transport.

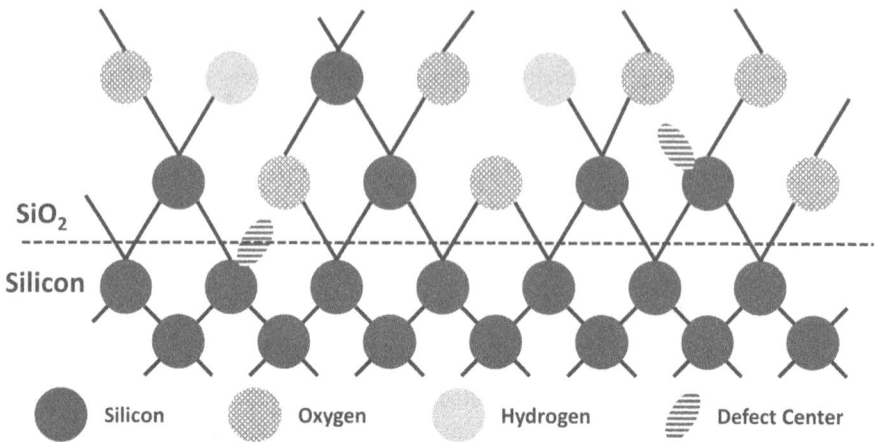

FIGURE 4.3 The silicon/silicon oxide interface has very low defect density because of the bonding of the surface silicon atoms with oxygen. It is further reduced by annealing with hydrogen that provides passivation to the remaining defect centers.

Silicon forms four covalent bonds with its closest neighbors to make the crystal lattice. However, this is not true for silicon atoms close to or at the surface of the wafer, because there are not enough closest neighbors for these atoms to form four bonds. This gives rise to unshared electrons because of the missing covalent bond which leads to "dangling bonds" at the surface. When a semiconductor/oxide interface is formed, say by annealing silicon wafer in the presence of oxygen, these dangling bonds are satisfied through covalent bonding with oxygen, as shown in Figure 4.3. This reduces the defect density at the semiconductor oxide surface, resulting in better MOS transistor performance. The formation of silicon dioxide, however, does not fully eliminate the defect centers. Among all low index silicon wafers, i.e., (100), (110), and (111), the defect density in silicon (100) wafers has been found to be the lowest. This is also the case when other high-κ oxides such as aluminum oxide or hafnium oxide are grown on the silicon surface. Hence, silicon (100) wafers are preferred as the workhorse of the CMOS industry. As a side note, a further reduction in the defect density is achieved by annealing the wafer after oxide growth in the presence of hydrogen. Hydrogen atoms form covalent bonds with the unshared electrons left after the oxide layer formation, thus, passivating them. In energy band terms, bonding with hydrogen essentially removes the energy state of these defects away from the silicon bandgap, thus, the defects are not able to behave as charge traps.

4.3 FLEXIBLE MONOCRYSTALLINE SILICON

Because silicon wafers are the de-facto standard in electronics production, it follows logically that they should be pursued for flexible electronics applications. However, a major issue with silicon is that it is rigid and brittle, i.e., its flexural rigidity is high. However, we know from Chapter 2, that every material can be bent to the desired bending radius, given the thickness of the material in the direction normal to the axis of bending is sufficiently reduced. The process of making silicon wafers flexible originates from this concept. These processes essentially reduce the thickness of the silicon wafer to obtain a certain level of flexibility in the electronics thus fabricated. It should be noted that the methods listed here are in reference to single-crystal silicon wafers. Because polycrystalline thin films can be deposited on any surface, including a flexible one, the process of obtaining polycrystalline flexible silicon is relatively straightforward. Whereas in the case of single-crystal silicon, the Czochralski and float zone processes (discussed in Chapter 3) give rise to ingots that are sliced into wafers. Thus, special process steps need to be followed to obtain flexible silicon pieces from these wafers.

It is fortuitous that the structure of the MOS-based transistors allows for silicon wafers to be thinned down because all the electronic activity is limited to the top few microns of a silicon wafer. The remainder of the wafer is only present to make it mechanically stable during production and packaging. In theory, the top surface of the wafer can be decoupled from the rest of the wafer and we can still obtain the same circuit performance. In fact, the circuit performance may improve slightly because of the reduction in leakage current to the bulk wafer. In the past decade, many processes have been introduced to thin-down single-crystal silicon wafers. These processes can be categorized as device-last approach and device-first approach. In the device-last approach, the silicon wafer is first thinned down, and transferred to a destination substrate before fabrication of electronic devices, thus, the devices are made last. On the other hand, in the case of the device-first approach, electronic devices are first fabricated on the silicon wafer using standard CMOS processes, following which the silicon wafer is thinned-down and flexed.

4.3.1 DEVICE-LAST APPROACH

The device-last approach is one of the most widely studied approaches in the flexible silicon field. The most common technique in the device-last approach is to transfer small pieces of single-crystal silicon (sometimes referred to as silicon microribbons or nanoribbons) from a silicon-on-insulator (SOI) substrate. The SOI substrate consists of a thin layer of single-crystal silicon on a layer of silicon dioxide, attached to a handle silicon substrate. The thickness of the top silicon varies from a few nanometers (3 nm) in the case of ultra-thin body SOI (UTBSOI) to 30 μm, depending on the fabrication process and the intended application. Electronic devices are fabricated using the top silicon layer. The oxide layer (also called the buried oxide layer or the BOX layer) shields the leakage of current from the electronic circuit to the handle substrate, thus increasing the efficiency of the system. The fabrication of a single-crystal silicon thin film is a very difficult process, hence, fabrication of SOI substrates is a cumbersome (and hence expensive) process. This leads to the overall substrates being many times more expensive than the regular (100) silicon substrates of the same size. One of the ways of fabricating SOI substrates is known as the SIMOX (separation by implanted oxygen) process. This was one of the first processes used to obtain SOI substrates. The idea is to implant oxygen atoms into a regular silicon wafer at the desired depth, followed by annealing to form bonds between silicon and oxygen leading to a buried silicon oxide layer. The depth of the buried oxide can be controlled using the implantation energy of oxygen ions, and the thickness of the layer can be controlled using the amount of oxygen implanted (or the implantation dose). As discussed in Chapter 3, when a substrate is subjected to implantation, it suffers surface damage due to the impact of higher energy ions. The damage from impact is generally higher for more massive and more energetic ions, and for a large dose of ions. In the case of the SIMOX process, because oxygen ions are heavy and a large dose of oxygen is needed to make SiO_2, the damage suffered by the top silicon layer is extensive. Annealing does help reduce the extent of damage, but the silicon top surface is not pristine after the SIMOX process. Because this is also the surface to be used for device formation, it is critical to recover this damage. One of the ways in which this can be done is using the epitaxy process. It is a specialized CVD process in which the deposition is carried out to form a single-crystal layer following the template of a seed layer. An epitaxial layer of silicon is grown using the damaged silicon as a template to hide the vacancy and other defects at the surface of the wafer. However, the defect density of the top surface is still higher than that of bulk silicon (100) wafers.

Another process that is commonly used to fabricate SOI substrates is called the Smart Cut process. In this process, hydrogen ions are implanted into an oxidized silicon wafer. The wafer is then bonded with another silicon substrate. When this structure is annealed, the hydrogen ions congregate to form microbubbles that cause the silicon substrate to split. The bonded silicon substrate then has an oxide layer and a silicon layer attached to it. The top layer is polished to create a pristine surface for electronic device formation. The thickness of the top silicon is again controlled using the implantation energy, while the thickness of the buried oxide is controlled through the oxidation of the initial silicon wafer. A key advantage of this process is the use of hydrogen ion implantation instead of oxygen ions. Because of the small size and mass of hydrogen ions, the lattice damage due to implantation is significantly lower in the Smart Cut process compared to the SIMOX process. Further, the surface exposed to the implantation, where the majority of lattice damage occurs, is the silicon oxide surface (that ends up being the lower silicon-oxide interface in the final SOI substrate). Thus, the top silicon surface of the final SOI substrate is of much better quality which can be further improved with epitaxy. However, this process makes the resultant SOI substrate more expensive because of the use of two silicon wafers to fabricate one SOI substrate. Another process involving the use of two wafers is the bond-and-etch-back SOI (BESOI) process. In this process, two silicon wafers (one of them oxidized) are bonded, followed by the back etching of one of the substrates. The etch-back and polish are used to control the thickness of the final top silicon, while the thickness of the buried oxide is controlled through oxidation of the initial silicon wafer. An advantage of this process is that it avoids the implantation process altogether, thus reducing the damage to the silicon surface. However, the etch-back is hard to control precisely over the large silicon surface, which can lead to an inconsistent thickness of the top silicon layer. The SIMOX and Smart Cut processes have been represented schematically in Figure 4.4.

The use of SOI substrates for the device-last approach has been widely successful because of the selective isotropic etching of silicon oxide using hydrofluoric acid

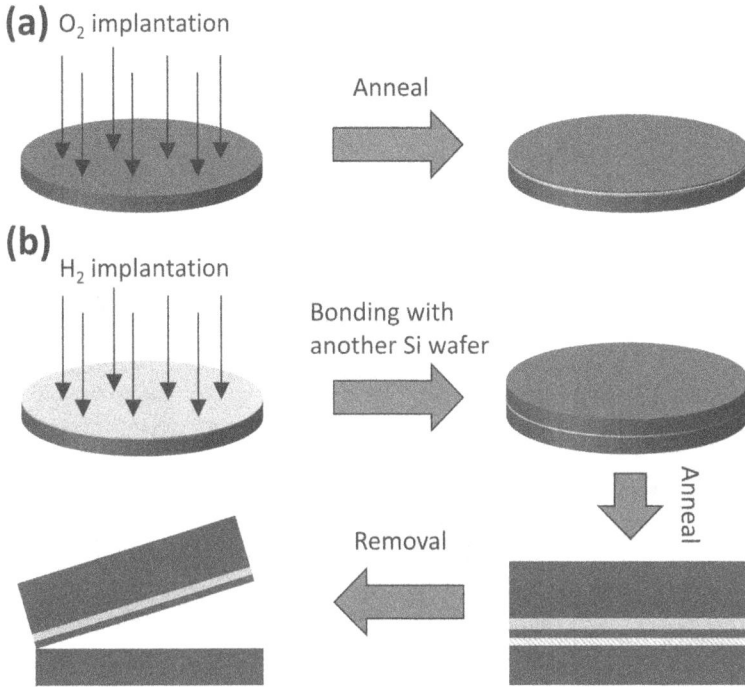

FIGURE 4.4 (a) The SIMOX involves the implantation of the silicon substrate with oxygen followed by annealing to obtain the SOI substrate. (b) The Smart Cut process starts with implantation of hydrogen in an oxidized substrate, followed by bonding with another silicon wafer, and annealing to tear away the SOI substrate.

(HF). The top silicon layer is patterned using lithography and then etched up to the buried oxide layer. This silicon etch is generally performed using directional reactive ion etching (RIE). The exposed BOX layer is then etched using an HF bath. Because this etch is completely isotropic, it removes the unexposed oxide from below the top silicon layer, causing the top silicon layer to become free-standing. Once the top silicon is unattached, or weakly attached to the buried oxide, it is transferred to a destination substrate. The transfer process usually involves a flat "stamp" typically made using a polymer such as poly-dimethylsiloxane (PDMS). The PDMS stamp is pressed against the weakly attached silicon after the oxide etch. This causes the silicon pieces to adhere to the stamp due to van der Waal's interactions. The stamp is then used to transfer the silicon onto destination substrates using various techniques such as difference in adhesivity, heat treatment, UV curing, and so on. This process has been reported many times in literature and is generally referred to as the "transfer printing" process. Figure 4.5 shows the illustration of the transfer printing process used to transfer silicon microribbons from SOI to a flexible substrate.

There are some advantages of using this process. The transfer process can be performed at room temperature and pressure using inexpensive equipment and materials. Also, because the silicon pieces are bonded to the destination substrates early in the circuit fabrication process, the integration of silicon-based devices with other

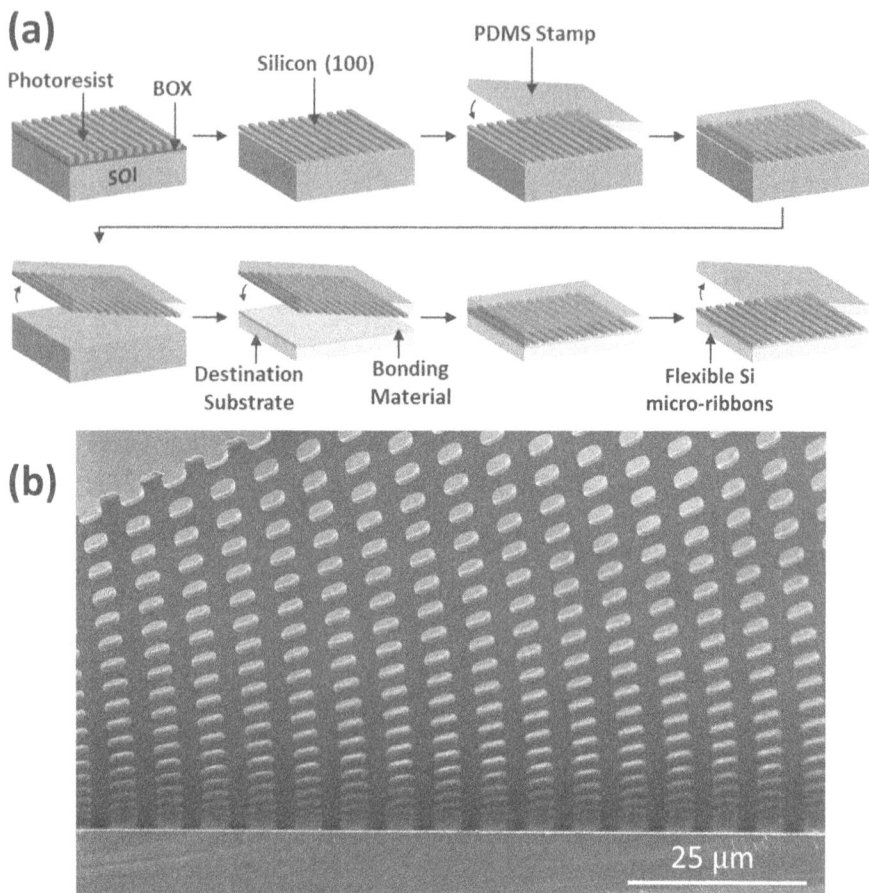

FIGURE 4.5 (a) The transfer printing process involves the etching of SOI substrates up to the BOX, followed by isotropic etching using HF and removal of silicon microribbons using PDMS stamps. (b) The scanning electron microscopy (SEM) image of the flexible silicon membrane after release from SOI substrate. (Adapted with permission from [2], © John Wiley and Sons, 2016.)

types of electronics (like organic electronics) is easier using this technique. The process is very versatile and can be used to transfer silicon pieces of any shape and size – the key limitation to size being the successful etching of buried oxide layer. This limitation arises from the fact that the buried oxide thickness is very small (few tens of nanometers) compared to the typical size of the silicon piece to be transferred (few tens of micrometers). Thus, reactant species such as the hydrogen and fluoride ions need to travel across the silicon piece to be undercut to reach the last remaining oxide support while the reaction products need to be ejected out in the other direction. This causes a gradient in the concentration of reaction species causing a slowing of the etching rate as the etch progresses. Thus, reaction time increases exponentially with the size of silicon pieces to be released. Further, because the edges of the silicon pieces are unsupported during the reaction, a section of the piece, if sufficiently large, can sag and stick to the silicon handle substrate. This may cause the transport of reaction species to completely stop, thus putting a limit on the size of the silicon piece that can be successfully undercut using lateral etching of SOI substrates. Generally, silicon pieces of a few hundreds of microns across have been reported in the literature using the transfer printing process. Thus, the SOI substrate is patterned into microribbon structures before being subjected to the HF etch. This increases the cost of the process because a percentage of the top silicon surface is lost to etching. Another disadvantage of the process is that there are several steps remaining in the process of converting the silicon pieces into working MOS devices. After the transfer printing, these steps must be completed while the silicon pieces are attached to a flexible substrate. This limits the temperature and overall thermal budget the follow-up processes can afford which might lead to a sub-optimal final device. Because of this limitation, an interesting variation of this approach involves the completion of some process steps before the silicon piece is transferred to the destination substrate. These generally include the doping and annealing steps because they require a very high temperature, and polymeric flexible substrates are not able to withstand it. Another challenge in processing the transferred flexible silicon pieces is that most semiconductor fabrication tools are designed for rigid substrates and cannot be used directly on flexible substrates without rigid support. This makes the processing more complicated and time-consuming.

There are several variations to the transfer printing process. In some cases, the intermediate stamp is replaced with the destination substrate itself, so that a single step transfer can be done. Further, the entire top layer of silicon can be detached from the handle SOI substrate using an HF bath. However, some unique adjustments are made to solve the mass transport problem encountered in the large distance lateral etch process. The top layer needs to be constantly pulled away from the handle substrate in order to make sure it does not sag onto the handle substrate thus closing the gap for further lateral etching. This can be done by attaching the top layer to the flexible destination substrate which is attached to a rolling cylinder. With time, the top layer is pulled away by the rolling cylinder, thus providing space for HF to continue etching the buried oxide. The same effect can be achieved by suspending a weight attached to the end of the top layer. As the buried oxide is etching and the top layer is detached, the weight pulls it further away from the handle substrate. Another technique reported in the literature utilizes the surface tension of the HF bath to push the top layer away from the substrate. These methods ensure that the lateral etch of the buried oxide continues and can be used to peel off large area flexible silicon pieces from SOI substrates.

Apart from the transfer printing process, another alternative is to produce thin and flexible silicon wafers and go for device fabrication on these. This is considered a device-last approach because the devices are fabricated on the silicon wafer after it is flexed. Many wafer manufacturers offer flexible, or ultra-thinned-down versions of silicon wafers. However, because these wafers are flexible, they have to be processed using specialized tools or have to be temporarily attached to a rigid backing, just as in the case of the transfer printed silicon pieces. An essential difference between the two processes is that silicon can withstand much higher temperatures and thermal budgets compared to polymeric substrates that are used in transfer printing, thus, theoretically, processes involving high temperatures should be possible with flexible silicon substrates. However, thermal cycling, particularly with many thin-film layers of different materials, can cause the flexible silicon wafers to warp. In fact, wafer bowing due to thermal cycling is a common problem encountered even in normal thickness rigid silicon wafers (the bowing is generally microscopic but can lead to process variations). In the case of flexible silicon wafers, the warping is exacerbated due to the low thickness of the substrate leading to low flexural rigidity.

4.3.2 Device-First Approach

The device-first approach refers to the flexing of silicon chips or wafers after the fabrication of the complete circuit. The key significance of this approach is that the circuit has already been completely fabricated using standard CMOS process technologies. This removes the constraints on the processes and tools that need to be used after flexing the substrate, thus enabling a truly state-of-the-art flexible silicon system. We can categorize the device-first approach into two methods, the top-down method and the bottom-up method. In the top-down method, the top silicon layer, containing the complete circuit, is removed while keeping the remaining handling substrate intact. In the case of the bottom-up process, the silicon wafer is removed starting at the bottom all the way up to the desired thickness, thus, eliminating the handling substrate in the process.

4.3.2.1 Top-Down Method

The top-down method involves the removal of the top silicon after the circuits have been completely fabricated. It should be noted that state-of-the-art circuits involve multiple layers of metal interconnects. Similarly, there are many layers of doping in the silicon substrate to fabricate a variety of device types (such as n-well, deep p-well, etc.). These doped regions along with the junctions they form are critical for the functioning of the circuit. Thus, the processes used to remove the silicon top layer should be designed in such a way that these layers of metal interconnects or doped regions are not damaged. Several processes have been reported in the literature for the top-down method. In this section, we will go through the processes along with their advantages and disadvantages.

One top-down method reported for fabrication of single-crystal silicon (100) based flexible circuits is the "trench-protect-etch-release" (TPER) process. As the same suggests, in this process, the silicon wafer is trenched, protected, etched, and then released. The process starts with patterning and etching deep trenches in the silicon substrate. This is done using a technique called Deep Reactive Ion Etching (DRIE), which is a variant of the RIE process discussed in Chapter 3. In this process, the wafer is maintained at -20°C, using liquid nitrogen and the etch is performed in cycles. Each cycle consists of directionally etching the silicon using SF_6, followed by passivation of all the surfaces using C_4F_8. In the next cycle, the passivation is removed by the directional SF_6 etch along with some more silicon, without damaging the passivation along the sidewall. In this way the trench grows deeper, passivated along the sidewalls, to increase the anisotropy of the etch. Thus, a very deep and high aspect ratio trench can be formed using DRIE. Once a set of trenches is dug into silicon, the wafer is subjected to atomic layer deposition (ALD) of a passivation layer to protect the sidewalls from the subsequent isotropic etching. In the case of the silicon wafer, a passivation layer of aluminum oxide (Al_2O_3) is used. The ALD process is ideal for this because it offers a conformal deposition of the passivation layer, thus completely covering the sidewalls. This step is followed by a directional etch (in the vertical direction) of the aluminum oxide layer to remove it from the bottom of the trench. This is necessary to expose the silicon at the bottom of the trench for the isotropic etch. Following this step, the only place with exposed silicon on the entire wafer is the horizontal surface at the bottom of the trenches. Finally, the wafer is subjected to isotropic etching of silicon using XeF_2 gas. The reactant and reaction products in this reaction are all gases, thus enabling a completely isotropic process. The etching itself is done in cycles as follows: introduction of XeF_2 in the chamber, wait time for the reaction to take place, and pumping out of the reaction products. The etching progresses in all possible directions starting at the bottom of the trenches. This creates a growing "bubble" of etched-out silicon deep inside the silicon wafer. The three-step etch cycle is repeated multiple times until the bubbles from adjacent trenches meet and the complete top layer of silicon is released from the substrate below. The process schematic, the resulting flexible silicon pieces, and their scanning electron microscopy (SEM) images are shown in Figure 4.6. The process can also be used for releasing flexible top surfaces of other substrates such as gallium arsenide (GaAs). In these cases, the isotropic etchants that etch and undercut the substrates may be different, which leads to a possible change in the protective sidewall thin-films deposited using ALD. However, the core philosophy of "trench-protect-etch-release" is preserved.

The TPER process has several advantages:

1. It is a device-first process, thus high-performance state-of-the-art circuits can be fabricated before they are flexed using this process.
2. The steps involved in the process do not subject the wafer to high temperatures or high thermal budgets. Because circuits have already been fabricated on the wafer, it likely consists of millions of transistors, p-n junctions and metal interconnects before being subjected to the TPER process. These components are temperature sensitive and can be damaged irreparably if the wafer is subjected to high temperature or thermal budget. The highest temperature in the TPER process is the ALD deposition of aluminum oxide (in the case of silicon substrate). This step can be carried out at a temperature of sub-100°C, which is an acceptable temperature for silicon wafers, post-circuit fabrication. Thus, the performance of the prefabricated circuit does not degrade due to the TPER process.
3. The process steps involved in the TPER process are compatible with CMOS fabrication technology. The materials, chemicals, tools, and techniques used in the process are already used in many steps involved in the CMOS fabrication process. Thus, the fabrication facility adopting the TPER process to fabricate flexible silicon circuits will not need additional tooling or infrastructure to carry out the process once the circuit fabrication is complete.

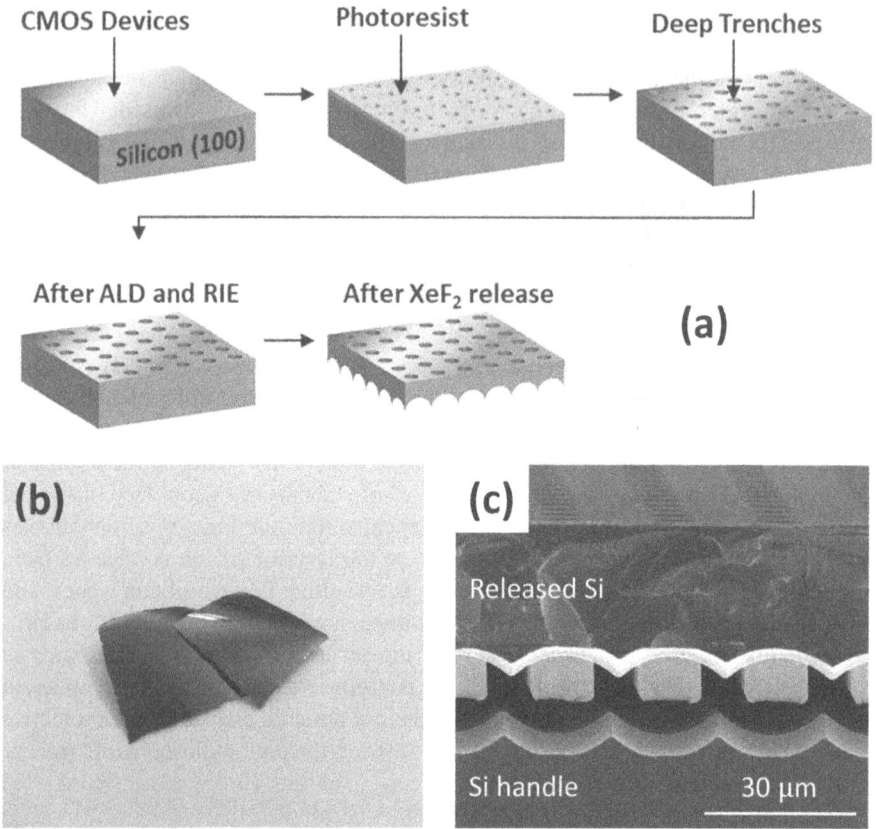

FIGURE 4.6 (a) The TPER process involves the development of trenches, followed by their sidewall protection and isotropic etching. (b) The image of flexible monocrystalline silicon after release from a bulk silicon substrate using The TPER process. The size is 2 cm × 2 cm. (c) Scanning electron microscopy (SEM) image of just released silicon using the TPER process clearly shows the convergence of etch bubbles from individual trenches. (Adapted with permission from [2], © John Wiley and Sons, 2016.)

4. The TPER process affords a high degree of control over the thickness of the top silicon released. This is achieved by varying the depth of the trenches inside silicon, which, in turn, can be controlled by changing the number of cycles of the DRIE process. It should be noted that the thickness of the released silicon is lower than the depth of the trench because the isotropic etch of silicon also progresses vertically upward. However, for a required final thickness of silicon, the depth of the trenches can be calculated based on mathematical modeling for the TPER process and empirical evidence.

5. In this process, the silicon substrate remaining after removal of the top silicon surface can be reused after polishing. The typical thickness of a silicon wafer is 0.5 mm. If the top 0.1 mm is lost during the release process (the thickness of the released piece will be less than 0.1 mm because of

losses in lateral etching etc.), the remaining 0.4 mm silicon substrate can be polished and reused as normal. This can be done several times depending on the thickness of the top portion removed. Eventually, the silicon substrate will be too fragile to undergo the complete CMOS circuit fabrication process and will have to be discarded. The cost of the silicon (100) substrate is low compared to the other substrates, such as SOI and III-V substrates, however, the reusability of the silicon substrate can still be an important cost-saving element for the industry and a way to reduce its carbon footprint.

6. The flexible silicon piece resulting from the TPER process has a unique characteristic – it has uniformly distributed microscopic through-holes throughout the silicon surface. This gives the silicon piece a semi-transparent characteristic. It has a transmittance of around 7% in the visible light region. An interesting advantage of these through-holes is that they can be used for heat dissipation or for fabrication of through silicon vias (TSVs) for 3D integration of flexible silicon pieces.

There are certain disadvantages to the process as well. A key disadvantage is the loss of the silicon area because of the placement of trenches on the silicon floor. These trenches are made after the completion of the circuit and metal interconnect fabrication; thus, their placement can place a major constraint on the placement and routing of the silicon circuit. On average, these holes consume about 20% of the silicon floor in most of the flexible devices reported in the literature. While this is a significant amount of real-estate area loss, the real loss in the usable area can be minimized by utilizing the inter-device separation gap for placing the trenches. However, this can significantly increase the time and effort required for placing and routing circuits on the silicon floor and will require modification of place-and-route algorithms and tools. Further, because the inter-device separation is very small in modern state-of-the-art devices, there will be a constraint on the size of the trenches that can be placed there. This will limit the depth of the trenches because DRIE processes can only provide a certain aspect ratio of trenching. This, in turn, will limit the thickness of the final flexible silicon piece obtained using this process. Thus, some loss of silicon real estate may be unavoidable if the TPER process is used. Another disadvantage of the process is that it produces flexible silicon pieces with semi-spherical "scallops" at the bottom, as shown in Figure 4.6c. These are caused because the etch bubbles of the isotropic etch progress at the same rate in all directions, including upward from the bottom of the trench. This causes more silicon to be removed near the trench bottom compared to other areas, leading to the formation of scallops. This uneven surface finish can cause problems when the flexible silicon pieces are transferred onto a flat surface. Further, while bending, the flexural stress may be distributed unevenly because of these structures, which may lead to cracking or fracture of the silicon pieces even before the minimum bending radius is achieved. However, this effect can be mitigated by creating thinner-than-required flexible silicon pieces so that the increased flexural stress can be adjusted against the lower flexural rigidity to obtain a specific bending radius.

An interesting variation to the TPER process can be used to obtain flexible silicon circuits using SOI as a starting substrate. As in all device-first approaches, the devices and metal interconnects are first fabricated on the SOI substrate using standard CMOS processing. Once complete, trenches are patterned on the substrate, and the top silicon surface is etched up to the buried oxide. The BOX is then etched using the same trench patterned, so that silicon from the handling substrate is exposed. The wafer is then subjected to conformal deposition of a protective thin film using ALD, followed by directional RIE to remove the thin film from the bottom of the trenches (same steps as TPER for bulk silicon). This is followed by XeF_2 based isotropic etching of silicon. The etch progresses into the handle substrate but cannot progress vertically upward from the bottom of the trench because XeF_2 does not etch silicon oxide. Thus, the top silicon, and hence all the devices and circuits, are protected by the BOX layer. Once the etch bubbles in the silicon handle substrate meet, the top silicon along with the attached BOX layer is released. This process has all the advantages mentioned for the TPER process. Additionally, there is no requirement of DRIE of silicon because the thickness of the top silicon layer is low, and the standard RIE process can be used. Another added advantage of this process is that it does not produce scallops at the bottom of the released flexible silicon. This is because the upward etch from the bottom of the trench is protected because of the BOX layer, thus leading to a smooth surface at the bottom. Finally, the thickness of the released silicon can be very low because it only consists of the top silicon layer, metal interconnects, and buried oxide layer. The lower thickness of the final flexible silicon results in lower flexural rigidity and better bending radius.

The TPER process uses chemical processes and the properties of directional and isotropic etching, to secure the release of the top silicon layer after circuit fabrication. The top layer can also be released using physical processes using the controlled spalling technique (CST). This process is also known as the Slim-Cut process. In the process, a stressor layer is deposited on the silicon substrate so that the crystal can be made to fracture along a plane parallel to the surface. The stressor layer is a thin film that is under stress because of the deposition technique used. Stress in a thin film can arise due to the mismatch in the lattice of the thin film and substrate, the difference in thermal expansion coefficient, or the use of non-equilibrium conditions during deposition. In the last case, the stresses are said to be internal stresses, because it is independent of factors outside the thin film. Generally, thin film deposition is carried out in such a way, so as to minimize the stress in the deposited thin film. Stressed thin films lead to wafer bowing which can cause problems during subsequent processing, particularly lithography because the photoresist focal plane is not fixed. However, in the case of CST, a large stress is required for the release of the top silicon surface, and stressful thin films are deposited by design. A stressor layer with tensile stress is used so that it can create an upward shear force on the wafer surface. This causes the top surface of the wafer to peel off under the stress of the thin film, thus creating a thin and flexible silicon circuit (Figure 4.7).

Initial experiments involving CST used metal thin films such as silver and gold as stress material. The wafer was also subjected to thermal cycling to increase the stress. However, in the case of the device-first approach, given that all the electronic circuits and their interconnects are already fabricated, the wafer cannot be subjected to thermal cycling. In this case, thin films with large internal stresses are used as stressor layers. Thus, stressor materials such as sputtered nickel or evaporated

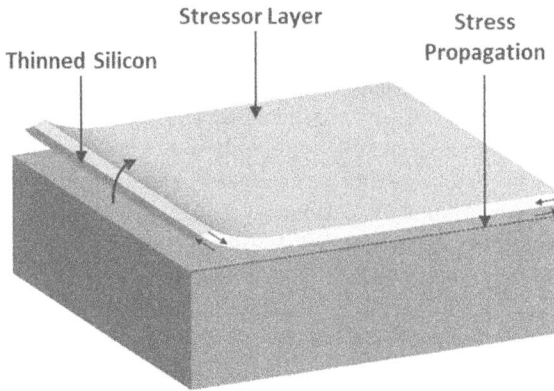

FIGURE 4.7   A schematic illustration of the controlled spalling technique (CST) used for removing the top layer of silicon after device fabrication (device-first approach). (Adapted with permission from [2], © John Wiley and Sons, 2016.)

aluminum have been used as stressor layers. The thickness of the flexible silicon piece is controlled using the thickness and internal stress of the stressor layer. An initial fracture can also be induced at the side of the wafer that can be propagated through the wafer using the stress from the stressor layer.

Advantages of the process include the ability to obtain flexible silicon chips of precise thickness without the loss of silicon material (say in lateral etching). The materials and processes used in this technique are compatible with CMOS fabrication processes, thus, can be included in the process flow for flexible silicon chip fabrication. Further, the back of the flexible silicon piece is crystallographically smooth, owing to the propagation of the crack along a cleave plane. However, the stress induced in the silicon because of the stressor layer can change the charge transport properties of the lattice. In any CMOS circuit, the balance between the on-state current carrying capacity of p-MOS and n-MOS transistors is tightly controlled to achieve desired circuit performance. A change in lattice properties affects these transistor types differently, thus causing a mismatch in the CMOS circuits. This can change the performance of the circuit if it is not accounted for at the circuit design stage. Even if the stress is taken into account in the CMOS circuit design, it is challenging to model the distribution of stress throughout the chip at a nanoscopic level. Thus, the uneven distribution of stress inside the lattice can cause the circuit to malfunction. This problem is somewhat mitigated by removing the stressor layer after the spalling process is complete. Selective wet chemical etching is a common process to remove the stressor layer. However, due to plasticity effects, the silicon lattice does not completely return to its original state after the removal of the stressor. This is known as stress memorization and may still lead to unexpected circuit behavior.

Apart from the TPER and CST, there are some techniques that are based on the precise epitaxial growth of silicon. One of these processes is the ChipFilm (and subsequently ChipFilm II) technology. The process starts with creating porous silicon on the substrate surface while keeping some areas (known as anchors) unchanged. This is done using a process called anodic etching. This process involves the use of electric current to activate the etching of a substance (similar to electrodeposition or electroplating, but reverse). The current is induced by providing a potential difference between the substrate and a reference electrode, immersed in an electrolyte bath. In the case of anodic etching of silicon, an aqueous solution of HF is used as the electrolyte. This process makes the top silicon surface porous. The degree of porosity can be controlled using the current density and the time of etching. Further, an interesting feature of the anodic etching of silicon is that it is highly selective between p-type and n-type silicon. While p-type silicon is rapidly etched, there is almost no etching of the n-type silicon. Thus, detailed microstructures can be defined on the silicon wafer simply by masked doping of the substrate. In the case of ChipFilm technology, small areas of the silicon wafer are n-doped to create anchors of solid silicon while the rest of the surface is made porous. This is generally done in two steps to create two layers of silicon with different porosities. A fine-porous layer of silicon is formed on top of a highly porous layer. The substrate is then subjected to annealing, which leads to micro- or nano-cavities forming in the fine-porous top layer, while a large cavity is formed in the highly porous bottom layer. The creation of a large cavity is supported by the fact that micro- and nano-cavities from the fine porosity silicon tend to diffuse toward the larger cavities in the layer below. The layer of silicon at the ceiling of the cavity and the layer at the top surface thus has less percentage porosity compared to the remaining layer. Thus, the result of these doping, etching, and annealing processes is a layer of fine porosity silicon detached from the handle substrate (because of the formation of a large undersurface cavity) and supported with solid silicon n-doped anchors.

Because pristine single crystal silicon is required for state-of-the-art CMOS circuit fabrication, the top porous silicon layer cannot be used. It is, instead, used as a seed layer to deposit single-crystal silicon using the epitaxial process. This process, which is a variation of the CVD process, involves the use of high temperatures to break down gaseous reactants into elemental silicon for deposition (pyrolysis). The process is carried out very slowly and precisely to mimic the underlying silicon lattice (from the seed layer) to result in a pristine defect-free silicon layer for CMOS device fabrication. Once this pre-fabrication module is complete, the silicon wafer is subjected to standard CMOS processes for fabricating the circuit. After the completion of the CMOS device and metal interconnect fabrication, the silicon chips are released by creating trenches through the top silicon surface up to the buried cavity. These trenches are patterned such that metal interconnects and device layers are not affected. This leaves the silicon chips weakly attached to the handle substrate through the n-doped silicon anchors. These silicon chips can then be detached from the substrate using a pick-and-place tool with a vacuum chuck. The tool also provides mechanical support to the flexible silicon chip before it is integrated with a flexible substrate. This is important because the flexible silicon chip may experience warping when it is completely detached from the substrate because of stresses in the silicon or the metal layers. The process is schematically illustrated in Figure 4.8.

The process described here is the basic process involved in the ChipFilm technology. Many variations have been reported in literature differing in process conditions and recipes in a few steps, however, the general approach remains the same. This technology has several advantages such as supporting state-of-the-art CMOS

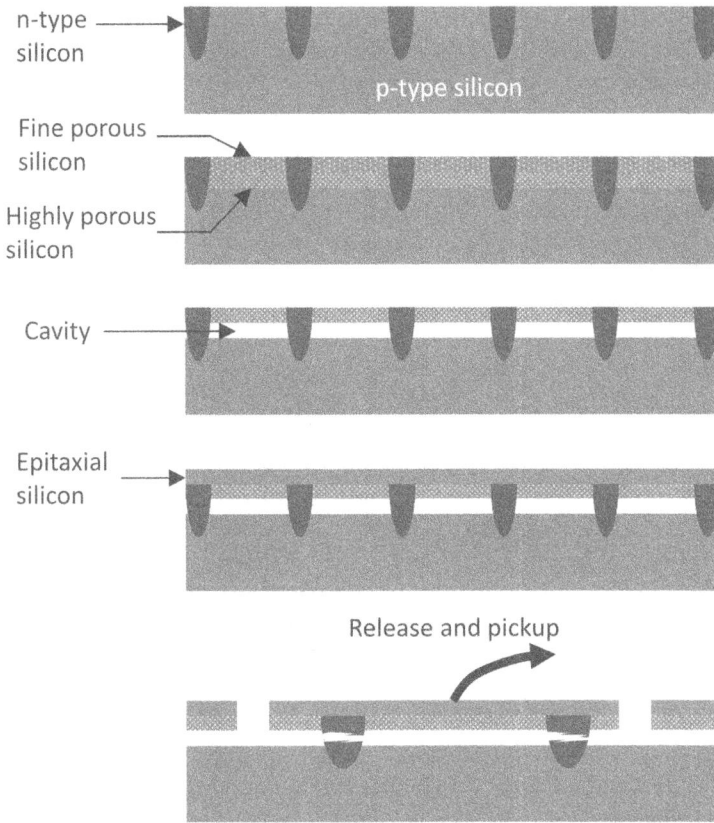

FIGURE 4.8 The ChipFilm process involves the development of n-type anchors, followed by anodic etching of Silicon to create porous Si. The top surface is made pristine again using epitaxy, followed by circuit fabrication using standard CMOS processing.

circuits (being a device-first approach), being CMOS fabrication compatible, providing thickness control (through thicknesses of fine porous silicon and epitaxial silicon layer), having a pick-and-place strategy in-built into the process, and having a smooth bottom finish for flexible silicon chips. However, silicon wafers are subjected to the complete CMOS fabrication cycle with the large undersurface cavities in place, which may lead to increased surface roughness and tightening of the process window in some steps. Further, the technology entails implementing an extensive pre-fabrication module, including an epitaxy process step, which can reduce throughput and add to the cost of the chips.

The process of formation of double-layer porous silicon followed by annealing and epitaxial growth can also be used to create a high-quality silicon layer weakly bonded to a handle substrate. This layer can be then be separated from the handle and transferred to a destination substrate for further processing and device fabrication. This device-last approach involving the use of porous silicon is called the ELTRAN process. It is generally used to fabricate SOI substrates by using the high porosity silicon layer to break the top silicon away from its handle substrate. However, if the receiving substrate is flexible, and the thickness of the top silicon is controlled, it can result in a single crystal flexible silicon piece.

4.3.2.2   Bottom-Up Method

As the name suggests, the bottom-up method is the process of reducing the thickness of the silicon substrate by removing material from the back of the wafer. The front end of the wafer consists of the CMOS circuits and metal interconnects. This side of the wafer is generally protected during the process by covering it with a photoresist layer or a thin-film laminate. The removal of material from the back end of the wafer can be done using physical or chemical processes. One of the simplest processes for this is called the "back-grinding" process. This process involves the use of abrasive grinding wheels to remove silicon. These wheels are usually made of materials that have more hardness (materials that can resist plastic deformation are "hard") than silicon so that they can remove silicon material effectively and can be used repeatedly. Tiny particles of these abrasive materials are generally embedded in a matrix to form the grinding wheel. The top surface of the silicon wafer is protected using laminate tape, while the bottom surface is exposed to the grinding wheel. The silicon wafer and the grinding wheel generally rotate in opposite directions around their central axes to provide areal uniformity in thickness which is measured using a parameter called total thickness variation (TTV), which is defined as the difference between the maximum and minimum thickness of the wafer and is expressed in units of distance (such as microns). It is important to have the TTV as low as possible because it shows that the wafer has not undergone uneven removal of material that can lead to a catastrophic failure later, particularly during bending.

The advantage of the back-grinding process is that it is highly scalable and provides a very fast process for thinning silicon wafers. This also reduces the effective cost of the process per wafer processed. Further reduction in the cost of operation is obtained from the longevity of the grinding wheels, particularly those made using synthetic diamond particles. However, there are several disadvantages of using back-grinding to obtain flexible silicon pieces. First, because the process is highly abrasive, it puts a substantial amount of mechanical stress on the silicon wafers. These wafers are already sub-millimeter thick, are routinely as big as 12-inch in diameter, are made of a brittle material like silicon, and have been subjected to several hundred process steps during CMOS circuit and metal interconnect fabrication. This can cause wafers to crack, particularly if the final thickness is very low which is a requirement for high flexibility. Second, because it is a relatively fast removal of material, it produces a significant amount of heat. The grinding process is generally accompanied by jets of deionized water to limit the rise in the temperature of the substrate. Third, the wafer can end up with microscopic scratches at the back surface because of the action of the abrasive particles in the grinding wheels, leading to high values of TTV. Owing to these disadvantages, the back-grinding process alone is not used to create flexible silicon wafers. It can be used as a coarse first step to remove large amounts of material quickly, followed by other less stressful and more precise processes. One of these processes is the chemical mechanical polishing (CMP) process.

The CMP process uses a combination of chemical reaction and mechanical abrasion to remove silicon from the substrate. The silicon substrate is applied a protective tape and mounted against an abrasive pad. The substrate and the pad are rotated against each other and a constant supply of chemical slurry is maintained during the process. The slurry reacts with silicon to form hydroxylated silicon on the surface which is then removed using mechanical abrasion. The combination of the force on the silicon substrate toward the polishing pad and the chemical reaction causes parts of the silicon surface that are higher than the general topography to be etched faster than the parts that are lower. This results in planarization of the surface. Further, the abrasive material is chosen such that its hardness is less compared to silicon lattice, but more than hydroxylated silicon. Thus, the action of grinding against the material removes only hydroxylated silicon and does not remove silicon, resulting in polishing of the silicon surface rather than roughening. The process is relatively low-stress, in fact, it helps reduce the stress and the excessive TTV caused by the back-grinding process preceding it. However, it is a slow process and involves the use of specialized chemical slurries, which makes it expensive.

Another process used for thinning down silicon substrate is the "spin etching" process. This is a completely chemical process that involves the use of etchants to react with and remove silicon. The etchant used in this case is called HNA which is an abbreviation for the three chemicals used in the mixture – hydrofluoric acid (HF), nitric acid (HNO_3), and acetic acid (CH_3COOH). While silicon can be etched using hydroxide-based compounds like potassium hydroxide (KOH) and trimethylaluminum hydroxide (TMAH), these etchants proceed anisotropically because etch rate is higher for certain lattice directions. In the case of HNA, silicon itself is not directly etched; it is oxidized by nitric acid to form silicon dioxide, which is then etched by HF. The process can be described using the following chemical reaction:

$$Si + HNO_3 + 6HF \rightarrow H_2SiF_6 + HNO_2 + H_2O + H_2$$

The presence of acetic acid reduces the probability of dissociation of HNO_3 and HF, causing the reaction speed to reduce. This results in a controlled and uniform etch of silicon. In the spin etching process, the etchant is poured onto the silicon substrate while it is placed upside-down on a spindle and rotated at high speeds. The rotation ensures uniform delivery of the etchant to all parts of the substrate and instant removal of reaction products. Uniformity can be further enhanced by controlling the etch rate using the relative percentages of acids and the chamber temperature. These processes are illustrated in Figure 4.9.

Dry etching can also be used to remove silicon from the backside of the substrate. In a process called the "soft etch back" process, the silicon wafer is flipped upside down and subjected to deep reactive ion etching (DRIE). The top surface of the wafer is protected using a thick layer of photoresist. However, there can be variations in the etch depth depending on the design of the chamber and the substrate loading. Because a significant percentage of the substrate needs to be etched, it is difficult to stop the etch at the precise time to obtain a specific thickness of silicon. Thus, this process is ideally used in the presence of an etch stop layer such as the buried oxide in the case of circuits fabricated on the SOI substrate.

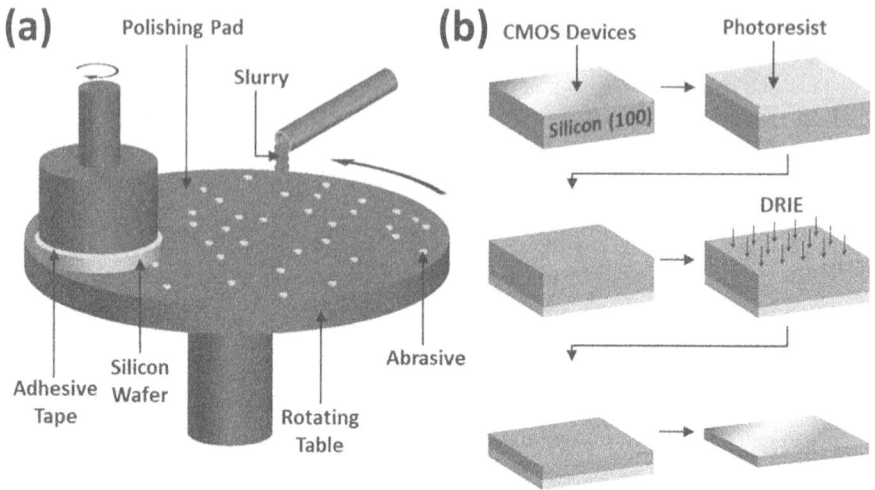

FIGURE 4.9 (a) Schematic illustration of the chemical mechanical polishing (CMP) process. (b) The process steps involved in soft back etch. (Adapted with permission from [2], © John Wiley and Sons, 2016.)

The processes described here for the removal of silicon from the backside of the substrate have their unique advantages and disadvantages. While back-grind is fast and cheap, it can cause substrate damage. On the other hand, methods such as CMP and chemical etching are relatively low stress but are slow and expensive. Hence, in practice, a combination of these processes is used to reduce the thickness of the substrate up to the desired value. Ideally, back-grinding is used as the first step to remove large quantities of material quickly, particularly when the substrate thickness is high, and it can withstand large mechanical stresses. This is followed by CMP to remove stresses and TTV created by the back-grinding process. The final step can be one of the chemical etching processes (wet etching or dry etching) so that there is no application of mechanical stress on the fragile flexible silicon piece. The various processes described in this section for the fabrication of flexible silicon chips can be summarized as shown in Figure 4.10. The processes are either device-first or device-last depending on whether CMOS circuits are fabricated before or after thinning down silicon. In the case of the device-first approach, the processes can be further subdivided based on whether the top surface of silicon is removed while keeping the remaining substrate intact (top-down processes), or the entire substrate is thinned down from the bottom part (bottom-up processes).

4.4 SILICON TFTs

In this chapter, the focus has been on creating flexible silicon pieces using monocrystalline silicon because it is the workhorse of the electronics industry. However, there are several applications where thin-film transistors (TFTs) can contribute significantly, a prominent example being the control electronics for display pixels. Because TFTs are based on thin films deposited using CVD or PVD processes, the underlying

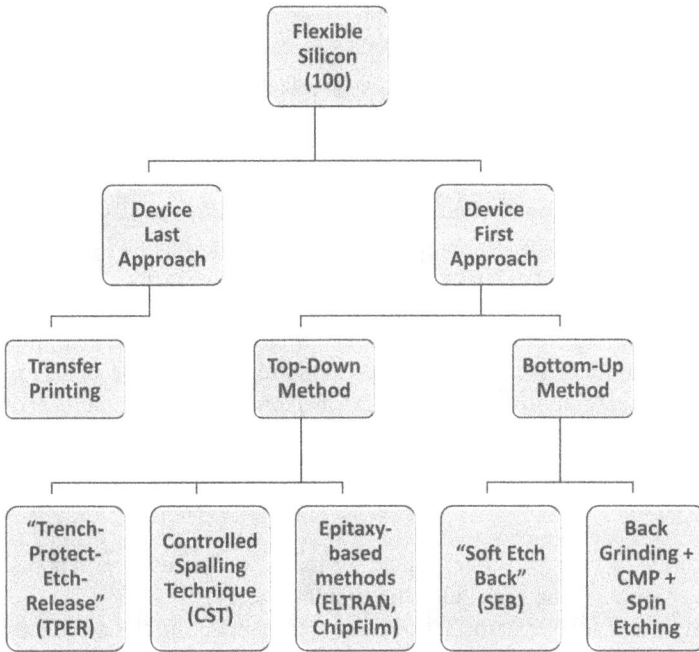

FIGURE 4.10 Flexible silicon tree showing some of the possible ways of obtaining monocrystalline flexible silicon pieces.

semiconductor is not monocrystalline, leading to lower performance in terms of current output and mobility compared to MOSFETs of similar size. However, ease of fabrication on any substrate makes TFTs an interesting option for flexible electronics applications. In particular, silicon TFTs have been extensively studied and deployed in commercial applications because of the proliferance of silicon processing infrastructure in the electronics industry, and because of the knowledge associated with silicon lattice, doping, contacts, and so on. To this effect, amorphous silicon TFTs are among the first TFTs to be reported in the literature. However, amorphous silicon has very high defect density leading to charge traps, which leads to poor TFT performance. This has been one of the reasons amorphous silicon TFTs took a long time before becoming feasible as switching devices. The breakthrough came in the 1970s with the discovery that annealing amorphous silicon thin films in hydrogen environment led to the passivation of dangling bonds. The resulting thin film, known as hydrogenated amorphous silicon (a-Si:H), had significantly improved defect density, leading to the possibility of TFT fabrication.

The performance of silicon TFTs can be further improved by crystallizing amorphous silicon thin films into polycrystalline silicon. This can be carried out by annealing amorphous silicon at high temperatures. The resulting polycrystalline film has significantly lower defect density and reduced carrier scattering, leading to better charge transport properties. Carrier mobility can be enhanced up to 100x by converting amorphous silicon into polysilicon thin films. However, the advantage of higher mobility comes at a cost of high-temperature annealing, which is not compatible with most flexible substrates. This problem is overcome using a process called excimer laser annealing (ELA). In this process, the amorphous silicon thin film is irradiated with an excimer laser beam (typically XeCl at 308 nm). The focus, intensity, and duration of the beam are fine-tuned such that the silicon at the surface melts and undergoes recrystallization, whereas the temperature of the underlying substrate remains relatively low. This process produces polycrystalline silicon, known as low-temperature polysilicon or LTPS, and is widely used in the display industry to create silicon TFTs. LTPS TFTs have much better performance compared to hydrogenated amorphous silicon TFTs allowing for smaller devices to produce the same drive current, leading to higher integration density of display pixels. However, a key disadvantage of this process is the areal non-uniformity in properties of the LTPS thin film created. This restricts the use of LTPS to small footprint, high-resolution displays such as smartphones, whereas large area displays are created using hydrogenated amorphous silicon.

EXERCISES

4.1. What is an SOI substrate? What are the processes used to fabricate SOI substrates starting with bulk silicon substrates?

4.2. List the advantages of the TPER process for fabricating flexible silicon from bulk silicon substrates, post-device fabrication. What are the disadvantages?

4.3. Compare the four processes for bottom-up device-first approach – 1. Back grinding, 2. CMP, 3. Wet etching, 4. Soft back etch – in terms of speed, stress on the wafer, and TTV obtained.

4.4. Consider the use of the TPER process to create flexible silicon circuits using an SOI substrate. We create circular trenches with 10 μm diameter, with their centers on a square lattice of 20 μm side. If one cycle of XeF_2 etch removes n moles of silicon, how many cycles are required to flex the SOI piece?

4.5. Name all the key methods that can be used to create flexible monocrystalline silicon. Mention their advantages/disadvantages in brief. Draw a tree indicating the classifications of all the methods.

5 Organic Electronic Materials

5.1 INTRODUCTION

The quest for obtaining a flexible electronic device can be seen from two different perspectives. We have high-performance electronic devices made using silicon
chips as the key active ingredient, we can try to make them flexible using manufacturing tricks, or we can take the already available flexible material systems such
as polymers, and try to induce electronic characteristics in them. The last chapter
was about the former approach while this chapter is about the latter. We see flexible
materials all around us. There are synthetic flexible and stretchable materials such
as rubbers and plastics, and there are naturally occurring ones like cotton fibers,
muscle fibers, skin tissues, and so on. A common attribute in all of these materials
is that they are made of chains of carbon atoms. In some materials, these chains
occasionally branch out and attach to other structures, or to each other to form
a giant 3-dimensional molecule. In most cases, the electrons in these chains are
tightly held in the molecular orbitals around the atomic nuclei, thus making it difficult for electronic conduction to take place. Thus, most of the polymeric material
systems are inherently non-conductive and thus, are not usable for creating flexible
electronic devices.

The problem of conduction (and semi-conduction) in organic molecules can be
solved using some interesting aspects of organic chemistry. The first discovery of a
conductive organic material consisted of the formation of complexes between polycyclic aromatic compounds and halogens (Perylene-Bromine) in 1954 by researchers
from the University of Tokyo. Even in that case, it was reported that these compounds are not stable and do not keep conductivity for long. In 1963, Australian
researchers B.A. Bolto, R. McNeill, and D.E. Weiss reported electrical conductivity
in a polymer, polypyrrole. Finally, in 1977, it was reported that the electrical conductivity of doped polyacetylene can be systematically controlled over 11 orders of
magnitude. Systematic control of conductivity is a key feature of silicon that enables
the fabrication of state-of-the-art electronic devices. Thus, this discovery proved that
organic materials can be used to fabricate electronic devices and paved the way
for research in organic electronics. The authors of the paper, Alan J. Heeger, Alan
G. MacDiarmid, and Hideki Shirakawa, were jointly awarded the Nobel prize in
Chemistry in 2000, "for the discovery and development of conductive polymers".
The conduction mechanisms in organic compounds are markedly different from
those in inorganic materials (like metals or silicon). This difference also gives rise to
some peculiar properties that are observed in organic electronic devices. The following sections discuss the mechanisms of conduction and the properties of electronic
conduction in organic materials.

DOI: 10.1201/9781003010715-7   **63**

5.2 CONDUCTION IN ORGANIC MOLECULES

In order to understand the conduction mechanism in organic molecules, it is important to reflect upon the conduction mechanism in inorganic lattices such as silicon and metals. In Chapter 3, we discussed the formation of the band structure in the silicon lattice and how it results in the observed electronic properties of silicon. The key realization here is that individual atoms, even if they are of a highly conductive element like silver, are not conductive on their own or as a collection. Electrical conductivity only arises in these lattices when multiple atoms come together in such a way that allowed energy states for electrons are shared across the lattice and electrons in these energy states can travel across the lattice. The probability of the existence of electrons at a particular point in the lattice and the distribution of their velocities then depend not only on the internal distribution of charged nuclei but also on external factors such as applied electromotive force. This phenomenon is referred to as electrical conductivity. In inorganic lattices, the shared energy states across the lattice, obtained because of the specific distribution of potential wells (nuclei), give rise to the observed electrical conductivity. In most organic molecules, the electrons associated with the molecules are tightly held in carbon-carbon bonding orbitals. Again, these molecular orbitals arise because of the solution to the distribution of electrons according to the potential wells of the nuclei of carbon atoms and other atoms participating in the molecule. These solutions permit electrons to have energy levels lower than those in the individual atoms (thus leading to a stable molecule). These molecular orbitals are called "bonding" orbitals. Electrons in bonding orbitals have a high probability of being between the two nuclei participating in the bond. The complete solution to the electron distribution within the molecule also results in higher energy molecular orbitals. These orbitals may permit electrons to have shared energy states across the molecule where intramolecular conduction may be possible, however, the energy of these "anti-bonding" molecular orbitals is very high for any practical conduction to take place.

As an example, let us consider the simple molecule of methane in which four hydrogen atoms bond with a carbon atom (Figure 5.1). The total number of electrons taking part in the reaction is eight (four electrons from four hydrogen atoms, and four valence electrons of carbon). We do not consider the electrons in the 1s orbital of carbon because they do not participate in the chemical reaction. The molecular orbitals for the methane molecule are formed such that there are four bonding molecular orbitals and four antibonding orbitals. Interestingly, contrary to intuition, all the four bonding molecular orbitals do not have the same energy – one of them has lower energy compared to the other three degenerate orbitals. The eight valence electrons are distributed in these four orbitals to form a stable methane molecule. A similar process occurs when an ethane molecule comes together. In this case, there are a total of 14 valence electrons (six from six hydrogens atoms and eight from two carbon atoms) distributed in seven bonding molecular orbitals, of which, four orbitals occur as two pairs of degenerate orbitals and others occur at different energy levels. In the case of ethene, however, the number of valence electrons is 12, whereas the number of bonding orbitals is only five. This results in the five molecular orbitals being occupied by ten valence electrons (the σ-bond framework), while

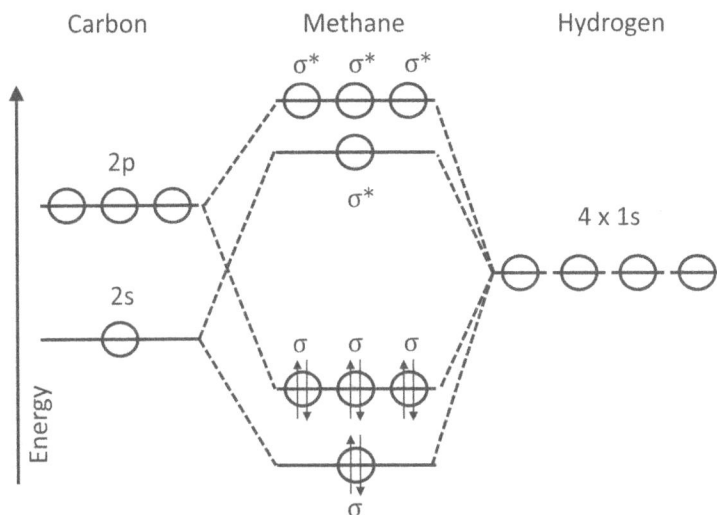

FIGURE 5.1 The molecular orbitals of methane consist of four σ orbitals occupied by electrons with four σ^* unoccupied higher energy orbitals. The difference in energy between the σ and σ^* orbitals is 12.57 eV. 1s orbitals of carbon have not been shown because they do not participate in bond formation.

the last two valence electrons occupy the π molecular orbital. The resultant π-bond is a result of interaction between two pure p orbitals (unhybridized) of the carbon atoms. The structure of these six orbitals is such that each one occupies a different energy level with the π-bonding orbital being at the highest energy. The next highest energy is associated with the π-antibonding orbital, which remains unoccupied at room temperature. The two π-orbitals are known as the "frontier orbitals" and dominate the chemistry of ethene and other aromatic organic compounds. The π-bonding orbital is known as the highest occupied molecular orbital, or HOMO, while the π-antibonding orbital is known as the lowest unoccupied molecular orbital, or LUMO. These frontier molecular orbitals can be thought of as valence band and conduction band in silicon lattice, respectively, and the energy difference between them is similar to the bandgap for inorganic semiconductor lattices.

In the case of ethene, the energy gap between HOMO and LUMO is 6.7 eV, which is high enough to make it an insulating material (Figure 5.2). However, π-bonds, in close proximity, can interact with each other to form more complex electron energy states. In particular, π-bonds in carbon chains containing alternate single and double bonds can interact to form a "conjugated" π-bond system across many carbon atoms. This system allows for lower energy orbitals, thus stabilizing the system while also allowing for electrons to be shared across the conjugated chain. The simplest example of such a system is 1,3 butadiene. The molecule is usually drawn as having double bonds between C1-C2 and C3-C4, while having a single bond between C2-C3. However, the four p-orbitals come together to form four π-orbitals with four different energy levels. Two of these orbitals are lower in energy compared to the original atomic orbitals and are occupied by the four p-electrons from the four carbon atoms

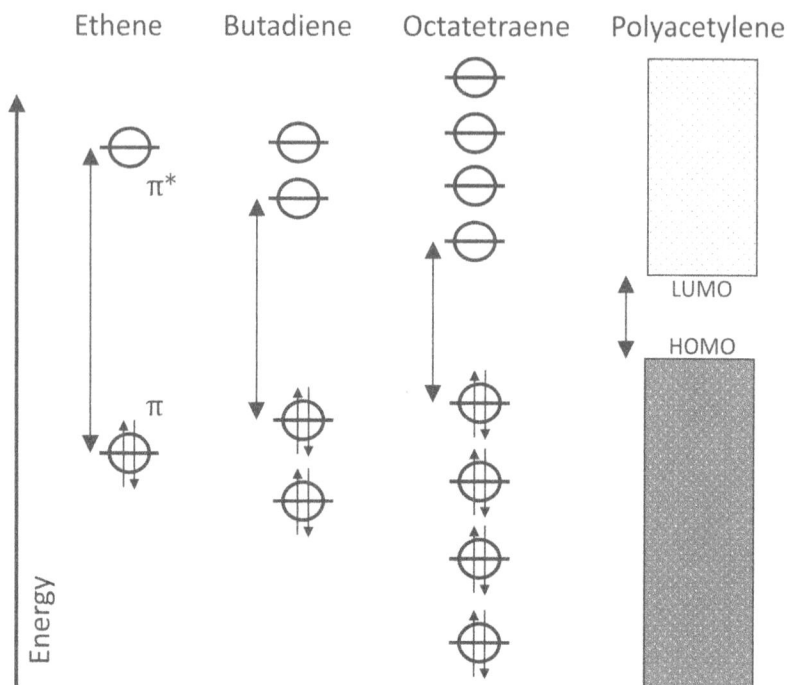

FIGURE 5.2   The gap between the HOMO and LUMO levels in the conjugated π-bond system reduces as the number of carbon atoms in the conjugated chain increases.

(π-bonding orbitals), while the other two are higher in energy and remain unoccupied (π-antibonding orbitals). The difference in energy between the HOMO and LUMO orbitals, in this case, is 5.8 eV. Thus, the energy band gap effectively reduces with an increase in the number of conjugating carbon atoms. An interesting effect of the formation of π-bonds is that all the four carbon atoms have similar bonding and electronic conditions. This is reflected in experimental results showing similar bond lengths and bond energy for all three carbon-carbon bonds. The conjugation of π-bonds in 1,3 butadiene is graphically represented by drawing a dotted line between the carbon atoms. The solid line refers to the stable σ-bond framework, while the dotted line represents the concept of delocalized π-bonds. It is important to note that the occurrence of alternate single and double bonds is essential for the delocalization of π-orbitals, and delocalization can be localized to specific parts of a molecule. For instance, the compound 1,3,6 heptatriene has four carbon atoms with delocalized π-bonds, but the π-bond between C6 and C7 is an isolated one because it is preceded with two single bonds (Figure 5.3).

If the length of the conjugated chain is increased, the number of delocalized π-bonding and π-antibonding orbitals increases. The difference between the HOMO and LUMO reduces and the number of energy states available for electrons starts to represent an energy band. In the case of the polymers which consist of a chain of alternating single- and double-bonded carbon atoms, the energy bands start to look like those in inorganic lattices. The difference between HOMO and LUMO behaves

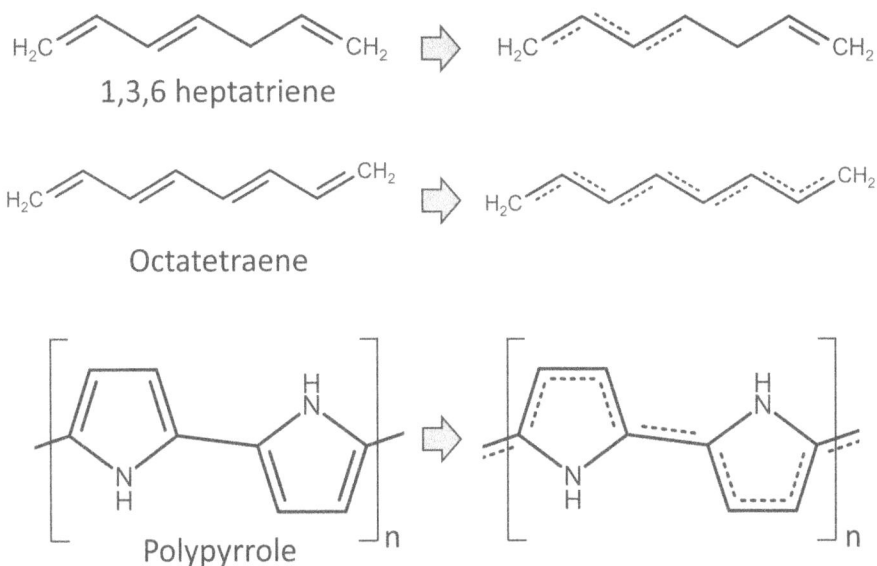

1,3,6 heptatriene

Octatetraene

Polypyrrole

FIGURE 5.3 Conjugation of π-bonds occurs in carbon chains with alternating single- and double- bonded carbon atoms, even for polymer molecules that have aromatic chains, such as polypyrrole.

as the bandgap in the case of inorganic semiconductors. Electrons in the π-bonding orbitals can be removed from their respective atoms and can participate in conduction through the molecule, if subjected to an electric field. This gives rise to electrical conduction in organic molecules. An example of this phenomenon is the electrical conductivity in the polymer polyacetylene as reported in the groundbreaking paper in 1977. Polyacetylene is a simple polymer with a single carbon chain of alternating single and double bonds. The energy bandgap in this case reduces to around 1.5 eV, depending on the number of carbon atoms in the chain. In the case of polypyrrole, as reported by Weiss *et al.*, the structure also consists of an aromatic cycle containing nitrogen atoms; however, it preserves the alternating single and double bond structure necessary for delocalization of π-electrons and conduction within the molecule (Figure 5.3).

5.3 PROPERTIES OF ORGANIC CONDUCTION

5.3.1 CARRIER MOBILITY

The interaction and delocalization of π-electrons in a conjugated carbon chain lead to electronic conduction in organic polymers. However, this phenomenon only translates into intra-molecular conduction, i.e., conduction through a single polymer chain. In order to achieve a conductive solid or thin film, it is important to have charge transfer between two adjacent polymer chains. The charge transport mechanism responsible for inter-molecular charge transfer is called hopping

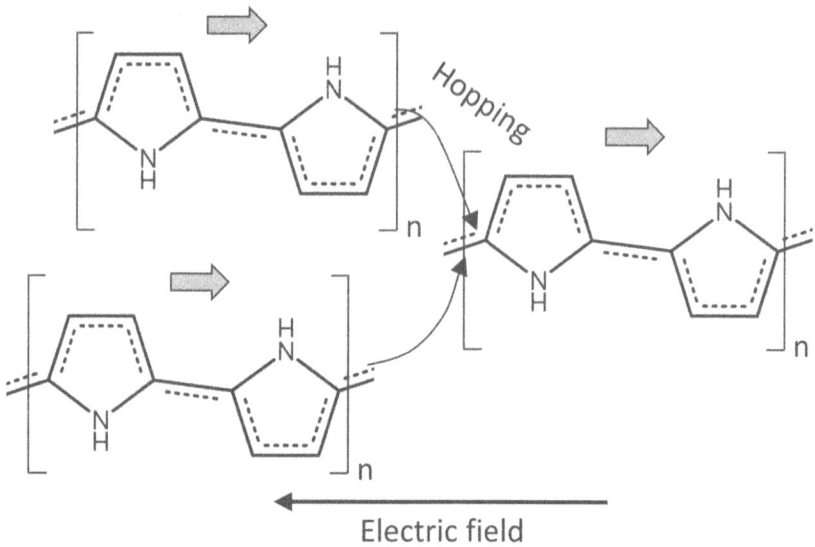

FIGURE 5.4 In the dispersion of a polymer, such as polypyrrole, the electron transport according to the direction of the applied field depends on the conduction of the electron through the polymer (band transport) and between two polymer molecules (hoping transport).

transport (Figure 5.4). The underlying principle of hopping transport is a quantum mechanical phenomenon known as tunneling. According to this phenomenon, there is a finite probability for a particle to tunnel through a potential barrier greater than the potential energy of the particle, without acquiring additional energy. Thus, electrons from one site can tunnel to another site across a potential barrier created due to the weak interactions between different polymer chains. The tunneling probability depends on the distance between the two localized sites, the energy of the electron, and the height of the potential barrier. Because of this hopping, electronic conduction can be observed throughout the material. It is important to note that tunneling probabilities for electrons reduce exponentially with barrier height and distance. Because of the low tunneling probabilities, the charge conduction because of hopping transport is much lower compared to band transport within the polymer chain.

To quantify the ease with which electrons can flow through a material, we define a parameter called mobility. In general, the average drift velocity (over and above the diffusion velocity) of an electron going through a medium is proportional to the applied electric field. The mobility of an electron is the constant of proportionality for this equation

$$v = \mu E$$

where, v is the average velocity of electrons through the material in the direction of the applied electric field E, and μ is the mobility of electrons through the material. This is a very important material property routinely reported in the literature to compare different semiconductor materials. In general, electrons can be accelerated to very high speeds (close to the speed of light) because of an applied electric field, however, because they collide with vibrating lattice atoms along the way, they are scattered, significantly reducing their average velocity in the direction of the field. These scattering interactions (also called electron-phonon interactions) play a major role in determining mobility in the case of band transport of carriers, such as that seen in inorganic lattices and single polymer chains. Further, organic materials lack the 3-dimensional order of inorganic semiconductor lattices.

In terms of band structures, perfectly ordered systems provide a continuous energy band for charge transport. In reality, no crystal is perfect, and localized crystal defects produce localized energy states inside the forbidden bandgap. In most inorganic crystals, these midgap states act as charge traps and produce some reduction in the number of charge carriers available for transport. However, because these systems have a very weak disorder, the midgap states are highly localized. With the increase in disorder, the number of midgap states increases. For amorphous materials or organic polymers, the disorder is large enough that no continuous energy band structure remains and hopping becomes the primary conduction mechanism. In the case of hopping transport, the probability of tunneling determines the mobility of a carrier; and because it is very low, it significantly reduces the mobility in organic materials. Thus, mobility values reported for materials with hopping transport are much lower than those with band transport. For example, single-crystal silicon, in which conduction relies on band transport, has an electron mobility of around 1300 cm²/V-s, whereas amorphous silicon, in which there are significant defects and scattering events, has an electron mobility of around 1 cm²/V-s, even though both materials are based on the same element. In the case of organic semiconductors, because hopping transport is involved for intermolecular charge transfer, the overall mobility is around a few cm²/V-s, at best.

5.3.2 ELECTRON-HOLE STABILITY

The stability of an electron-hole pair in a semiconducting medium is a critical factor determining its electronic properties. In essence, the number of electrons and holes available for conduction is determined by the rate of creation of these pairs and the stability of these pairs with time. The number of charge carriers available for conduction, in turn, determine the total conductivity of the material as follows:

$$\sigma = nq\mu$$

where n is the number of free charge carriers, q is the electronic charge and μ is the mobility of electrons through the material. These electron-hole pairs are created because of thermal or photoelectric effects and usually last for only a few picoseconds in most materials. They recombine to release energy in the form of heat (nonradiative recombination) or photon (radiative recombination). When an electron-hole pair is formed, they are generally linked to each other through electrostatic attraction between the negatively charged electron and the positively charged hole. The strength of this interaction determines the stability of the electron-hole pair. Weakly bonded pairs can be separated to account for separate electron and hole conduction through the material, while strongly bonded pairs remain together and eventually recombine without contributing to net electrical conduction. The strength of the electrostatic

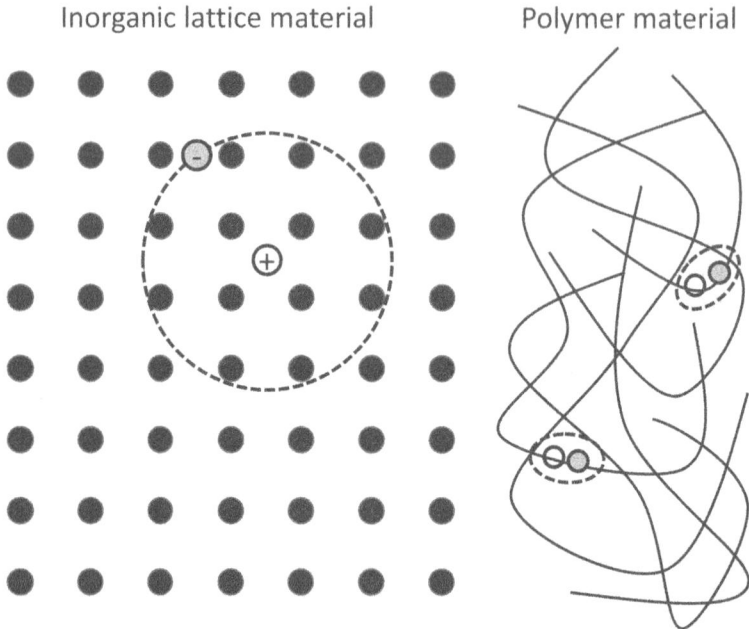

FIGURE 5.5 The electron-hole pairs (excitons) in an inorganic lattice are more loosely bound compared to excitons in organic materials because of Coulombic screening.

interaction between an electron and hole is determined by the electric field screening as a result of the dielectric properties of the material. In inorganic lattices, such as silicon and germanium, the dielectric constant is large, leading to higher electric field screening and lower Coulombic interaction between the electron-hole pair. This leads to weakly bonded electron-hole pairs, also referred to as Wannier-Mott excitons, with binding energy in the range of a few tens of meV. In the case of organic materials, the dielectric constant is low, hence, the Coulombic interactions between electrons and holes are higher, leading to strongly bonded excitons, referred to as Frenkel excitons, with binding energy around 0.1–1 eV (Figure 5.5). The strength of the electron-hole interaction in organic compounds leads to several important properties. Because exciton movement does not contribute to net current, the number of free carriers available for conduction is lower in organic compounds leading to lower conductivity values compared to metals. However, because electron-hole pairs are strongly bound, they are also highly likely to eventually recombine, thus leading to a higher probability of radiative recombination in organic light-emitting diodes (OLEDs).

5.3.3 POLARON TRANSPORT

The energy spectrum of an electron moving along an energy band through a lattice or polymer chain is defined by the positions of the atomic nuclei (or the periodic potential wells) that the electron encounters. Conversely, the positions of the nuclei are disturbed because of the presence of the access negative charge of the electron in their vicinity. This disturbance manifests itself in the form of vibration of atoms around their equilibrium positions and contributes to the overall atomic vibrations in the material already present due to thermal effects. The additional phonons generated due to electrons also interact with the electrons that are responsible for their creation leading to higher electron-phonon interactions, electron scattering, and lowering of electron mobility. The charged particle and the polarization it induces in the material move together and can be modeled as an independent particle called polaron. The resulting quasi-particle has a higher effective mass compared to an electron in the energy band, thus having lower mobility. Theoretically, polarons exist in organic as well as inorganic materials, however, the polarization of organic materials is higher leading to a larger loss of energy and hence a higher reduction in mobility. In the case of inorganic lattices, the strong covalent bonding and high mass of atomic nuclei prevent the strong coupling of electrons with lattice deformation. This results in a very small difference in the effective mass of carriers because of polarization, and the polaron model is generally ignored.

5.4   ORGANIC SMALL MOLECULES

An organic small molecule substance is a low molecular weight organic compound, generally considered to have molecular weights lower than 1000 g/mol. In the case of the polymer chain scenario discussed in the previous section, a clear electronic band structure is created because of the presence of conjugated π-bonds and their interactions to form molecular orbitals. This facilitates the conduction of carriers through the chain, while inter-molecular conduction follows the hopping transport regime. Because of the possibility of π-π interactions across molecules and the probability of carriers tunneling through, there are some organic compounds that are conductive/semiconductive without having long polymer chains. These organic small molecule electronic materials also, however, have some conjugated π-bonds in the molecule. The best example of such a system is pentacene ($C_{22}H_{14}$), which is a polycyclic aromatic hydrocarbon (particularly, acene) with five benzene rings "linearly-fused" together (Figure 5.6). It has a high degree of conjugation with 22 π-electrons (one from each carbon atom) interacting with each other. It has a relatively large energy gap between the HOMO and LUMO, indicating semiconducting

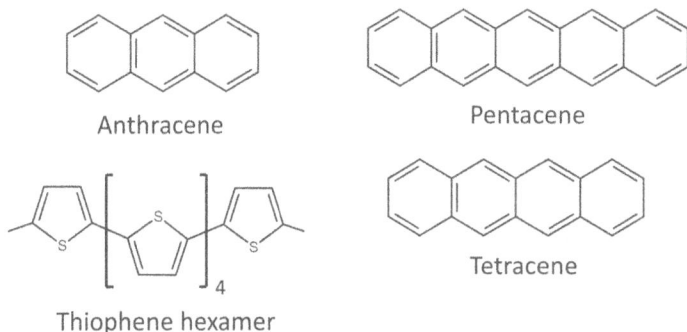

Anthracene

Pentacene

Thiophene hexamer

Tetracene

FIGURE 5.6   Examples of commonly used conducting/semiconducting organic small molecule materials include anthracene, pentacene, tetracene, thiophene oligomers, and the members of the phthalocyanine family.

behavior. The field-effect mobility as high as 2 cm²/V-s has been reported in the literature for pentacene-based devices. A key disadvantage of pentacene is that it rapidly oxidizes on exposure to air, as is the case with many organic conductors. It also has low solubility in many organic solvents, thus in most cases, thin films of pentacene have been deposited on a substrate using evaporation. To overcome these problems and to preserve high carrier mobility, many researchers have studied derivatives of pentacene by adding one or more organic moieties. Pentacene is one of the most studied organic compounds for electronic applications and other small molecules of the acene family, such as anthracene (three benzene rings), tetracene (four benzene rings), and hexacene (six benzene rings) that also show π-conjugation and are being studied for their electronic and semiconducting properties.

5.5   ORGANIC POLYMERS

The phenomenal aspect of conduction in organic materials is that because of the vastness of organic chemistry, almost limitless molecules can be designed and synthesized. Various properties of these materials such as conductivity, absorption/emission spectra, Young's modulus, solubility, thermal stability, and so on, can be controlled by changing the moieties attached to the primary carbon chain. This method of tweaking electrical, mechanical, chemical, and thermal properties of electronic materials does not exist for their inorganic counterparts. There are, of course, tradeoffs arising out of fundamental laws of physics in competing desirable properties. It is exciting, nonetheless, to witness the power of synthetic organic chemistry being used to develop new age conducting and semiconducting organic materials with tailor-made electronic properties. Some of the most commonly used polymers for the creation of organic electronics are polyacetylene, polycyclic aromatic hydrocarbons (such as poly p-phenylene), polypyrrole, polyaniline, polythiophenes, and so on (Figure 5.7). Some of these polymer bases

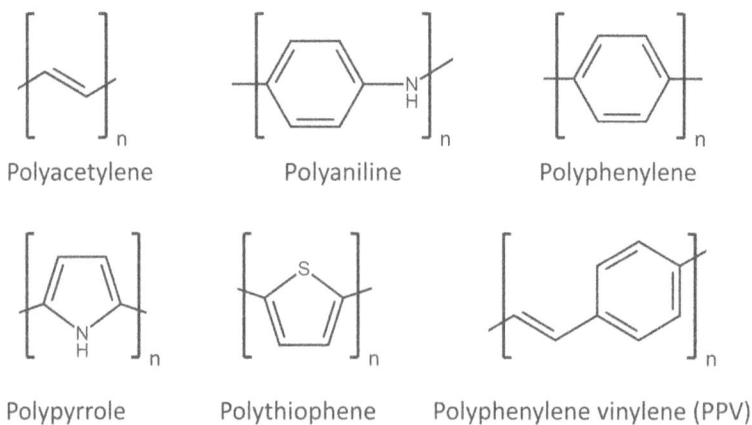

Polyacetylene   Polyaniline   Polyphenylene

Polypyrrole   Polythiophene   Polyphenylene vinylene (PPV)

FIGURE 5.7   Examples of conductive organic polymers commonly used in electronic applications include polyacetylene, polyaniline, polyphenylene, polypyrrole, polythiophene, and PPV.

can be coupled with a particular organic moiety to change some of the electronic, chemical, mechanical, or thermal properties of the material. Further control over the material and thin film properties can be exercised by changing the polymerization process to influence the average molecular weight and the polydispersity index of the material. However, the influence of these parameters over the material properties of organic thin films can create problems with reproducibility. Hence, it is important to study the effects of such parameters on properties so that a process window can be established for batch-to-batch reproducibility in the case of large-scale production. Another advantage of using organic polymers is that polymers have very low vapor pressure and diffusion constants because of their large size and molecular weights. This provides for easier integration in a process flow with multiple stacked thin films and several thermal cycles. Finally, polymeric thin films exhibit robust mechanical properties compared to small molecular organic thin films leading to the possibility of integration in a roll-to-roll production paradigm.

5.5.1 COPOLYMERS

An interesting trick used in creating organic electronic materials is the use of copolymers. This is yet another trick in the synthetic organic chemistry playbook for tweaking properties of the organic thin films to create a perfect material for a specific application. A copolymer is an organic polymer derived from two or more monomers. There are several ways in which these separate monomers combine into forming a copolymer material (Figure 5.8). If the monomers are attached to the polymer chain in an alternating fashion, it is called an alternating copolymer. For example, with monomers A and B, the copolymer chain (-A-B-A-B-A-B-A-B-) is an alternating copolymer. If the monomers are randomly distributed inside a single polymer chain, the copolymer is called a random copolymer. With monomers A and

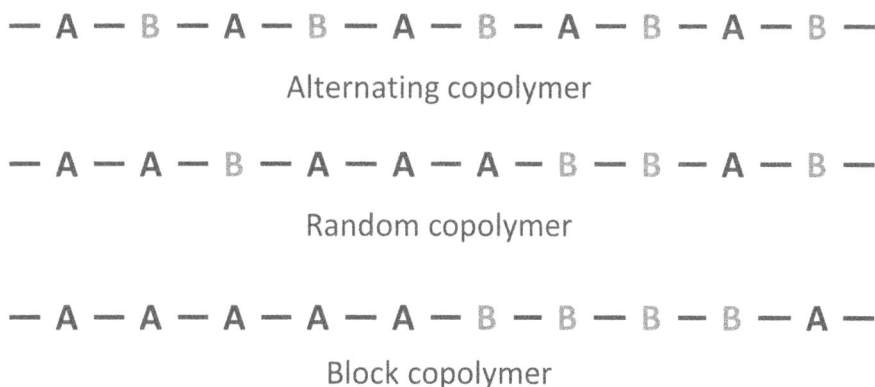

— A — B — A — B — A — B — A — B — A — B —

Alternating copolymer

— A — A — B — A — A — A — B — B — A — B —

Random copolymer

— A — A — A — A — A — B — B — B — B — A —

Block copolymer

FIGURE 5.8 (a) Copolymers are classified as alternating, random, or block depending on the distribution of monomers in the polymer chain. (b) PEDOT:PSS is a mixture of two separate polymers – poly(3,4-ethylenedioxythiophene) (PEDOT), and polystyrene sulfonate (PSS).

B, the copolymer chain (-A-B-B-A-A-B-A-B-B-A-A-B-A-B-) is a random copolymer. The most commonly available commercial example of a copolymer is ABS, which is a copolymer thermoplastic made of acrylonitrile (C_3H_3N), butadiene (C_4H_6), and styrene (C_6H_5-C_2H_3) monomers. Random copolymers are formed by polymerization reactions consisting of the two monomers, and the probability of finding a specific monomer at a specific point in a copolymer chain can be calculated based on the reaction kinematics. The addition of a particular monomer in the chain depends on the ratio of the reaction rate for the addition of that monomer to the reaction rate for the addition of the other monomer, also known as the reactivity ratio. Copolymerization can also be achieved such that there are smaller blocks of polymers of specific monomers attached to form the larger copolymer chain; the copolymer is then called block copolymer. For example, the copolymer chain (-A-A-A-A-A-B-B-B-B-B-B-A-A-A-A-A-) is a block copolymer. These copolymers are synthesized by polymerizing one monomer, followed by polymerizing the other from the ends of the existing polymer chains. The synthesis conditions are maintained such that the reactivity ratio for a particular monomer is much higher for a given time so that only that monomer is added to the chain. Copolymers are generally used to obtain properties that are intermediate between the properties of polymers obtained from individual monomers. For example, the glass transition temperature (the temperature at which an amorphous material transitions from hard/brittle to soft/viscous state) for a copolymer is between the glass transition temperatures for polymers with individual monomers and depends on the mass fraction of the monomers in the copolymer chain. In the case of organic electronics, copolymers can be used to fine-tune properties such as light emission color for OLEDs.

Polymers can also be used in the form of mixtures, as in the case of PEDOT:PSS, which is a mixture of two polymers, poly(3,4-ethylenedioxythiophene), PEDOT, and polystyrene sulfonate (PSS). The polythiophene PEDOT is in itself a conjugated polymer showing reasonable electronic conductivity. It is a relatively stable polymer with high transparency, opening up avenues for use in transparent electronics applications, apart from flexible electronics. The addition of PSS makes the conductive polymer stable in aqueous and other organic solvents, increasing its processability. PEDOT:PSS finds commercial usage as a transparent conductor in photographic films and OLED fabrication. For commercial applications, PEDOT:PSS is synthesized by oxidative polymerization of the EDOT monomer in presence of PSS.

5.6   ORGANIC THIN FILM DEPOSITION

There are several ways of obtaining a uniform thin film for both small molecule and polymeric organic semiconductors, such as evaporation (PVD), low-pressure chemical vapor deposition (CVD), spin-coating, spraying, and so on. For flexible electronic applications, it is important to focus on low-temperature processes so that the flexible polymeric substrates can withstand the process. In the case of vacuum deposition, such as evaporation and low-pressure chemical vapor deposition, the thin film is formed by condensation of vapors of the material on the substrate. The deposition is carried out in a chamber under vacuum (hence the name vacuum deposition). In the case of evaporation, the material vapor is obtained by evaporating the source material using a heating element. For some volatile organic compounds, sublimation can be used to form vapors at room temperature. While the process is similar to the thermal evaporation of inorganic and metal thin films, there needs to be more careful control over process parameters in the case of organic sources. In particular, complex organic structures can have low bond strength that may result in the decomposition of the structure upon application of heat. Further, the evaporation rate, and hence the rate of deposition, increases exponentially with temperature. Hence, very accurate control of temperature is required for the thermal evaporation of organic semiconductors. Vacuum is maintained to reduce contamination of the source material vapor with air or other gases, and to increase the mean free path of the molecules to obtain a uniform thin film. The level of vacuum depends on the acceptable level of contamination in the thin film. Low-pressure CVD process is performed with the evaporated or sublimated material being carried to the substrate using an inert carrier gas. The substrate temperature is maintained such that condensation can take place on the substrate. In both evaporation and CVD, deposition rates are generally monitored using thickness monitoring systems. These are generally based on the variation in the resonant frequency of a quartz crystal because of the mass variation per unit area which changes with the thickness of the deposited thin film. An important advantage of evaporation is the possibility of the use of shadow masks for patterning the deposited thin film. This is particularly important because most organic thin films are unstable in the atmosphere and may not be able to withstand the chemicals used in standard lithography processes.

In the case of organic polymers, the boiling point can be high and vapor pressure can be low because of large molecular weights and internal interactions between polymer chains. Thus, solution-based techniques are used for deposition. A preferred method of thin film formation, in this case, is the spin-coating method. In this process, a small amount of polymer solution is dropped on a rotating substrate. After a fixed duration, the substrate is subjected to annealing to remove access solvent, leaving the polymer thin film behind. The speed of rotation, the duration of the process, and the viscosity of the solution determine the thickness of the final thin film. Another solution-based process is spray-coating, which involves the formation of a jet of spray droplets of the solution containing the organic semiconductor directed towards the substrate. This is a probabilistic method, mainly used in cases where uniformity of the thin film across the substrate is not critical. Thickness control in spray coating depends on the mechanism of droplet formation, rate of spraying, and deposition time. As in the case of evaporation, shadow masks can be used for patterning the deposited thin film.

5.7   ORGANIC SINGLE-CRYSTALS

We have discussed the differences between transport in organic materials and inorganic crystals, however, organic electronic materials can themselves be crystalline and can be formed into single-crystal solids. Research interest in organic single-crystal materials has been traditionally low because they are very difficult to form into stable thin films because of fragility and handling difficulties. However, with new developments in fabrication and integration techniques, there is a renewed interest in single-crystal organic electronic materials. This is because single-crystal materials, in general, offer a charge transport regime based on band transport, resulting in much higher mobilities compared to polycrystalline or amorphous materials where hopping transport is also involved at the grain boundaries. As in the case of inorganic lattices, organic single-crystal materials provide a nearly defect-free system for charge transport and provide a reproducible platform for device fabrication. Further, from a theoretical point of view, it is important to study organic single-crystals because charge transport mechanisms in organic crystals can be better understood owing to less interference from structural defects.

In order to study organic crystals and fabricate electronic devices on them, it is important to have a process to grow macroscopic crystals. We have seen that silicon single crystals are grown using a technique called the Czochralski process. However, this technique is not applicable for most small-molecule organic semiconductors because of their low melting point (high vapor pressure). Organic crystals can be grown using a vapor phase or solution phase approach. In the vapor phase process, an organic source is heated to the point of sublimation; the vapor is carried using an inert gas and is allowed to cool and form single-crystal structures. In the case of the solution-based approach, the polycrystalline organic material is dissolved in a high boiling point solvent by heating the mixture. This is followed by cooling, which results in the precipitation of single-crystal organic material. This technique is, of course, not universally applicable, and is highly dependent on the choice of the organic material and the choice of the solvent. The solution-based approach is particularly useful for organic materials with very large molecular structures. These materials typically dissociate into constituent components before vaporization, thus the vapor-phase technique is not applicable. Solubility can also be increased by increasing the pressure of the mixture. A simple example is the dissolution of sugar crystals in hot water and their subsequent crystallization upon cooling. From an organic electronics point of view, we can take the example of hexathiophene which can be dissolved in benzyl phenyl sulfide sealed in a glass tube. Cooling of the system results in small platelets of hexathiophene.

In both solution-based and vapor-based processes, the properties of the crystal can be controlled by varying process parameters such as temperature, pressure, cooling rate, materials involved, and so on. A key advantage of the vapor-phase approach is that control of the temperature of sublimation of the source organic material can also lead to the high purity of the resulting crystals. Thus, the same process can be used for both purification and crystallization. Contrary to inorganic single-crystal materials such as silicon, germanium, etc. that have atoms as the base units, organic materials have complex and highly structured molecules as the base. Further, in the case of electronic organic compounds with conjugated π-bonds, there is no degree of rotational freedom around the bond, reducing the number of ways in which these materials can be incorporated into a crystal lattice. The complex structures and the organic moieties associated with these materials require energy to reorient into crystalline lattice and form crystals with low-symmetry unit cells. Thus, the mechanical and electronic properties of these crystals are highly anisotropic (directional). A practical ramification of this is the crystallization of rubrene (Figure 5.9), which is the organic semiconductor with the highest reported carrier mobility (around $20\ cm^2/v\text{-}s$ for holes). Vapor-phase process can be used to fabricate Rubrene crystals,

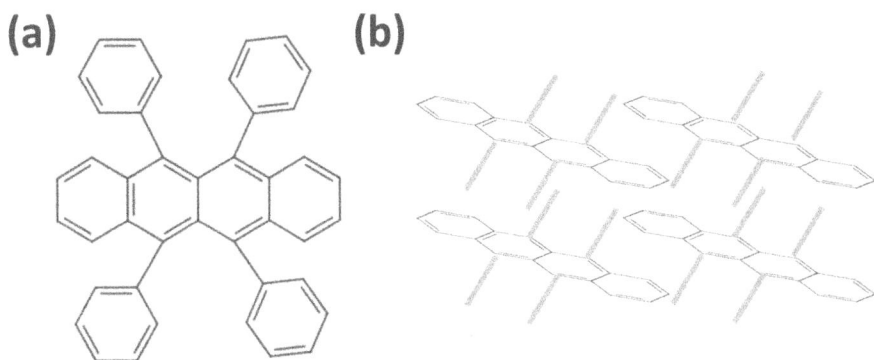

FIGURE 5.9 (a) A single molecule of 5,6,11,12 tetraphenylene tetracene (rubrene). (b) A single crystal of rubrene in the monoclinic polymorph. The grey lines represent phenyl groups perpendicular to the paper.

however, crystallization does not occur at low temperatures because molecules do not have enough thermal energy to reorient and position themselves in the crystal lattice. Another problem with complex molecule crystals is that there can be multiple polymorphs for a given organic molecule. Thus, organic crystals need to be grown within very tight process windows for the fabrication of crystals with reproducible properties and their subsequent integration into large-scale electronic systems.

With refinement in crystal growth methods, organic crystals of large size (hundreds of micrometers in dimensions) have been formed. These have been used to create field-effect transistors (FETs) on the surface of the crystal similar to the use of silicon to fabricate transistors. However, from a flexible electronics point of view, organic single-crystal substrates tend to be brittle, just like their inorganic counterparts, thus restricting the minimum bending radius possible. Further, creating large-area ultra-thin single-crystal organic substrates remains a challenge that will require more understanding of the growth mechanisms and additional fine-tuning of the growth processes. For example, in the case of Rubrene, thin large-area crystals have been reported with increased temperature of sublimation (to increase Rubrene vapor formation) and increased flow rate of the inert carrier.

5.8 ORGANIC FIELD-EFFECT TRANSISTORS (OFETs)

Field effect transistors are the workhorse of the modern electronic industry. Particularly, the use of complementary metal oxide semiconductor (CMOS) FETs has resulted in a revolution in the silicon-based electronics industry. Ideally, with the growing interest in organic single-crystal materials, these advances and know-how can be used to fabricate FET structures on organic substrates for specific applications such as displays and photovoltaics. However, even with the best growth techniques, the biggest crystals of organic electronic materials are of the order of a few centimeter square areas. Further, the large-area crystals that we are able to produce using organic materials tend to be extremely unstable or fragile or are not conducive to atmospheric exposure. On the other hand, state-of-the-art silicon substrates are

12-inches in diameter, only around a millimeter thick, and are mechanically, chemically, and thermally stable. It is clear that there is still a long way to go for organic crystals to be used as a substrate for batch fabrication of field-effect transistors for electronic systems. However, there are some advantages of making single FETs on these substrates and studying their electronic properties. Detailed study of electronic properties of FETs and the metal oxide semiconductor (MOS) structure can provide vital information about the behavior of the semiconductor substrate. For example, studying MOS capacitors and FETs based on a specific semiconductor can provide information about its electronic band structure, Fermi levels, carrier concentrations, mobility, charge transport regime, surface charge trap density, and so on. Further, studying these properties at low temperatures can provide additional insight into the semiconductor's behavior. Hence, it is desirable to fabricate FETs based on the large-area organic single crystals that have been successfully grown so far. This method allows for the study of organic materials in their pristine form, without interference from defect centers generally associated with polycrystalline or amorphous thin films. However, we need to create metal contacts and insulator thin films to form the MOSFET structure. Because these crystals are still relatively small and fragile, they cannot be used for thin-film processing such as deposition, lithography or etching, directly on their surface. Thus, some innovative techniques have been developed to fabricate FETs based on organic single-crystal materials.

One of the techniques used for fabricating FETs using organic crystals is the use of silicon or plastic substrate for metal contact and dielectric layer formation. In this method, the source/drain contact and dielectric layer are deposited and patterned on silicon or plastic substrate using standard thin-film processing techniques. The organic single crystal is then pressed against the prefabricated FET structure to complete the fabrication. This process allows for the use of any metal or dielectric thin film and any known process of deposition because the substrate for prefabricated structure is silicon. However, a major challenge is the formation of contact between the organic semiconductor and the metal lines. This is particularly important if the structure is to be used for analyzing the properties of the semiconductor because contact resistance at the source and drain can significantly reduce device performance. An interesting application of this technique is when there is no gate dielectric deposited, and the airgap between the source/drain terminals from the sides, gate metal from the bottom and organic substrate from the top, is used as the dielectric (Figure 5.10). In this configuration, the FET structure can be exposed to the external

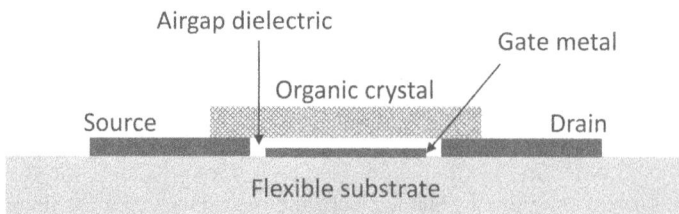

FIGURE 5.10 Organic FET formed by pressing an organic single-crystal material against the source, drain and gate metals prefabricated on a flexible substrate.

environment and can be effectively used as a gas sensor depending on the chemistry of the organic substrate used. Another technique for the fabrication of FETs is the direct deposition of metal and dielectric on the organic substrate. The metal deposition is done using techniques that are room-temperature and soft on the substrate, like evaporation or spraying coating. Shadow masks are typically used for patterning the metal lines. The dielectric layer is then deposited using evaporation or spin coating, followed by gate electrode deposition.

In conclusion, the use of organic materials for flexible electronic applications seems to be inevitable. There are many challenges yet to be overcome – such as low mobility compared to inorganic materials, processability, air stability, integration with current electronic systems, and so on. However, I use the word "inevitable" because the best computing system known to man – the human brain – is organic. With all the limitations organic materials have, we only need to look around in nature to see their true potential. Comparing a human brain to a computer, a cheetah to a car, or a pigeon to a drone, shows that organic materials, when used right, have the potential to revolutionize the world as we know it. Further, if the complete potential of designing systems with organic materials is harnessed, there can be electronic devices that are not only flexible and stretchable, but also grow with time and heal themselves when damaged. It is only our limitation that is keeping us from realizing a drone as efficient as an eagle.

EXERCISES

5.1. Draw the conjugated π-bond structure for the following organic polymers:
   a. Polyacetylene
   b. Polyaniline
   c. Polyphenylene
   d. Polypyrrole
   e. Polythiophene
   f. Poly(p-phenylene vinylene)
5.2. State the key differences in electronic conduction properties of organic electronic materials compared to their inorganic counterparts.
5.3. Why are organic materials better suited for the fabrication of LEDs compared to photovoltaics?
5.4. What is one method of fabricating organic field-effect transistors without exposing the organic material to the fabrication process steps?
5.5. Draw the structure of ABS thermoplastic, given that it is an alternating copolymer of the monomers, acrylonitrile (C_3H_3N), butadiene (C_4H_6), and styrene (C_6H_5-C_2H_3).

6 Metal Oxide Semiconductors

6.1 INTRODUCTION

Transition metals belong to the d-block of the Periodic table and are characterized by a partially filled d sub-shell, tendency to form complex coordination compounds, variations in oxidation states, catalytic properties, and compounds with vivid colors. In some cases, the oxides of transition metals such as zinc, copper, tin, etc. exhibit semiconducting properties with relatively large bandgaps (compared to inorganic lattice semiconductors) and high carrier mobility (compared to organic semiconductors). Many of these oxide semiconductors can be deposited as a thin film on a substrate for electronic device formation. In terms of carrier mobility, transition metal oxides (TMOs) are midway between organic semiconductors on the lower end and crystalline inorganic lattice semiconductors on the higher end. However, because these materials do not need to be crystalline, their deposition and use as a thin film open up several avenues of application. These are particularly suited to flexible electronic applications because of their tendency to form high-performance thin films on flexible substrates. There has been growing interest in metal oxide semiconductors since the 1960s when transistors based on zinc oxide were first studied and reported to have relatively high mobility. Further investigations into the charge transport mechanisms and properties led to the realization that many transition metal oxide semiconductors exist. Subsequent interest in transition metal oxide thin films has led to the discovery of many metal oxide materials with varying bandgaps, electronic properties, and deposition techniques.

6.2 METAL OXIDE SEMICONDUCTOR PROPERTIES

In order to understand the electronic properties of metal oxide semiconductors, it is important to know the mechanism of carrier formation and conduction in these materials. Transition metal oxides can be binary, with only one transition metal (such as ZnO, CuO, SnO, etc.), ternary, with two transition metals (such as $ZnSnO$, $InZnO$, etc.), or quaternary, with three transition metals (such as $InGaZnO$). Further, the ratio of the individual metals in the ternary and quaternary compounds can be different depending on the metals involved. Thus, a large variety of transition metal oxide semiconductors are possible, each with its unique electronic and material properties. In order to understand the conduction mechanisms involved in these materials, we will take the example of zinc oxide, because it forms the basis of many ternary and quaternary metal oxides and it is one of the most well-studied metal oxide semiconductors. Prior to its use in the electronics industry, ZnO had many industrial uses as a pigment, coating, and even as a medicine in ancient times. ZnO is a wide bandgap

DOI: 10.1201/9781003010715-8

semiconductor with a direct bandgap of 3.3 eV at room temperature, which makes it a transparent thin film (ideal for display-related applications). ZnO crystallizes in wurtzite structure (named after the naturally occurring ore, ZnS), and can be grown into large, stable, and transparent single crystalline structures.

Elemental zinc consists of 30 electrons distributed as $1s^2\ 2s^2\ 2p^6\ 3s^2\ 3p^6\ 4s^2\ 3d^{10}$ and oxygen has 8 electrons distributed as $1s^2\ 2s^2\ 2p^4$. In zinc oxide, the electronic states for the conduction band (lowest unoccupied molecular orbital) are determined by the 4s states of zinc, whereas those of the valence band (highest occupied molecular orbital) are determined by the 2p states of oxygen. The involvement of the 4s states of Zn in the formation of the conduction band is significant because s-orbitals are non-directional. Thus, the conduction of electrons in ZnO remains isotropic. Further, in the case of polycrystalline or amorphous thin films of ZnO, the non-directional nature of the 4s orbitals creates a higher probability of overlap with any nearest neighbor molecule, without the consideration of the angle of incidence, leading to a lower occurrence of scattering. Thus, electronic conduction and mobility remain relatively high in polycrystalline or amorphous ZnO compared to conventional semiconductors such as silicon or germanium. This is also the case for metal oxides with other metals such as copper (4s conduction) and indium (5s conduction).

For electronic conductivity, it is also important to have high free carrier concentration along with high mobility. Because the bandgap of ZnO is relatively high (~3.1 eV), the carrier concentration in the conduction band because of thermal excitation is low at room temperature. However, free carriers can be generated in ZnO at room temperature through native point defects or unintentional hydrogen presence. The native point defects can be in the form of oxygen vacancies, i.e., missing oxygen atom at a lattice site where it should be; or zinc interstitials, i.e., presence of zinc atom inside a lattice interstice. In some cases, interstitial hydrogen can also be responsible for free carriers. These hydrogen atoms can be a result of unintentional doping during the thin film or crystal formation. In any case, the control of conductivity through the control of the concentration of charge carriers is an important feature of any semiconductor. This is conventionally done by introducing dopants into the semiconductor lattice. In the case of ZnO, conductivity can be controlled using substitutional dopants or by intentionally introducing native point defects.

Because of the native point defects and some unintentional doping during formation, ZnO is generally highly n-type. While this suffices and the use of dopants to control conductivity provides an important tool toward the formation of some electronic devices such as thin-film transistors (TFTs), in order to truly unlock its potential in all areas of electronics, it is important to form p-type ZnO. This is particularly important where a junction of p and n-type materials is required, such as in the case of CMOS processors, LEDs, photovoltaics, lasers, and so on. It is relatively easy to increase the n-type conduction of ZnO by using deficient oxygen or excess zinc during deposition. This increases the probability of O-vacancies or Zn-interstitials. Further, doping with aluminum and gallium has been shown to increase n-type conductivity in ZnO. However, obtaining p-type ZnO is a challenge, and only recently some promising results have been achieved. As in the case of silicon, the substitution of group III element leads to p-type conduction, in ZnO, the most promising candidates for p-type dopants are the elements from group V. These elements when introduced as a substitutional impurity for oxygen, promise to provide free holes for p-type conduction. However, most candidate dopants for p-type conduction in ZnO create deep acceptor levels leading to low levels of ionization at room temperature. In particular, nitrogen-doped ZnO thin film has been deposited using various methods and has been extensively studied for p-type conduction, however, p-type conduction is not comparable to n-type ZnO that can be obtained using various techniques. Some metal oxide thin films that are predominantly p-type as-deposited include CuO_x and SnO_x. However, the high formation energy of metal cations limits the generation of free holes, leading to lower free carrier concentration. Further, because the main conduction pathway for holes, the valence band maximum, consists of zinc 3d states and oxygen 2p states, it is anisotropic. This leads to directional dependence on carrier conduction leading to lower mobility in polycrystalline and amorphous thin films because of grain boundary effects. Even so, metal oxides can truly break into traditional CMOS roles only when the n-type and p-type nature of an individual thin film can be selectively and significantly controlled using doping, which is yet to be reported in the literature.

6.3 TERNARY AND QUATERNARY METAL OXIDES

Transition metal oxides with two or more metal cations can provide a means to tune some of the properties of this material system. Ternary metal oxides are formed using two metal cations and can be represented by the formula ABO_x, while quaternary metal oxides are formed using three metal cations and can be represented by the formula $ABCO_x$. The letters A, B, and C each represent a particular transition metal such as Zn, Cu, V, Bi, Sn, In, Mn, and so on. Further, for a given combination of metal cations, many ratios of individual metals are possible, however, not all ratios are stable at room temperature. These combinations are generally represented using a phase diagram – a triangle in the case of quaternary metal oxides. The vertices of the triangle represent the three binary metal oxides from which the compound is formed. The three lines connecting the three binary metal oxides represent the ternary metal oxides. The area inside the triangle represents the quaternary metal oxide. The proportion of the three metals depends on the relative distance of the point from the vertices. For example, consider a quaternary metal oxide phase diagram for the system represented by ABCO, as shown in Figure 6.1. It will have AO_x, BO_y, and CO_z as the vertices. The line connecting AO_x and BO_y represents the compounds of the ternary system ABO. Similarly, the line connecting BO_y and CO_z represents the BCO system, and the line connecting AO_x and CO_z represents the ACO system. The mid-point of line AB represents the ternary compound $ABO_{(x+y)}$ with equal quantities of A and B, whereas the point at a two-thirds distance from A represents the ternary compound $AB_2O_{(x+2y)}$. Similarly, the distance of a point on the lines from a particular vertex represents the ratio of the metal in the ternary compound. All the points inside the triangle represent the quaternary system ABCO. In general, for a point at a distance p, q and r from vertices A, B and C respectively represents the compound given by, $A_{\left(1-\frac{p}{a}\right)}B_{\left(1-\frac{q}{a}\right)}C_{\left(1-\frac{r}{a}\right)}O_{\left(1-\frac{p}{a}\right)x+\left(1-\frac{q}{a}\right)y+\left(1-\frac{r}{a}\right)z}$, where a is the

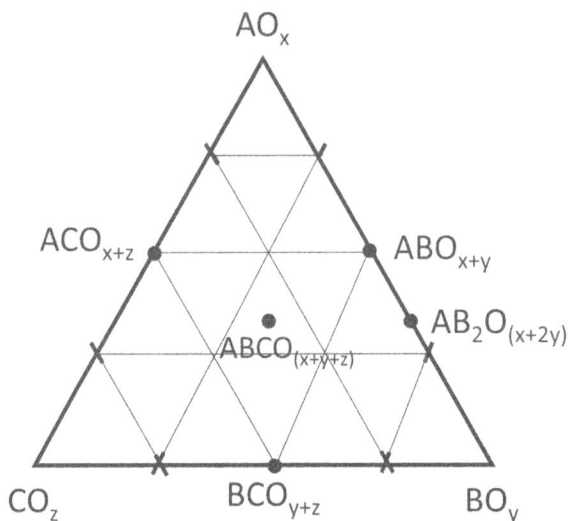

FIGURE 6.1 The graphical representation of the quaternary oxide system helps develop trends of physical properties of the compounds according to their relative concentration in the compound.

side of the triangle. The properties of a compound of such a system vary according to the ratio of the constituent cations and can be tuned by changing the relative ratios of cations. However, it should be noted that not all the points on the lines and inside the triangle are achievable due to lattice stability considerations.

There are many examples of ternary metal oxide semiconductors being studied for their possible applications in flexible electronics. Some of the well-studied ternary systems include the indium-gallium, indium-zinc, zinc-tin, and zinc-gallium systems. An oxide of the indium-tin system, Indium tin oxide (ITO), is one of the most studied metal oxide thin films because of its use in commercial applications as a transparent conductive oxide (TCO) thin film. Although it has a large bandgap (around 4 eV), the free electron concentration at room temperature, because of the presence of point defects, is sufficient for high electrical conductivity. The large bandgap also leads to the thin film being transparent. Consequently, it is widely used as a conductive thin film for touch screen applications in mobile phones and other electronic displays. The conductivity obtained in ITO thin films depends heavily on the composition of the thin film, the deposition process, and the post-processing, however, it should be noted that the conductivity, even in the best case, is several orders of magnitude below metal thin films. Hence, ITO thin films are used as conductors only for applications with low current requirements, and where transparency is of prime importance.

An example of a phase diagram of a practical quaternary system is the IGZO system, composed of oxides of metals zinc (ZnO), gallium (Ga_2O_3), and indium (In_2O_3). The IGZO system is one of the most studied quaternary transition metal oxide semiconductor systems. The compound with equal concentrations of all the constituent cations, i.e., $InGaZnO_4$, has been commercially deployed in modern television displays. The bandgap of this semiconductor is 3 eV, which makes it a transparent thin film, ideal for display device applications. This compound lies at the circumcenter of the IGZO phase diagram because it is equidistant from all the vertices.

6.4 DEPOSITION TECHNIQUES

As in the case of organic semiconductors, there are two main routes through which thin films of metal oxide semiconductors can be obtained – vacuum deposition and solution processing. However, there are several considerations to be made specifically for metal oxide deposition. In this section, we will discuss the processes most commonly used to obtain metal oxide thin films.

6.4.1 VACUUM DEPOSITION

Vacuum deposition involves the use of a vacuum chamber, and for metal oxide deposition, can take the form of sputtering, atomic layer deposition (ALD), or pulsed laser deposition (PLD). One of the most common modes of deposition is the use of sputtering, in which, a plasma of gas is targeted at a source and the ejected material is deposited on the substrate (Figure 6.2). The deposition of metal oxides, such as zinc oxide, can be achieved by bombarding a zinc oxide target with a plasma of an inert gas, such as argon. However, we can also achieve the deposition of oxides using a process called reactive sputtering. In this case, an oxygen plasma is created and targeted at a pure metallic source. The metal atoms ejected because of ion bombardment react with the oxygen plasma to produce metal oxide and deposit onto the substrate as a thin film. This process provides control over the amount of oxygen saturation in the resulting thin film by controlling the flow of oxygen, the pressure, and the temperature of the chamber. As discussed previously, the role of O-vacancies

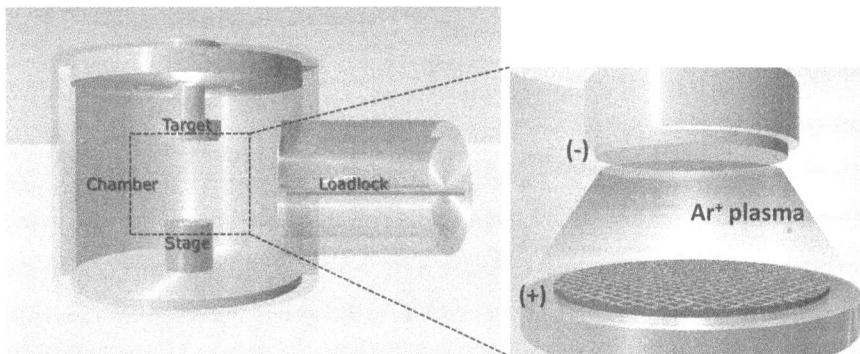

FIGURE 6.2 An illustration of a sputtering tool with a load-lock to facilitate quick loading and unloading of the wafer, without changing the base pressure in the chamber. The plasma is created using RF power sources and a DC field is maintained to accelerate the ions toward the target.

can be important in determining the electronic properties of the oxide thin films. In practical scenarios, a mixture of an inert gas (such as argon) and oxygen is used for reactive sputtering to afford more control over oxygen saturation, deposition rate, thin-film morphology, and so on. To obtain ternary or quaternary compounds, a technique called co-sputtering can be used. In single-target sputtering, the target is usually placed directly above the substrate to obtain a uniform thin film deposition. This configuration cannot be used for multiple source metals simultaneously. In the case of co-sputtering, the target is positioned at a tilted angle relative to the substrate surface. Because this configuration changes the distance profile between the substrate and the target, the target is moved off the rotational center of the substrate to obtain acceptable thin film uniformity. With this configuration, multiple metal targets can be installed to achieve simultaneous deposition of oxide thin film. This method is useful for creating ternary or quaternary thin films because it provides precise control over the ratios of the cations to produce the desired thin film. However, because this is a physical deposition process, the metal cations do not completely interact with each other, leading to possible variations in properties across the thin film. The problem can be somewhat mitigated by annealing the thin film after deposition. Complex metal oxide thin films are also obtained by simple sputtering of targets that are ternary or quaternary oxides themselves. An example of this is the sputter-deposition of ITO using an ITO target and argon plasma.

A variation of this process involves the use of lasers to eject material from a source target. The process is called pulsed laser deposition (PLD) and is also a form of vacuum deposition of metal oxide thin films. The incidence of a high-energy laser beam on the target increases the temperature of the local area causing evaporation. Along with this, there are other effects like exfoliation of a chunk of material because of intense lattice vibration, ejection of electrons due to the photoelectric effect, and the formation of ions because of electron ejection. All these effects cause a plume of material, which includes atoms, clusters, and ions, to eject from the target toward the substrate and deposit on the substrate as a thin film. In the case of metal oxides, the deposition is carried out in the presence of oxygen gas so that the stoichiometry of the thin film is maintained. Like sputtering, PLD can be used for depositing ternary and quaternary metal oxide thin films by using multiple targets simultaneously. The laser is pulsed sequentially over multiple targets to obtain the desired cation composition.

Atomic layer deposition is another vacuum deposition process that has become popular for metal oxide thin film depositions. The process works by injecting precursors into the chamber one pulse at a time, followed by an inert gas purge. The purge removes the precursor gas from the chamber, except for the molecules that have adsorbed on all the available surfaces, such as the substrate surface, chamber walls, and so on. This layer of material then reacts with the incoming precursor to form an atomically thin film of the material to be deposited. The process is widely used in the semiconductor manufacturing industry for the deposition of high-κ dielectrics (that are also transition metal oxides like HfO_2). The precursors used for metal oxide ALD are generally organo-metallic complexes for the transition metal cation and water vapor, oxygen, or ozone for the oxide part. Because the ALD process produces a single atomic layer of the thin film in one cycle, it is the most controllable technique in terms of deposition thickness. Further, because the process involves surface adsorption of precursors in every cycle, the deposited film is extremely conformal. Deposition can even be obtained on the underside of microstructures on the substrate surface using ALD. Furthermore, there is a high degree of crystallinity in the deposited thin film because the adsorption and deposition of subsequent atomic layers keep following the template of the underlying layers. Usually, the substrate is kept at a mildly elevated temperature (100–200°C) for the reactions to take place. The temperature is not too high to discourage the use of flexible plastic substrates for ALD deposition.

As an example of the process, let us consider the deposition of ZnO thin films using ALD. The most commonly used precursor for zinc is diethyl zinc ($[C_2H_5]_2Zn$), while water is used as the source of oxygen. Diethyl zinc (DEZ) is a highly reactive organo-metallic reagent, that ignites spontaneously when it comes in contact with air (pyrophoric). It is used in many synthetic organic chemistry reactions as a source of the ethyl group, however, in this case, it is used as a source of zinc atom. During the deposition, a single cycle consists of pulsing of the DEZ precursor, followed by a purge with nitrogen. This is followed by injection of water vapor, followed by another nitrogen purge. The four-step process deposits a single atomic layer of ZnO on all the exposed surfaces inside the chamber. During the process, the use of water vapor keeps the surfaces inside the chamber hydroxylated. Some of these oxygen molecules remain as water molecules adsorbed to the surface while others are in the form of hydroxyl (-OH) groups. The zinc atom of the incoming DEZ molecule from the next cycle of deposition preferably adsorbs to the water molecule. The ethyl group attached to this molecule is removed in the form of ethane in a ligand-exchange reaction with the neighboring hydroxyl group. The ethane molecule thus formed is removed from the chamber through the vacuum pump. The remaining ethyl group accepts a hydrogen atom from the water molecule attached to the zinc atom, in a second ligand-exchange reaction, creating another ethane molecule. The zinc atom thus bonds with two oxygen atoms and the next injection of water vapor creates the hydroxylation of the surface for the next cycle to start. The process is schematically illustrated in Figure 6.3. It should be noted that the ALD process is a complex chemical process that is set up for the specific precursors and the reaction mechanisms involved. This is a key disadvantage of the ALD process, that it is not a universal process that can be used for any transition metal oxide thin film, particularly for complex thin films like ternary and quaternary films. Further, because only a small fraction of the precursor injected into the chamber is actually used to create the thin film, ALD is an expensive process overall. This is particularly problematic because organo-metallic precursors used in the process can be very expensive, particularly those involving rare transition metals such as indium and gallium.

For the growth of ternary or quaternary metal oxides using ALD, we can use the concept of a "supercycle". For example, the indium-zinc oxide (IZO) thin film can be grown by depositing one atomic layer of ZnO, using the four-step process previously discussed, followed by a layer of In_2O_3, using a separate four-step cycle. The precursors in the case of indium oxide deposition are [1,1,1-trimethyl-N-(trimethylsilyl) silanaminato]-indium (InCA-1) for indium and hydrogen peroxide (H_2O_2) for oxygen. The entire eight-step process is then called the IZO supercycle. Hydrogen peroxide

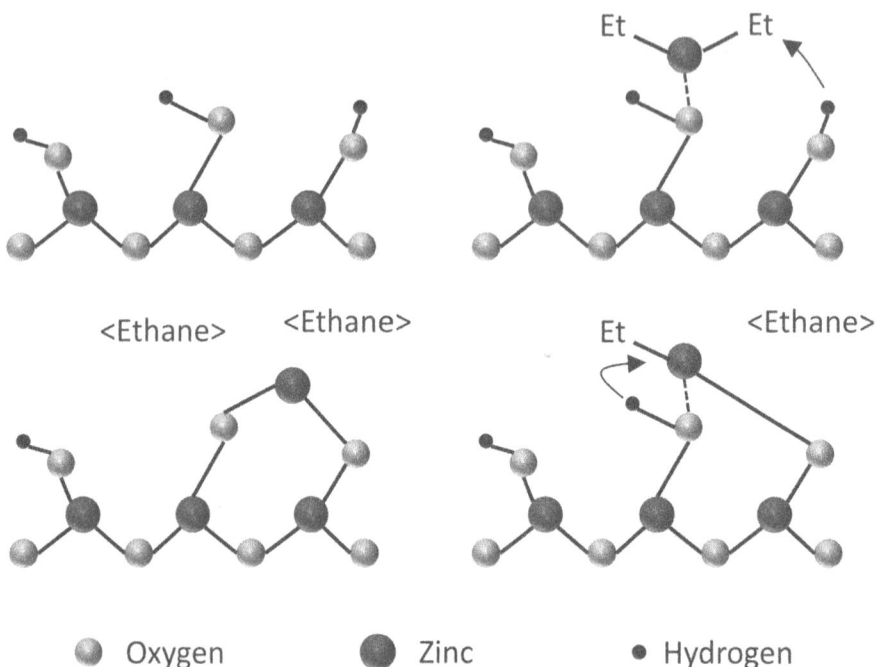

FIGURE 6.3 The ALD deposition of ZnO starts by adsorption of ZnO on the hydroxylated surface, followed by two ligand exchange reactions leading to the creation of two Zn-O bonds and two ethane molecules.

can also be used with DEZ for ZnO deposition, thus, it is used as the sole source of oxygen for the IZO supercycle to simplify the setup. IGZO, which is an important quaternary metal oxide, can also be deposited using a supercycle including the same four-step process for the deposition of gallium oxide using trimethylgallium (TMGa) as a precursor. The relative amounts of In, Ga, and Zn in the final IGZO thin film can be controlled by creating a larger supercycle consisting of a variable number of individual four-step cycles for each metal oxide deposition.

6.4.2 SOLUTION PROCESSING

Solution processing is a general term used for any process involving the deposition of material using a solution, such as a spray deposition, spin coating, inkjet printing, sol-gel process, and so on. Solution processing can be a low-cost approach for obtaining metal oxide thin films because it eliminates the requirement of a vacuum chamber. Further, because it can have much higher throughput, it offers a more scalable solution for mass manufacturing. It also leads to less wastage of precursor material compared to vacuum deposition processes such as ALD, which is significant while using rare and expensive materials such as indium. However, for the specifications of electronic devices to fall inside a particular performance window, it is important to realize all the thin films inside a given process window. The narrower the performance window, the narrower the process window. Thus, the mathematical modeling and control of solution processing techniques are extremely important to achieve the desired thin-film properties. The general process of deposition of a solid thin film using solution processing involves the creation of a liquid solution containing the thin-film material, and the subsequent removal of the solvent, usually through evaporation. This is true for most of the processes used such as spray coating, spin coating, ink-jet printing, and doctor blading. The challenge in these processes is to obtain a solution of the material with the right viscosity and surface tension for the deployment of a particular technique.

The most common method of obtaining a solution or an "ink" for a metal oxide is to have nanoparticles of the material suspended in a solvent. When the solvent is evaporated, the suspended nanoparticles agglomerate to form a thin film. The advantage of this process is that it can be carried out at low temperatures (depending on the solvent used). It should be noted that the process of creating nanoparticles can be a high-temperature process, such as chemical vapor deposition, however, in most cases, it can be scaled to produce the required metal oxide nanoparticles in large quantities. The actual deposition of the liquid droplets on a substrate can be carried out using spraying coating, spin coating, or ink-jet printing. Once the solvent has evaporated, the nanoparticles condense into a thin film. To obtain good thin-film characteristics, it is important to have good contact between the individual nanoparticles. This is sometimes done by annealing the thin film to make sure that the solvent has completely evaporated. This process of deposition is particularly useful for metal oxide thin films because of the 4s or 5s band-based conduction which increases the probability of electronic conduction between adjacent nanoparticles. For example, commercially available nanoparticle ink of ITO from Ulvac has a 4-nm average particle size. The sheet resistance of a 200 nm film formed using this ink has been reported as 350 Ohm/sq, after annealing at 230°C for 60 minutes. The ink can be used for ink-jet printing patterns on flexible substrates followed by vacuum annealing to obtain ITO thin films. Similarly, ZnO nanoparticle ink with 10–15 nm particle size and isopropyl alcohol as the solvent is commercially available. While the nanoparticle solution-based processes are very versatile, they produce a nano-crystalline/amorphous thin film with many grain boundaries and defects because of the lack of interaction of adjacent nanoparticles. The resulting thin film is, after all, only a collection of particles, not a homogenous thin film.

To overcome this disadvantage of nanoparticle inks, the sol-gel process is used. The process involves the precipitation of solid particles or a dense network from a precursor-containing solution. In the case of metal oxide thin films, the precursors are metallic compounds like alkoxides, nitrides, hydroxides, and so on. The choice of precursor and the solvent depends on the metal oxide to be deposited and the required morphology of the deposited thin film. In general, the sol-gel process can be used to obtain nanoparticles, fibers, thin films, and solid structures as well. The process starts with a solution of the metal precursor compound in a solvent, such as water or alcohol. In some cases, a catalytic reagent is added to speed up the process. The metal compound undergoes hydrolysis (the process of undergoing a bond rupture due to the presence of water), followed by polycondensation to form a metal oxide dispersion, called the "sol". Further condensation leads to the conglomeration of individual particles, leading to the formation of a wet gel. The gel can then be heated to remove the residual water and moisture leading to a dense ceramic. In the case of thin films, the solution or the sol can be deposited on the substrate using one of the coating methods like spin-coating, spray-coating, or printing. The gel-formation and densification can then be done in-situ. Because the chemical decomposition and subsequent conglomeration lead to fewer barriers to particle-particle interactions, the properties of the thin films produced using sol-gel are better compared to the ones produced using nanoparticle inks. An illustration of the sol-gel process for thin film formation is shown in Figure 6.4.

The final densification step, in the sol-gel process, can require high temperatures, commonly above 300°C, but ideally more than 500°C. This can be detrimental to the use of sol-gel with flexible substrates. An approach to reduce the final anneal temperature is to control the anneal environment. It has been reported that the use of water vapor or an oxygen-rich environment during the final anneal, allows for a lower temperature. This also leads to thin films with lower defect density and carbon/hydrogen residue from the solvent. Besides this, other forms of excitation like microwave radiation or UV radiation can be used for the densification step, leading to a lower temperature requirement. Another problem with the sol-gel process, particularly for transition metal oxides, is that the control of the rate of hydrolysis for a uniform film deposition can be difficult given the instability of some of the transition metal precursors. However, for any given metal oxide thin film, there can be many precursors, solvents, catalysts, and stabilizers that can be used to obtain the desired film properties. Apart from these, the reaction temperature and deposition rate can provide further control over thin-film properties.

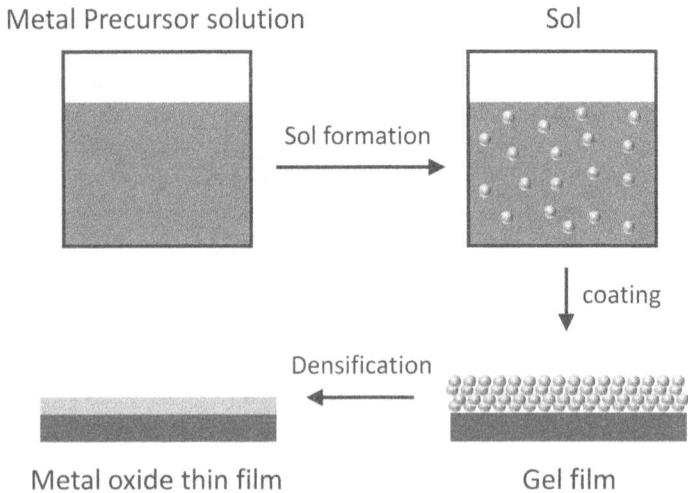

FIGURE 6.4 The sol-gel process for thin-film fabrication involves the creation of a sol using metal precursors dissolved in an appropriate solvent. The sol is then coated on a substrate using a process such as spin-coating, followed by densification of the film to lead to the final metal oxide thin film.

To take an example of the sol-gel process, let us consider the deposition of ZnO thin films using the sol-gel process. One of the simplest ways of depositing ZnO using the sol-gel process is to use zinc acetate ($[CH_3COO]_2Zn$) as the precursor and ethanol as the solvent. The precursor undergoes hydrolysis in the presence of water and alcohol. This is followed by the condensation of two precursor molecules to form a Zn-O-Zn bond, subsequently leading to ZnO particle formation. The solution can then be spin-coated on a substrate followed by annealing at 300°C for film densification. The sol-gel process can also be used for obtaining ternary and quaternary metal oxide thin films, however, the control of reaction rate of individual metal precursors and obtaining a uniform film composition throughout the film become major challenges. Given the versatility and scalability of the process, some efforts have been made to fabricate thin films of ternary and quaternary metal oxides using the sol-gel process. An example of this is the deposition of zinc tin oxide (ZTO) thin films. Stannic chloride ($SnCl_2$) is used as the precursor for tin, whereas zinc acetate or zinc chloride ($ZnCl_2$) can be used as the precursor for zinc. These precursors are separately dissolved, followed by the mixing of the two precursor solutions for the formation of zinc stannate (Zn_2SnO_4). The sol can then be used to coat a substrate for gel formation, densification, and thin film formation. The sol-gel process has also been reported for quaternary compounds such as indium gallium zinc oxide (IGZO). In this case, nitrates of all the metals are used as precursors, ie, indium nitrate, gallium nitrate, and zinc nitrate, while 2-methoxyethanol (2-ME) can be used as the solvent. The precursor solutions are then mixed to form the IGZO sol, followed by spin coating and film densification anneal. The ratio of mixing can vary depending on the required film properties. For instance, an excess of indium can lead to high electron concentration, whereas the presence of excess gallium suppresses free electron concentration. In some cases, the sol-gel process is enhanced by adding prefabricated nanoparticles to the solution. This leads to embedded nanoparticles inside the dense sol-gel film leading to improvement of the overall properties of the thin film. The embedded nanoparticles can be used to change the stoichiometry of the film, particularly in the case of ternary and quaternary metal oxide thin films, leading to the enhancement of specific film properties.

6.5   METAL OXIDE TFTs

We have studied the concept of TFTs in Chapter 3. We know that silicon (amorphous and poly-) and organic electronic materials can be used to fabricate TFTs on flexible substrates. Similar techniques can be used for the fabrication of metal oxide TFTs both on rigid and flexible substrates using the deposition techniques for metal oxide thin films discussed in the previous section. The performance of metal oxide thin films (in terms of carrier mobility) lies between amorphous silicon and LTPS (low-temperature polysilicon). However, the key advantage provided by these thin films for TFT application is the scalable processes used for deposition of decent quality metal oxide thin films, particularly compared to LTPS films, without requiring complicated processes such as excimer laser annealing. A key cause of concern, however, has been the instability in threshold voltage of the TFTs based on metal oxide thin films, particularly in the long term and over repeated usage. A variation in threshold voltage can cause serious problems in the operation of the overall system, leading to higher-than-expected power consumption or, in the worst case, total system failure. Compared to the information available regarding the physics of silicon-based TFTs, the understanding of the underlying physics of threshold voltage instability in metal oxide thin films is limited. As a result, reliability issues and their causes have been the major focus area for metal oxide thin film research over the last decade, leading to some physical reasons for threshold voltage instability being identified, such as molecular absorption, charge traps, semiconductor/oxide surface defects, oxygen vacancy defect states, and so on. These insights have led to the development of optimized processes for deposition, device architecture, integration, and fabrication of metal oxide TFTs, paving the way for commercial deployment of systems based on metal oxide TFTs.

6.5.1 WAVY CHANNEL TFTS

The length and width of a transistor are important parameters that determine its electrical response. The output current of a metal oxide semiconductor (MOS) transistor depends on the width-to-length ratio (W/L ratio), the mobility of the channel material, and the capacitance (per unit area) of the gate. While the mobility and capacitance depend on the thin film used and the integration process, the W/L ratio can be controlled for an individual transistor for a given process. The control is generally exercised using the lithography process to pattern the thin film itself, or to pattern the gate stack, such that a specific area of the transistor is affected by the gate field. An increase in the width of the transistor leads to an increase in the saturation current (maximum output current). Higher current leads to more charge flow (per unit time) toward the output capacitance, which causes the output voltage to be more responsive to the gate field. This leads to a "faster" transistor. Further, transistors with higher current carrying capacity can be used to drive larger loads. For example, if used to control the pixels in an LED display, the TFTs with higher current capacity can lead to a brighter LED response. Thus, increasing the saturation current in TFTs is the underlying focus area when thin films with higher mobility and gate dielectrics with a higher dielectric constant are studied.

The goal of increasing the W/L ratio for higher current capacity in TFTs can also be reached using novel transistor architectures. One such unique device architecture is known as the "wavy channel" TFT architecture. In this approach, the substrate is patterned into fins before the deposition of the channel and gate stack material (Figure 6.5). This leads to a larger operational width of the transistor compared to coplanar transistors with the same substrate footprint. The increase in the width of the channel depends on the number of fins and their height, width, and pitch. The dimensions and pitch of the fins are chosen such that they can be reliably patterned and fabricated on the substrate while maximizing the operational width of the resulting transistors. Once the substrate is patterned according to the required fin structure, the metal oxide thin film, the gate dielectric, and gate metal are deposited on the substrate. For reliable device performance, it is important to have completely conformal coverage of the fins with the channel material and the gate dielectric. This is achieved using the ALD process for channel and gate dielectric thin films.

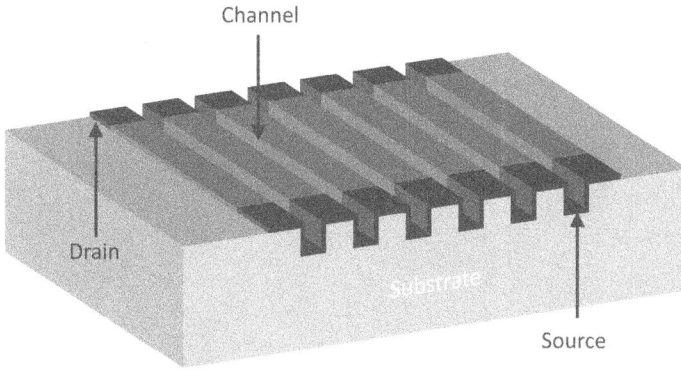

FIGURE 6.5 The wavy channel TFT comprises fins etched into the substrate before the deposition of the semiconducting channel, leading to a larger width of the TFT compared to a planar TFT with the same substrate footprint.

For example, a wavy channel transistor can be constructed using ZnO as the channel material and Al_2O_3 as the gate dielectric, both deposited using ALD. The process can be carried out in the same vacuum chamber without breaking the vacuum to obtain better semiconductor/oxide interface properties. The process, if performed at low temperature, can also be done on flexible substrates to make flexible wavy channel TFT arrays. However, the requirement of complete conformality and the use of ALD for channel material restricts the wavy channel architecture to metal oxide TFTs.

In this chapter, we discussed the potential for metal oxide thin films as a potential candidate for the fabrication of TFTs for flexible electronic systems. There are several transition metals with semiconducting oxides with a variety of electronic, optical, mechanical, and thermal properties. Further, the combination of transition metals into ternary and quaternary compounds leads to additional fine-tuning in film properties. These thin films can be deposited on the substrate of choice using various processes such as ALD, sputtering, sol-gel, nanoparticle ink, and so on. With high mobility, film stability, myriad compositions, and a variety in deposition processes, metal oxides are a promising material class for future flexible electronic system design.

EXERCISES

6.1. What is the fundamental reason for the isotropic conduction of electrons ZnO thins?

6.2. Draw the phase diagram for the IGZO system and indicate the following compounds on it: $InGaZnO_4$, ZnO, In_2GaO_x, $In_2Ga_3Zn_4O_x$, $GaZnO_x$.

6.3. Describe the steps for the sol-gel process for deposition of ZnO and zinc tin oxide (ZTO) thin films.

6.4. In the wavy channel architecture, we create fins with 5 μm width, 10 μm depth, and with fin pitch of 10 μm. What is the percentage enhancement in the width of the transistor compared to a planar transistor of the same substrate footprint, for a 10 fin structure?

7 III-V Semiconductors

7.1 INTRODUCTION

So far, we have discussed the semiconductors from group IV of the Periodic table (silicon) and the combination of materials from group II and group VI (such as zinc oxide). We can also form semiconductors by combining elements from group III and group V of the Periodic table. These material systems are very important for modern electronic systems, particularly from the point of view of displays and LEDs. These materials are known as III-V semiconductors. Group III of the Periodic table consists of boron (B), aluminum (Al), gallium (Ga), and indium (In), whereas group V consists of nitrogen (N), phosphorous (P), arsenic (As), and antimony (Sb). The four elements from group III and group V can be combined in diatomic compounds in 16 possible combinations (Figure 7.1). A stoichiometrically equal combination of elements from these two groups leads to the creation of III-V compound semiconductors. Examples of III-V semiconductors include gallium arsenide (GaAs), gallium nitride (GaN), indium phosphide (InP), and so on. Further, as in the case of transition metal oxide semiconductors, these material systems can also be combined to form ternary and quaternary systems such as indium gallium arsenide (InGaAs), and indium gallium arsenide phosphide (InGaAsP). In any case, the elements are combined in such a way that the total moles of elements from group III and group V in a compound are equal. Thus, for a ternary system with two elements from group III, the system can be generalized as $III_{(1-x)}III_xV$, for example, $In_{0.53}Ga_{0.47}As$. In the case of quaternary compounds, the generalization can be written as $III_{(1-x)}III_xV_{(1-y)}V_y$. Thus, there are limitless combinations of III-V compound semiconductor systems that can be created using a variety of fabrication processes. A change in the proportion of different elements brings about a change in the electronic, chemical, mechanical and thermal properties of the material, thus providing a way to tune the properties of a system in the desired way.

7.2 PROPERTIES

As always, we look at the structure of the semiconductor lattice to determine its electronic properties. III-V compound semiconductors form covalent bonds with their neighboring atoms and crystallize into a lattice. The lattice structure can be the diamond lattice (or zinc blende lattice), or the hexagonal lattice (wurtzite lattice). The electronic properties of the lattice depend on its band structure, which can be determined by the solution to the Schrodinger equation for the periodic distribution of the atomic nuclei of group III and group V elements in the lattice structure. As in the case of silicon, the lattice periodicity is used to obtain the solutions to the possible energy states available for electrons to occupy (Bloch theorem). The occupation probability of these states is then calculated for a given temperature using Fermi-Dirac statistics. Finally, the bandgap, the free electron concentration at room

DOI: 10.1201/9781003010715-9   **95**

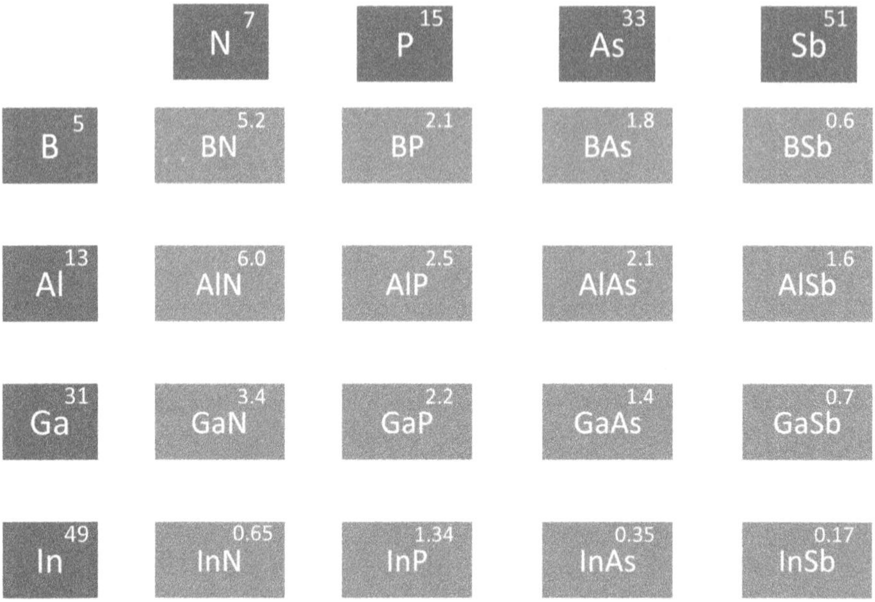

	7 N	15 P	33 As	51 Sb
5 B	5.2 BN	2.1 BP	1.8 BAs	0.6 BSb
13 Al	6.0 AlN	2.5 AlP	2.1 AlAs	1.6 AlSb
31 Ga	3.4 GaN	2.2 GaP	1.4 GaAs	0.7 GaSb
49 In	0.65 InN	1.34 InP	0.35 InAs	0.17 InSb

FIGURE 7.1 III-V combinations for the elements in the group III and group V of the Periodic table. The values at the top corner indicate atomic number of the elements, and the approximate reported bandgap for the compounds (in eV).

temperature, and the effective mass of carriers (and hence their mobility in the lattice) can be calculated. We will not delve into the details of these mathematical analyses in this book, but some interesting insights can be obtained by observing these properties for III-V compound semiconductors.

Let us start with the bandgap. The key difference between the III-V material system and the group IV materials (and their alloys) is that the bandgap of most III-V materials is "direct", whereas that of group IV materials is "indirect". This property of the bandgap comes from the structure of the top of the valence band and the bottom of the conduction band and is best represented through an E-k diagram, which represents the allowed energy states for electrons in a lattice corresponding to the allowed wavevector values. According to quantum mechanics, these E-k values are the only states in which electrons are allowed to exist. Consider the E-k diagram shown in Figure 7.2. The top of the valence band aligns with the bottom of the conduction band, i.e., there is no change in the wavevector between these two energy states. Such a bandgap is referred to as a direct bandgap. Conversely, if there is a change in the wavevector going from the top of the valence band to the bottom of the conduction band, then the bandgap is referred to as an indirect bandgap. This property of a semiconductor can have a significant impact on its behavior, especially pertaining to its optical properties. We know that an electron can be excited from the top of the valence band to the bottom of the conduction band, possibly giving rise to free carriers (electron and hole). This excitation is generally done by providing energy equal to the bandgap of the semiconductor. The requisite energy can be sourced from phonons, which are

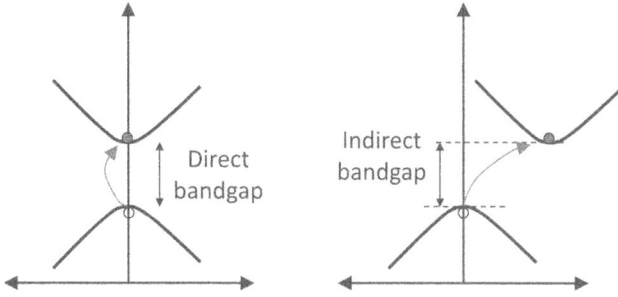

FIGURE 7.2 E-k diagram for direct and indirect bandgap shows the large change in the wavevector for an electron transition from valence band to conduction band in case of indirect bandgap.

lattice vibrations arising out of, say, the temperature of the lattice (thermal phonons), or photons from an external light source. However, there is a key difference between these two sources. While both can have the energy required to excite an electron into the conduction band, the momentum associated with both of these particles is very different. Electron-phonon interactions typically have a higher change in momentum associated with them compared to electron-photon interactions. Because both energy and momentum need to be conserved for any particle interaction, the same is true for electrons being excited from the valence to the conduction band. Thus, in the case of direct bandgap semiconductors, electron excitation and relaxation have a higher probability of occurrence because of electron-photon interactions (low change in wavevector), whereas in indirect bandgap semiconductors, electron excitation and relaxation have a higher probability of occurrence because of electron-phonon interactions (large change in wavevector). Thus, in direct bandgap semiconductors, electrons from the excited state have a high probability of radiating a photon to reach the ground state. This makes direct bandgap semiconductors, i.e., III-V material systems, the basis of very efficient light-emitting devices. On the other hand, established group IV materials such as silicon are not usable for light emission because of their indirect bandgap.

Another interesting feature of the III-V material systems is the tuning of the bandgap with stoichiometry in ternary and quaternary systems. This is important because the bandgap of a semiconductor determines the energy of the photons emitted, which determines the color of the light emitted. Thus, by changing the bandgap, we can theoretically design light-emitting devices of any required color. In many ternary systems of the form $A_x B_{(1-x)} C$, the bandgap of the semiconductor forms a quadratic equation with the molecular fraction x:

$$E_g = x E_{AC} + (1-x) E_{BC} - b \, x \, (1-x)$$

where E_{AC} and E_{BC} are the bandgaps of the III-V compounds AC and BC, and b is called the bowing parameter. The value of b depends on the three elements in the lattice and can be theoretically estimated using mathematical modeling or can be experimentally determined. Thus, the bandgap of a III-V compound can be precisely engineered by controlling the relative molecular ratios of the elements. However, a change in molecular ratios brings about a change in all the properties of the material. In particular, one key parameter to consider is the lattice constant of the resulting crystal. The change in the lattice constant of a ternary material is approximately given by the linear weighted mean of the lattice constant of individual components (Vegard's law). Thus, the lattice constant of materials of the form $A_xB_{(1-x)}C$ is given by:

$$a = xa_{AC} + (1-x)a_{BC}$$

where a_{AC} and a_{BC} are the lattice constants of compounds AC and BC respectively. However, in many cases, the quadratic bowing formula discussed for bandgaps is also applied as a general case of the Vegard's law for lattice constants. The lattice constant of a III-V material system is important because III-V compounds are usually grown using epitaxy processes that require lattice matching with the substrate for the deposition and growth of a high-quality, defect-free thin film. If the lattice constant of the deposited film is more than that of the substrate, the film tends to contract laterally to "fit" on top of the substrate lattice structure, leading to compressive strain in the thin film and tensile stress in the substrate. Conversely, if the lattice constant of the deposited thin film is less than that of the substrate, the film develops a tensile strain leading to compressive stress in the substrate (Figure 7.3). The degree of strain can be calculated from the lattice constant mismatch as:

$$e = \frac{a_{sub} - a_{layer}}{a_{sub}}$$

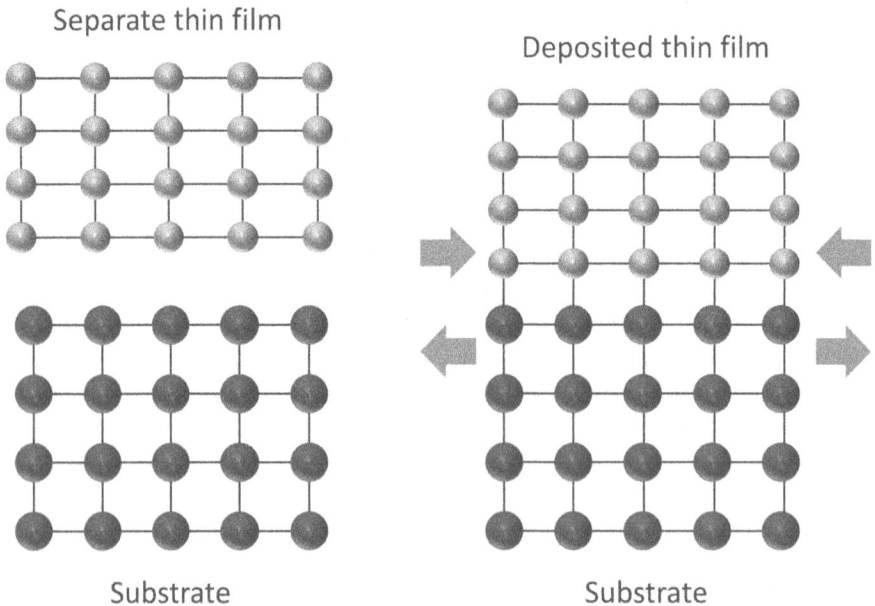

FIGURE 7.3 The deposition of thin films that are mismatched compared to the substrate lattice constant creates strain in the deposited thin film and stress in the substrate.

where, e is the strain developed in the layer, a_{sub} and a_{layer} are the lattice constants of the substrate and the layer respectively. However, this equation only holds for small amounts of lattice mismatch, because large mismatches can lead to local defects such as dislocation loops leading to some strain relaxation. However, these defects can lead to the formation of midgap electronic states which can be extremely detrimental to device performance.

Thus, it is important to keep the lattice constant of the crystal in consideration for determining the correct molecular ratios in ternary and quaternary compounds. The need to tune the bandgap along with the specific requirement of lattice leads to two possible solutions. One solution is to choose the elements in the III-V material system such that both parameters are satisfied. This can be done with the help of the bandgap versus lattice constant plot of various III-V systems, as shown in Figure 7.4. The points on the diagram represent the bandgap and the lattice constant of specific binary III-V compounds, while the solid lines connecting two points denote the ternary compounds formed using the elements from the binary compounds. However, if the elements in the III-V system are fixed due to other constraints such as availability or cost, another solution to reconcile both bandgap and lattice constant is to use a quaternary system. The two molecular ratio variables x and y in a quaternary system, such as $III_{(1-x)}III_xV_{(1-y)}V_y$, can be tuned such that a particular bandgap and lattice constant are obtained. However, it should be noted that there are practical limitations in engineering the stoichiometry of material systems, and not all molecular ratio systems will be stable after deposition.

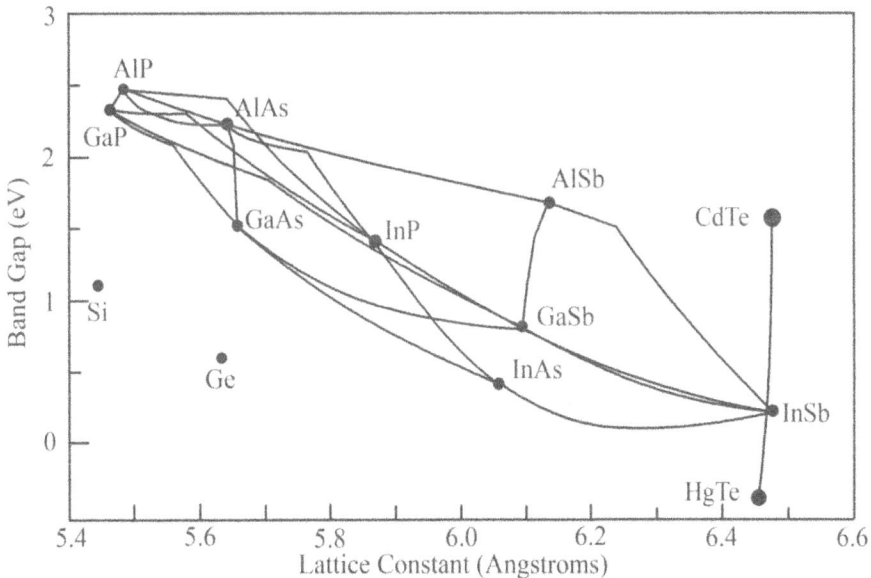

FIGURE 7.4 Bandgap versus lattice constant for various III-V semiconductor materials. The solid line dictates the variation of bandgap and lattice constant while moving from one compound to another in the ternary system. (Reprinted from [3] under the Creative Common CC BY license.)

III-V semiconductors exhibit very high electron mobility (approximately 10× that of silicon) and electron concentration at room temperature. This property, and the fact that high-quality III-V compound thin films can be grown, leads to the creation of a special type of transistor architecture called high electron mobility transistors (HEMTs). These transistors consist of a junction of two III-V compounds (heterojunction) as the channel, instead of a doped semiconductor in conventional MOSFETs (Figure 7.5). The two materials are chosen such that there is a difference in their bandgap. The material with the wider bandgap is doped to produce excess electrons that migrate to the undoped narrow bandgap material, leading to the formation of a junction. These electrons are then used to transport current through the channel. The accumulated electrons at the edge of the junction are confined to the heterojunction interface, leading to the quantization of energy states. Because these electrons resemble a two-dimensional sheet and are free to move at the junction interface, they are referred to as two-dimensional electron gas (2DEG). An interesting aspect of HEMTs is that because the conduction takes place in an undoped material, the electron mobility obtained is very high, owing to the lack of scattering from dopant atoms. The HEMT structure is particularly important from the III-V system point-of-view because the interface between two semiconductors (wide and narrow bandgap) can be seamlessly created using epitaxial growth. Doping can be achieved in-situ as well. The most commonly reported example of a HEMT structure

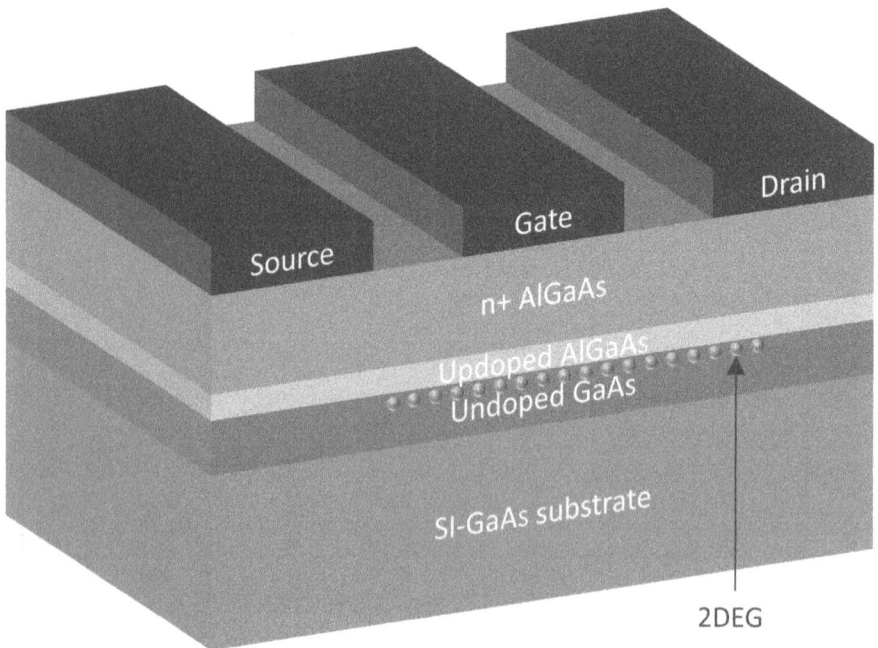

FIGURE 7.5 The high electron mobility transistor (HEMT) structure on a semi-insulating GaAs substrate, with the formation of 2-dimensional electron gas (2DEG) at the junction of the intrinsic GaAs and AlGaAs layers.

consists of gallium arsenide (GaAs) and aluminum gallium arsenide (AlGaAs). A semi-insulating GaAs substrate is used for the fabrication of the device. A layer of undoped GaAs (bandgap = 1.44 eV) is deposited as the channel material for the formation of 2DEG. This is followed by the deposition of a thin layer of undoped AlGaAs spacer (2–3 nm), followed by n-doped AlGaAs. The undoped spacer is deposited to make sure there is insulation between the 2DEG in the GaAs layer and the doped AlGaAs layer. The gate, source, and drain structures are then created to carry out transistor operations.

The electron mobility of III-V materials is very high compared to silicon, leading to very high-speed HEMT devices. However, to create CMOS circuits, the semiconductor should also exhibit high hole mobility. In the case of many III-V materials, the hole mobility is very low compared to silicon, leading to a large mismatch in the performance of n-type and p-type transistors. Over the years, many research efforts have focused on enhancing the hole mobility in III-V compound semiconductors or in search of materials that can be heterogeneously integrated with III-V materials to provide a high hole mobility transistor. Even if these efforts fructify, there remains one challenge that will restrict the use of III-V materials as a replacement of silicon for general purpose electronic applications – cost. Most commonly studied III-V systems rely on gallium and indium as the group III material. Both elements are rare in the earth's crust (indium more so than gallium), thus leading to an increase in the cost of fabrication. This is particularly true if the complete wafer is fabricated using a III-V material. Further, the techniques used for the fabrication of large silicon wafers are not easily applied for III-V systems, leading to the wafer sizes being limited. All of these practical limitations restrict the use of III-V materials for conventional electronics. At this moment, the only application that utilizes III-V material systems on a large commercial scale is the fabrication of light-emitting devices, because of their direct bandgap.

7.3   III-V WAFERS

Gallium arsenide (GaAs) is the most commonly used III-V material for wafer fabrication. In fact, it is the second most used wafer material after silicon. The most common uses include the growth of other III-V systems for the fabrication of light-emitting devices, such as LEDs and laser diodes, light-absorbing devices such as photodiodes and solar cells, or fast electronic devices (HEMTs). Other commercially available III-V wafers include GaN, GaP, GaSb, InP, InAs, and so on. Fabrication of single-crystal GaAs wafers (and of other III-V compound semiconductors) is done by first synthesizing high purity GaAs (polycrystalline), followed by recrystallizing the material using a seed crystal. This is similar to the process followed for silicon wafers, wherein, the Siemens process is used to obtain pure polycrystalline silicon, followed by the Czochralski process for single crystal silicon ingots. High purity polycrystalline GaAs is obtained by annealing high purity elemental gallium and arsenic in a sealed quartz furnace. The polycrystalline material can be recrystallized into single crystal wafers using one of these processes: vertical gradient freeze (VGF) process, liquid encapsulated Czochralski (LEC) process, or horizontal gradient freeze (HGF) process.

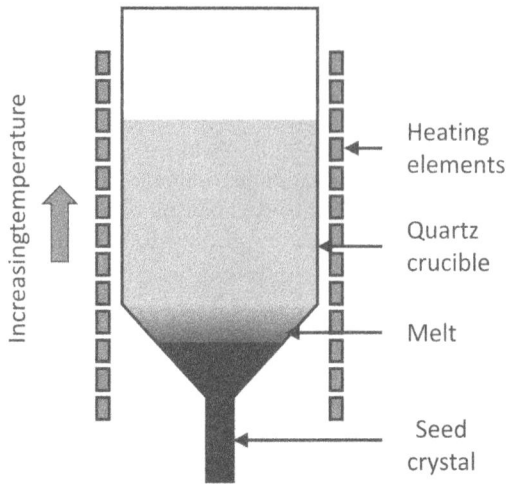

FIGURE 7.6   The vertical gradient freeze (VFG) process involves the use of a quartz crucible to melt the polycrystalline III-V material. The material recrystallizes based on the lattice of the seed crystal.

The vertical gradient freeze is the most commonly used process to fabricate GaAs wafers. The specially designed furnace for this process consists of a quartz chamber tapering at the bottom into a seed crystal (Figure 7.6). The polycrystalline material is placed into the quartz chamber and the temperature is set such that the material at the bottom melts first. The furnace is designed such that the temperature distribution in the furnace can be moved vertically by mechanical movement of the chamber or heating elements, or by electronic control of the heating elements. Thus, the heating is moved away from the bottom of the crucible allowing the bottom melt to recrystallize according to the seed template. The heating is moved vertically upwards leading to gradual melting, and solidification of the polycrystalline material into a single crystal mapping the seed design. This is similar to the float zone process discussed for silicon wafers. The process can also be carried out horizontally with one side of the crucible containing the seed crystal (horizontal gradient freeze process). Another process for obtaining single crystal III-V compound semiconductors is the LEC process. This process is similar to the Czochralski process used for silicon wafers, with the difference that the melt is encapsulated using a liquid layer to protect the melt from sublimation and evaporation. In the case of GaAs, a pellet of boron trioxide (B_2O_3) is inserted in the crucible to act as an encapsulation layer when the solution is melted. In LEC, the pressure in the chamber is generally kept high to further discourage evaporation of the melt material. When the melt is ready for recrystallization, a seed crystal is introduced in the melt through the liquid encapsulation layer. As in the case of silicon, the seed is slowly withdrawn from the melt to obtain crystallization of the melt material on the seed surface. The crystal puller is rotated while being drawn for better radial uniformity of deposition.

For GaAs wafers, a common option is the use of semi-insulating substrates (SI-GaAs). These substrates are "doped" GaAs wafers, however, they exhibit conductivity and carrier concentration close to those in intrinsic GaAs wafers. This is possible by choosing the right dopant species and concentration so that the Fermi level in the resulting crystal ends up being pinned at the center of the bandgap. This helps reduce the carrier concentration at room temperature, resulting in the semi-insulating nature of the substrate. Although they have similar electronic properties compared to intrinsic GaAs wafers, semi-insulating substrates are important because they are easier to manufacture compared to perfect GaAs crystals. A common approach is to manufacture GaAs wafers in the presence of excess arsenic leading to arsenic antisite defects (presence of an arsenic atom where a gallium atom should have been). These defects pin the Fermi level at the center of the bandgap leading to low carrier concentration and semi-insulating properties.

7.3.1 EPITAXY

One of the most important processes for obtaining single-crystal thin films of III-V materials is the epitaxy process. It is a process by which a monocrystalline thin film is grown on a monocrystalline substrate. It is generally divided into two groups: homoepitaxy, which is the growth of the same thin-film material as that of the substrate; and heteroepitaxy, which is the growth of a thin film different than the substrate. For example, the growth of GaAs on GaAs substrate is homoepitaxy, whereas the growth of AlAs on GaAs substrate is heteroepitaxy. Epitaxy can be performed with source materials in solid, liquid, or gaseous state, and there are many different approaches for epitaxial growth of III-V materials such as liquid-phase epitaxy (LPE), vapor-phase epitaxy (VPE), molecular beam epitaxy (MBE), and so on. In liquid-phase epitaxy (LPE), a supersaturated solution of the constituent materials is brought in contact with the seed substrate. As the temperature of the solution is reduced, the excess material precipitates on the seed substrate in the form of a single crystal thin film. The advantage of this process is the high throughput for the fabrication of commercial devices. The disadvantage is the precise control of temperature and stoichiometry required to obtain the correct deposition. Also, introducing dopants into the thin film is challenging with LPE because of the fine liquidus-solidus equilibrium to be maintained for a solution with multiple constituents.

Vapor-phase epitaxy entails the use of gaseous sources to perform epitaxial growth of thin films. It is a form of chemical vapor deposition, in which several gases (depending on the material to be deposited) are introduced in a heated chamber simultaneously. The temperature of the chamber depends on the energy required for pyrolysis (thermal decomposition of the gases). The gases undergo pyrolysis and the resulting subspecies react to form the desired material that is then adsorbed onto the substrate. The source gases are chosen such that the by-products of the reaction are gaseous and are carried away by the vacuum system. The rate of growth is a key parameter in obtaining high-quality single-crystal thin film and is controlled by regulating the flow rate of gaseous sources and the temperature of the chamber. If the rate of growth is too fast, the adsorbing molecules do not have enough time to rearrange according to the template of the seed substrate, leading to dislocation defects, and in the worst case, a polycrystalline thin film. If the source gases are organometallic compounds (especially for the group III component), the process is called metal organic chemical vapor deposition (MOCVD). Weakly bonded organometallic compounds are generally used as precursors to reduce the chamber temperature required for pyrolysis. For example, trimethyl indium ($[CH_3]_3$ In) is used as a source gas for indium while trimethyl gallium ($[CH_3]_3$ Ga) is used for gallium. The group V constituents are typically introduced as respective hydrides. For example, phosphine (PH_3) for phosphorus, ammonia (NH_3) for nitrogen, and arsine (AsH_3) for arsenic. A major advantage of the MOCVD process is the ability to dope the deposited thin film in situ. This is done by introducing a source gas for the dopant atom into the chamber. For example, if silane (SiH_4) is introduced into the chamber with $[CH_3]_3Ga$ and AsH_3, we can obtain silicon-doped GaAs thin films. Further, it is relatively easy to control the concentration of the dopant species which depends on the relative flow rate of the source gases. A disadvantage of the process is the low throughput because of the slow growth required for high-quality films and the wastage of expensive source gases because of deposition on chamber walls and in the vacuum system.

Molecular beam epitaxy is another process commonly used to obtain single-crystal III-V thin films. In this case, ultra-pure sources, in elemental form, are heated in separate effusion cells. This causes the elements to sublimate and travel to the substrate where they get adsorbed on the surface and react with other atoms to form the desired material and are deposited as a thin film. Surface mobility is important for the adsorbed atoms to react and subsequently form single-crystal thin films. Very high levels of vacuum (typically 10^{-12} torr) are maintained inside the chamber to make sure there is no source of contamination for the highly reactive atoms. This leads to very large mean free paths for the atoms, thus, creating a molecular "beam" going from the effusion cells towards the substrate. In this case, as well, the deposition rates are kept low and the substrate temperature is kept high to provide adsorbed atoms and molecules enough time and energy to travel across the surface and reach an appropriate lattice site, leading to a monocrystalline deposition. An advantage of MBE is that there is very little loss of expensive materials such as gallium and indium. The system also provides finer control over deposition rates and thicknesses using the shuttering mechanisms on effusion cells, and control over effusion cell temperature. The system can be used to dope the thin films being deposited in-situ, using the appropriate source cells.

7.4 FLEXIBLE III-V SEMICONDUCTORS

Bulk III-V compounds are generally brittle, as is often the case with crystalline covalent compounds. GaAs has Young's modulus of around 85 GPa, while the elastic strain limit has been reported to be 0.1%. This is close (in order of magnitude) to the values observed for silicon. However, we know that a good approach to obtaining flexibility in any material is to reduce the thickness of the material. Many of the processes discussed in the flexible silicon case are, in fact, applicable to III-V systems with some modifications in etch chemistry. For example, an LED stack can be deposited on a GaAs substrate using AlGaAs as the sacrificial layer. The etch rates for AlGaAs vary with the proportion of aluminum and the concentration of HF used, however, etch selectivity of over 100 with GaAs has been reported. The general

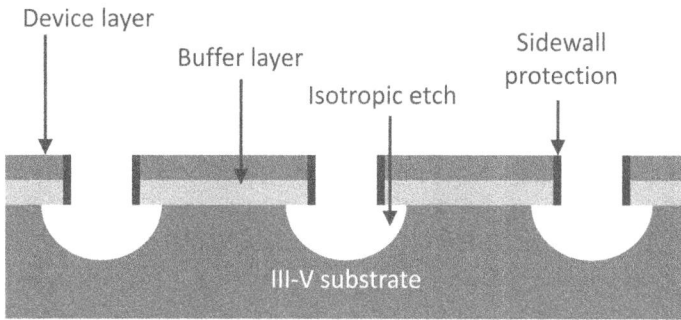

FIGURE 7.7 The trench-protect-etch-release (TPER) process discussed for flexible mono-crystalline silicon can be employed for III-V with a change in the sidewall protection layer material and the isotropic etch chemistry.

process for fabrication of flexible devices with this approach is to start with the epitaxial growth of the AlGaAs sacrificial layer on a GaAs substrate, followed by the growth of device layers. The structure of the device is then defined in terms of doping, metal interconnects, patterning, and so on. This is followed by the removal of the AlGaAs sacrificial layer with wet etching in HF, which undercuts the sacrificial layer and releases the top device from the substrate underneath. Thus, a thin and flexible device can be obtained. A process similar to the trench-protect-etch-release (TPER) in silicon can also be imagined where the III-V substrate is etched and covered with a protective coating on the sidewall, before being subjected to lateral etching using an isotropic etchant (Figure 7.7).

The process of epitaxy in III-V material systems provides some unique opportunities for producing flexible single-crystal thin films. Because the process is repeatable, a stack of device layers and intermediate sacrificial layers can be grown on the same substrate. The entire stack can then be subjected to wet etching resulting in potentially thousands of flexible single-crystal films floating in the wet etch bath. Thus, the process of forming high-quality flexible thin films can be scaled in a way that is not possible for silicon or other material systems. Further, after the process is complete, the wafer can be cleaned, polished, and reused multiple times significantly reducing the fabrication costs. Also, because the thickness of the flexible membranes is determined by epitaxy and their lateral dimensions are determined by photolithography, the overall dimensions of all thin films produced can be repeatable within a certain process window, allowing for reliable industry-scale production.

The epitaxial growth and subsequent removal of III-V single-crystal thin films using the sacrificial layer technique provides for opportunities to tune the electronic properties of the resulting thin films. With epitaxy, the deposition of very thin layers (< 10 nm) of high-quality III-V compounds is possible. This means the resulting flexible single-crystal thin films are so thin that they exhibit quantum confinement of carriers. This means that the solution to the solution to Schrodinger's equation for these thin films leads to different electronic energy bands compared to the bulk material. This typically leads to an increase in the bandgap leading to emission and absorption spectra different from the bulk material. Further, the intensity of light traveling through a material is given by:

$$I = I_0 \ e^{-ax}$$

where x is the distance traveled by light, I_0 is the original incident intensity, and a is the absorption coefficient. Thus, as the thickness of the thin film reduces, the transmittance increases exponentially. This has important repercussions for flexible transparent electronics applications.

7.5 III-V THIN FILMS

While single crystal III-V materials are most commonly used for applications such as light emission, solar cells, photodiodes, high-speed transistors, and so on, the processes required to fabricate single-crystal materials are complex and expensive. In some applications, III-V semiconductors have shown promise in polycrystalline or amorphous form. These thin films are easier to fabricate and scale-up and can offer better performance compared to other semiconductor thin films like silicon or metal oxides. This is particularly interesting for flexible electronic applications because the choice of substrate is very restrictive for single crystal III-V thin films. The use of polycrystalline III-V thin films opens up the opportunity to use flexible polymer substrates for the fabrication of electronic systems because processes such as sputtering can be carried out at much lower temperatures.

The most common method for fabricating III-V polycrystalline thin films is the use of CVD. This is similar to the MOCVD process discussed earlier, however, instead of having a seed substrate so that lattice matching and single crystal growth are obtained, we can use any substrate that can withstand the chamber temperature. Commonly used substrates are glass, quartz, graphite, and alumina. The source gases and the equipment for this process can be the same as those for MOCVD, i.e., organo-metallic compounds for group III constituents, while hydrides for group V. The thin films deposited using this process are polycrystalline with the grain size depending on the choice of substrate and chamber temperature. The electronic properties of these films are strongly dependent on the grain size and distribution because grain boundaries play a fundamental role in electron transport. The electrically active impurities tend to segregate preferentially at the grain boundaries, creating trap states and recombination centers. This can reduce the carrier concentration available for conduction leading to lower conductivity. Further, grain boundaries act as scattering centers leading to lower overall carrier mobility. While these disadvantages can reduce device performance, the non-requirement of high-quality and expensive seed substrates for deposition is a major advantage of the use of polycrystalline III-V thin films.

Another widely used non-epitaxial process for deposition of III-V material thin film is magnetron sputtering. In this case, an argon ion plasma is focused on a target (say GaAs) to eject material from it and deposit it onto the desired substrate. The temperature of the substrate and that of the chamber are typically much lower than those required for CVD deposition, however, higher temperatures do lead to better quality film deposition. Sputtering can also be used for ternary compounds such as InGaAs by co-sputtering two separate targets (say In and GaAs). A key advantage of the sputtering process for polycrystalline III-V deposition in flexible electronic applications is the ability to use lower substrate and chamber temperature. However, because these are multicomponent compound targets, the bombardment of ions can lead to compositional changes in the ejected material and the stoichiometry of the resulting thin film can be changed. Hence, it is important to model the ion-solid interaction process and the transfer of energy from ions to target, to obtain a thin film of the correct composition. Given the ease of use, scalability, and significance of sputtering for flexible electronic applications, there is great interest in developing a stable and repeatable polycrystalline III-V deposition process.

III-V compound semiconductors form an integral part of the electronics ecosystem, particularly because of their commercial usage in light-emitting devices such as LEDs, displays, lasers and because of potential use in solar cells and high-speed transistors. Their unique properties – direct, tunable bandgap and high electron mobility – ensure their continued significance in the broader electronics landscape. In the case of flexible electronic applications, there remain significant challenges, however, it is important to find ways to integrate these material systems with conventional electronic materials so that a complete flexible electronic system can be fabricated.

EXERCISES

7.1. What is meant by a direct bandgap semiconductor? Why is it important to increase the probability of radiative recombination of carriers?

7.2. We decide to create a ternary III-V material from InAs and GaAs that have bandgaps of 0.354 eV and 1.43 eV respectively. If the bowing parameter is 0.436 eV, what is the ratio of InAs required to create a semiconductor that can produce infrared LEDs of wavelength 900 nm?

7.3. For the material in question 7.2, what will be the lattice constant if the lattice constants of InAs and GaAs are 0.605 nm and 0.565 nm? If we try to epitaxially deposit the material on a GaAs substrate, what will be the resulting strain in the thin film?

7.4. Describe the working principle and structure of a high electron mobilty transitor (HEMT) based on the GaAs, AlGaAs material system.

7.5. What can be the advantages if a trench-protect-etch-release (TPER) process is developed for III-V material systems?

8 Nanostructured Materials

8.1 INTRODUCTION

The discussion related to all the key electronic material systems so far has been considering bulk wafers or thin films. However, with advancements in processing technology, imaging, and characterization techniques, we can now sculpt many of these materials into nanostructures. These structures allow us to manipulate the electronic and mechanical behavior of these materials, depending on their dimensions and the processing conditions. This is important from a flexible electronics point of view because a reduction in the form factor of electronic materials allows for flexible systems with complex geometrical shapes, not just in thin-film form. We look at the nanostructured material systems in terms of the number of significant spatial dimensions they possess. The bulk material or thin films discussed until this point had three significant dimensions, length, width, and thickness. Certain material systems are such that they naturally occur in atomically thin sheets weakly bonded together. We refer to them as two-dimensional atomic crystal structure materials (2DACS) or simply two-dimensional materials. These materials only have two significant dimensions – length and width. The most common example of such a material is graphite, which naturally organizes into atomically thin sheets of carbon, called graphene, weakly interacting with each other. Because the sheets are not bonded together, they can easily slip with respect to each other upon application of force. This has resulted in graphite being used as a solid lubricant. However, the atomically thin carbon sheets have some remarkable electronic properties as well. Then, there are materials that can be structured as nanowires or nanotubes. These only have one significant dimension – length. The thickness or radius of these materials also affects their properties, however, because it is much smaller compared to the length, we do not consider it as a significant dimension. Finally, we have some materials that can be structured as nanoparticles. In this case, all the dimensions are constrained (under 100 nm) resulting in the so-called zero-dimensional material systems. In this chapter, we will discuss the properties, fabrication processes, and possible applications of all these materials.

8.2 TWO-DIMENSIONAL MATERIALS

Two-dimensional material systems are atomically thin sheets with two significant dimensions – length and width. There are certain materials that form these sheets natively, i.e., in their natural form, for example, molybdenum disulfide (MoS_2). In other cases, these are allotropes of materials, i.e., one of the forms they naturally occur in is the 2D form, for example, graphene. In yet other systems, 2D sheets can be synthetically formed through precise micromachining techniques, for example, silicene. In either of these cases, the 2D atomic crystal structure materials offer some remarkable properties as electronic conductors and semiconductors. In fact,

DOI: 10.1201/9781003010715-10   **109**

they are quite suitable as a material system for flexible electronics because their angstrom-scale thickness allows for very small bending radii (nanometer scale). Further, because most of the atoms in 2D materials reside on the "surface", their surface-to-volume ratio is incredibly high. This provides some unique properties to this material class. In particular, the properties of these materials are very sensitive to surface chemistry, for example, adsorption of a specific molecule on the surface, leading to sensor systems being a possible application for these materials. Hence, they have been extensively studied in the last decade and many processing techniques and integration strategies have been developed to create complete electronic systems using these materials. We start, of course, with the most popular of these materials – graphene.

8.2.1 GRAPHENE

In 2004, K. S. Novoselov, A. K. Geim, and others reported the electronic properties of a remarkable material they had been working on. Nobody before them had successfully isolated the material on an insulating substrate (silicon substrate covered with SiO_2) and realized the powerful electronic properties the structure can entail. In 2010, K. S. Novoselov and A. K. Geim won the Nobel prize in physics for "groundbreaking experiments regarding the two-dimensional material graphene". They did not "discover" graphene, because its existence as a thin sheet of carbon atoms has been long known. In fact, the extraordinary electronic properties of graphene were also predicted decades ago. However, the experiments conducted by the duo were incredibly simple to perform and replicate, which laid the foundation for hundreds of research groups around the world to experiment and study various properties of graphene. They used ordinary scotch tape to peel layers off graphite to obtain single-layer graphene sheets. This simple and ingenious technique, called mechanical exfoliation or the "Scotch Tape" technique, was of course time-consuming and labor intense. Over the years, other processes to obtain single-layer graphene were discovered. Ten years later, we have an entire ecosystem of small-sized manufacturing industries producing graphene with academic research groups being their primary consumers. The material system indeed has some truly remarkable properties; however, no commercially viable application has been showcased. To this day, graphene remains a solution looking for a problem. Regardless, the study of graphene is important for flexible electronics because the atomically thin structure provides some unique opportunities.

As in any other material system, the key to understanding the properties of graphene is in the arrangement of atoms and the bonding between them. Graphene is an all-carbon material, hence, is considered as an allotrope of carbon. The carbon atoms are all sp^2 hybridized giving rise to the carbon-carbon double bonds and the existence of $2p_z$ electrons interacting to form π-orbitals. This is similar to the structure of organic electronic polymers, only, in this case, the structure is spread in 2D space. This is made possible because of the "flat" structure of the three σ-bonds distributed at an angle of 120° in a single plane. They bond with other carbon atoms with the same σ-bond distribution in the same plane resulting in a potentially infinite lattice of carbon atoms in a single plane. This gives rise to the stable atomically

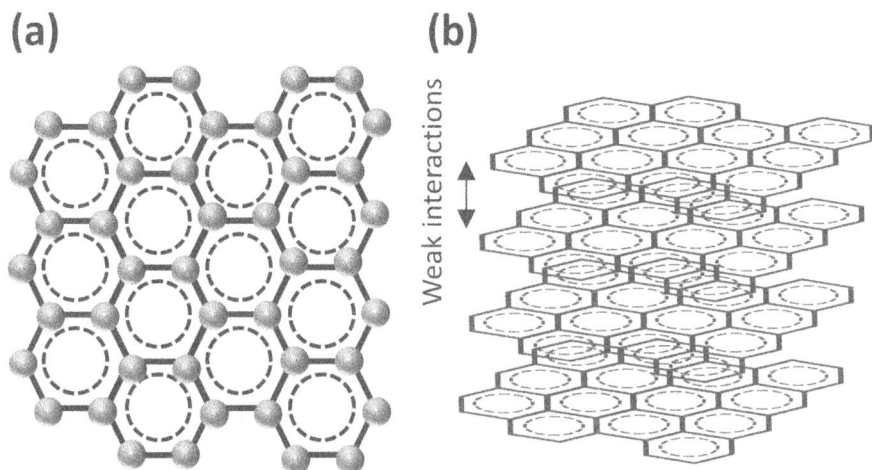

FIGURE 8.1   (a) The structure of graphene is an infinite tessellation of carbon atoms in A two-dimensional hexagonal lattice structure (honeycomb structure). The entire structure has conjugated π-bonds. (b) These layers stack together to form the bulk material graphite, with weak interactions holding the individual layers in place.

thin structure of graphene. The carbon atoms are placed at the vertices of a hexagon, similar to the structure of the benzene ring. Every alternate pair of carbon atoms forms a double bond. Further, because the double bonds are conjugated, the π-orbitals extend throughout the material structure allowing for seamless electronic conduction. Graphene sheets stack on each other, without any chemical bond, to form graphite (Figure 8.1). The free slippage of graphene sheets with respect to each other is the reason for the lubricating properties of graphite. The distance between two graphene sheets is 3.3 Angstrom (0.33 nm).

The unique structure of graphene bestows some unique electrical, mechanical, and thermal properties on graphene. Graphene is a zero bandgap semiconductor, i.e., it is not a metal that would have the conduction and valence bands overlap at multiple points in the E-k space; however, it is also not a semiconductor where there is a non-zero bandgap between the valence and conduction band. In the case of graphene, the valence and conduction bands meet at a single point in the E-k space, at six different points in the Brillouin zone (Figure 8.2). This implies that graphene can have free carrier concentration at any temperature above absolute zero, however, the number of these carriers is limited to the number of electronic states in the energy range at that temperature. Experimental results show that intrinsic carrier concentration in graphene is proportional to the square of temperature, as predicted theoretically. However, the most interesting electronic property of graphene is the behavior of electrons as a massless two-dimensional gas of Dirac fermions, thus requiring the use of the relativistic Dirac equation for accurately predicting their behavior, as opposed to the non-relativistic Schrodinger equation. The massless nature of the charge carriers can be predicted from the conical shape of the electronic band at the Dirac point. For other materials, the effective mass of charge carriers is inversely

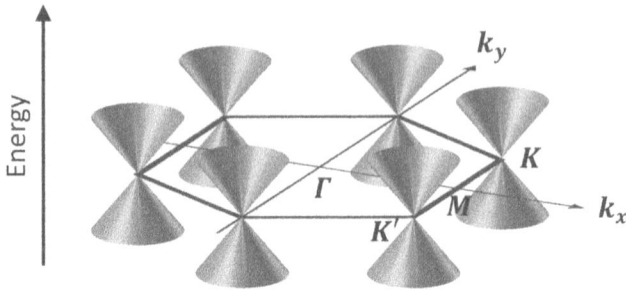

FIGURE 8.2 The first Brillouin zone of graphene showing the Γ-point along with K, K', and M points. The density of states is zero at the Dirac point, but extended in both directions of the energy axis in the form of a cone.

related to the curvature of the energy band with respect to the wavevector. However, in graphene, we have a linear dispersion dependence of energy on momentum, as in the case of photons, with a group velocity of 1/300 the speed of light. This theory, while seemingly outlandish, has been backed by several experiments.

Another extraordinary property of electron transport in graphene pertains to the mobility of charge carriers in the lattice. Dirac fermions are insensitive to external barriers, which would normally scatter electrons obeying the non-relativistic Schrodinger equation (Klein paradox). The scattering of charge carriers leads to the reduction in their average velocity, which leads to resistivity and reduction in mobility. However, because electrons behave like Dirac fermions, the diminished scattering in graphene results in much higher mobility. The speed of transport of electrons in graphene is then only limited by relativistic effects, leading to what is called the "ballistic" transport of electrons. However, these considerations are only true for an ideal graphene sheet with no defects, impurities, and other electron scattering mechanisms. Even so, experimentally reported values of mobility of electrons in graphene are in the excess of 15,000 cm^2/Vs. The phenomenal mobility and ballistic transport have inspired the use of graphene in ultra-high-speed devices for terahertz (THz) communication applications. The unique structure of graphene and the behavior of electrons as Dirac fermions offer many more unique properties in terms of thermal conductivity, optical properties, ambipolar transport, anomalous quantum Hall effect, and so on. These properties make graphene an exciting candidate for a variety of electronic device applications. However, deposition, processing, integration, and long-term reliability remain key challenges for the deployment of graphene devices.

One of the simplest processes for obtaining single-layer graphene is the micro-mechanical cleavage of graphite. In this process, flakes of single-layer graphene are extracted from a piece of graphite by overcoming the weak interactions between the individual layers using mechanical energy. The simplest way to perform exfoliation is by repeated stick-and-peel using tape until a monolayer flake is obtained. The flake can then be transferred to the desired surface using an adhesion layer. This process, known as the "Scotch tape" process was described by Novoselov and Geim in their groundbreaking paper. It can provide a very low-cost and effective solution to obtaining high-quality single-layer graphene sheets. However, it is not a scalable and repeatable process and involves a certain trial-and-error approach. Further, the size of the flake obtained is also limited to a few square millimeters at best. Thus, this process, while very useful and popular for research on graphene, cannot be used at an industrial scale. For large-scale applications, other mechanical exfoliation processes are deployed. These include using ultrasonication energy to separate graphene sheets in a solution, followed by transfer to a substrate. The process allows for very large throughputs because large tanks are used for sonicating large quantities of graphite, however, it does not provide control over the size of the resulting graphene sheets. In particular, high energy sonication can cause graphene sheets to break, while low energy may not be sufficient to completely separate each layer of graphene from bulk graphite. This trade-off results in a solution with a mixture of thick multi-layers sheets and broken graphene flakes. Further, because graphene is hydrophobic, the process requires the use of surfactants to stabilize the solution in polar solvents such as water or alcohol. This can increase the cost of processing and reduce the purity of graphene obtained after solvent evaporation. Graphite can also be exfoliated into graphene sheets using chemical processes. Chemical exfoliation involves the intercalation of certain materials into bulk graphite, leading to the weakening of interactions of graphene sheets with each other. In many cases, compounds containing alkali ions such as sodium (Na^+) and potassium (K^+) are added to the solution for intercalation into graphite. The exfoliation process is generally expediated using mild heating or exfoliation to completely dissociate the graphene sheets.

Another process to produce graphene sheets is to oxidize graphite to form graphite oxide (or graphene oxide) followed by reduction to reduced graphene oxide (rGO). Oxidation of graphite is usually carried out using the Hummer's process, which involves the use of potassium permanganate, sodium nitrate, and sulfuric acid to oxidize graphite. Graphite oxide (GO) is also a layered compound consisting of graphene sheets decorated with oxygen-containing functional groups such as hydroxyl (-OH), carbonyl (-CO-), carboxyl (-COOH), and so on. GO is soluble in water and most polar solvents, however, it is non-conducting owing to the disruption of the π-conjugate structure because of the presence of the functional groups. Thus, it is important to reduce GO into rGO to obtain the electronic properties of graphene. The reduction process is carried out using reducing agents such as hydrazine (N_2H_4). The reduction of GO leads to the removal of the functional groups attached to graphene sheets, restoring the original structure of graphene. However, this restoration is only partial, and many oxygen-containing functional groups remain in rGO after the reduction process. The carbon to oxygen atomic ratio of around (12-14):1 is generally reported for even the best reduction processes. Further, the reduction of GO also reduces its hydrophilicity, which can cause rGO to clump together and precipitate. Another downside of the oxidation process to produce graphene is that some carbon atoms remain sp3 hybridized even after the reduction process, leading to the creation of electron traps and loss of conduction. Even with all these disadvantages, the oxidation process is being widely studied and optimized because it is extremely scalable and can produce industrial quantities of rGO at very low costs.

The key limitation of the exfoliation and oxidation processes is that even in the best-case scenario, the size of the final graphene sheet is dependent on the size of the initial graphite piece. Thus, arbitrarily large graphene sheets cannot be obtained using these methods. For this reason, chemical vapor deposition (CVD) of single-layer graphene sheets is the most popular process for producing large-area, high-quality, single-layer graphene sheets. The process involves the use of a transition metal, such as nickel or copper, to produce sheets of graphene from gaseous hydrocarbon sources. As in any other CVD process, the source hydrocarbons undergo decomposition because of the temperature of the chamber (pyrolysis) followed by surface adsorption, surface mobility, and bond formation. The temperature used for such synthesis is around 1000–1200°C, while the source gases can be methane along with hydrogen and argon as carrier gases. Initially, nickel foils were reported for graphene CVD; however, adsorbed carbon atoms have low mobility on nickel surface leading to the growth of independent graphene flakes simultaneously on the nickel surface, leading to polycrystalline graphene. Further, because the solubility of carbon is high in nickel at the elevated temperatures of CVD, the dissolved carbon atoms precipitate upon cooling, leading to multilayer graphene in many places. Because of these problems, nowadays, copper is the most commonly used substrate for graphene CVD. Owing to the low solubility of carbon atoms in copper and high surface mobility, large-area single-layer graphene films are obtained on flexible copper foils. This is particularly exciting for flexible electronics applications and large-scale roll-to-roll production. However, a problem with growing graphene on copper is that copper itself is highly conductive, hence, it is not possible to use the graphene on copper directly for thin-film transistor (TFT) or any other electronic device fabrication. This requires graphene to be transferred to an insulating destination substrate for device fabrication. The transfer of graphene from copper involves the dissolution of copper using wet chemical etchants such as ferric chloride ($FeCl_3$) or ammonium persulphate ($[NH_4]_2S_2O_8$), followed by transfer of the graphene to an insulating substrate. During this process, the graphene sheet is supported by attaching it to a poly-methyl methacrylate (PMMA) thin film formed by spin-coating and curing. The PMMA thin film is then dissolved to release the graphene sheet. The graphene transfer process, however, can introduce impurities and residues into the graphene sheet, compromising its electronic properties.

With the CVD and transfer process, very large area (of the order of a few inches), high-quality graphene can be transferred to any desired substrate for further processing. Because of its high conductivity and transparency (97.4%), one of the most promising uses for graphene is as a transparent conductor. For semiconducting applications, a bandgap can be opened in graphene sheets by patterning them into nanoribbons. Graphene nanoribbons (GNRs) are graphene thin films with one or more dimensions constrained to a few nanometers (less than 50 nm). The constraint on dimensions causes changes in the electronic structure of the bands, leading to a bandgap dependent on the dimensions of the nanoribbon and the geometry of the edge.

8.2.2 Transition Metal Dichalcogenides (TMDs)

Transition metal dichalcogenides are, as the name suggests, dichalcogenides of transition metals. Dichalcogenides are compounds containing two atoms from the chalcogenide group, i.e., group VI of the periodic table, which includes oxygen (O),

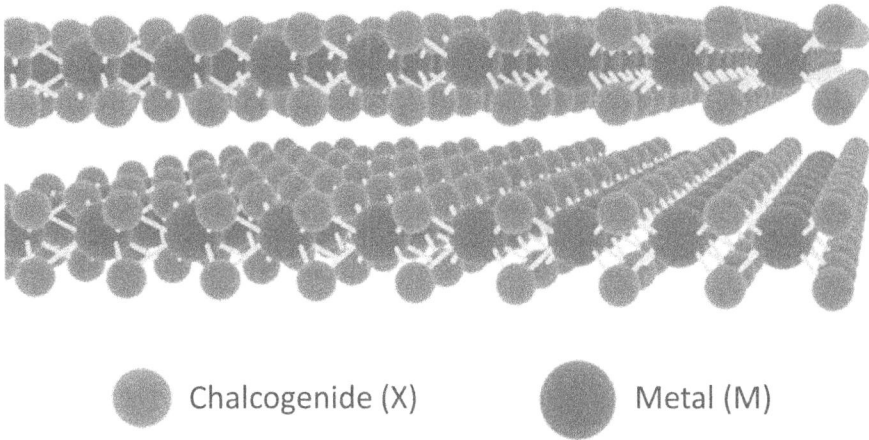

Chalcogenide (X)   Metal (M)

FIGURE 8.3   Each layer of transition metal dichalcogenides (TMDs) consists of a sandwich structure of the form X-M-X, where M is the metal atom and X is the chalcogenide atom.

sulfur (S), selenium (Se), tellurium (Te), and the radioactive element polonium (Po). As seen in Chapter 6, oxides of some transition metals are semiconducting, but their structure is not atomically thin sheets. When we discuss TMDs, as a class of atomically thin electronic materials, we consider sulfur (S), selenium (Se), and tellurium (Te) as the chalcogenides, and some transition metals such as molybdenum (Mo), tungsten (W), niobium (Nb), and so on. The general stoichiometry of these compounds is MX_2, where M is the transition metal and X is the chalcogen. A single layer of TMD consists of a layer of hexagonally arranged metal atoms sandwiched by layers of chalcogen atoms from the top and the bottom (X-M-X), as shown in Figure 8.3. The oxidation state of the metal is typically +4 and that of the chalcogen is -2. As in the case of graphene, the bulk TMD material consists of single layers bound together through weak interactions. Thus, just like graphene, they have found usage in the dry lubricant industry, typically as an additive for graphite-based lubricants. A single layer of TMDs generally has a thickness of a few angstroms (6-7), for example, MoS_2 single layer thickness is 6.5 Å. It is typically more than graphene (3.3 Å) because of the sandwich structure of the material. TMDs are a promising material class for electronic applications because of several reasons. First, TMDs are semiconductors with a bandgap in the range from 1 to 2 eV, thus making them ideal for several electronic device applications. In some cases, the bandgap goes from being indirect in bulk material to direct in the monolayer. This, coupled with the bandgap in the near-visible range, opens up applications in light emission (LED, lasers, etc.) and light absorption (solar cells, photodiodes, etc.) devices. Second, the reported carrier mobility for many TMDs is in the hundreds of $cm^2/V\text{-}s$. Thus, TMDs offer a potential replacement for silicon technology with their carrier mobility and bandgap being similar. Third, their atomically thin structure implies that the material cost of production can be low compared to other bulk semiconductors such as germanium or III-Vs. Their thin structure also makes them ideal for flexible electronic applications.

For the synthesis of single-layer TMD thin films, micromechanical exfoliation or the Scotch tape method is widely used. The process involves the use of bulk TMD material and repeated use of mechanical cleavage to obtain single-layer flakes. It provides a simple, cost-effective method to obtain small flakes for experimentation for determining the electronic, mechanical, and other properties of the materials. Exfoliation in the liquid phase using mechanical energy (sonication) as well as chemical intercalation have also been studied for TMD materials. Water is a natural choice for solvent, but its large surface energy restricts the exfoliation of 2D materials. Thus, it is necessary to use surfactants or non-polar organic solvents for exfoliating TMDs and restricting conglomeration. Even so, liquid-phase exfoliation has a low yield of single-layer flakes. This is a major issue because some of the unique properties of TMDs are only observed in monolayer form (such as direct bandgap). For this reason, CVD is considered to be the best process for synthesizing large quantities of high-quality TMD thin films. In the case of metal sulfides, the process simply involves the reaction of sulfur vapors with metal thin films at high temperatures. In most cases, sulfur powder and metal thin film deposited on a silicon substrate are introduced in a chamber and heated up to 750°C. The sulfur vaporizes above 444°C and reacts with the metal to produce MS_2 compounds. This can be used to produce industrial quantities of MS_2 materials on large substrates. The metal thin film can be deposited using physical vapor deposition (PVD) techniques such as evaporation or sputtering. However, the resulting TMD materials with this process are polycrystalline and multilayer, the number of layers depending on the initial thickness of the metal. This can be solved by using a precursor for transition metal instead of the metal thin film. An example of such a process is the sulfurization of molybdenum trioxide (MoO_3) to produce MoS_2. In this process, MoO_3 and sulfur powders in separate vails are heated in a chamber along with the substrate. This produces thin films of MoS_2 on the substrate, however, the growth of single-layer MoS_2 with this method requires pretreatment of the substrate with rGO to create nucleation sites. Even so, the resulting single-layer thin-film nucleates simultaneously at all the nucleating sites, resulting in a thin film that is not monocrystalline. This is in contrast with graphene, which produces a single crystal monolayer with CVD on a copper substrate. Because the existence of grain boundaries can significantly impact the electronic performance of any material, many research groups around the world are looking for processes to reliably synthesize single-crystal monolayers of various TMD materials.

8.2.3 Hexagonal Boron Nitride (H-BN)

Boron nitride (BN) is an interesting material. It consists of an equal number of boron and nitrogen atoms bonded together through covalent bonds. It can exist in many forms that are similar to carbon structures, such as cubic crystal lattice (analogous to diamond), hexagonal BN (analogous to graphene), and nanotubes (analogous to carbon nanotubes, CNTs). Its most stable form is hexagonal BN, in which boron and nitrogen atoms are arranged in a hexagonal structure in a single 2D plane (Figure 8.4). In bulk h-BN, these 2D layers are connected through

(a)   **(b)**

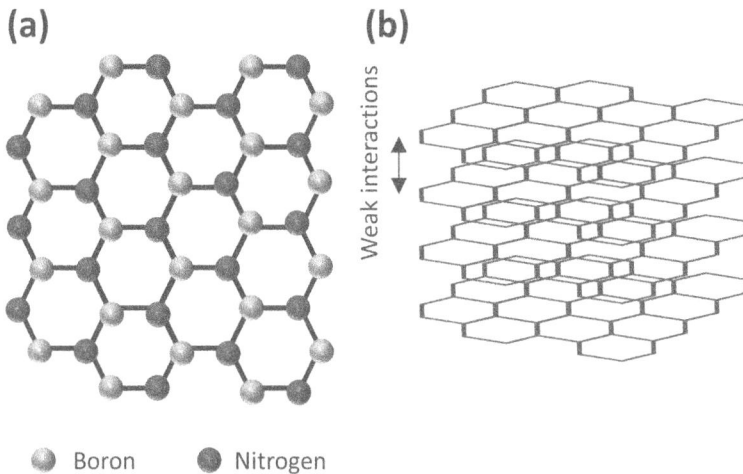

Weak interactions

⬤ Boron   ⬤ Nitrogen

FIGURE 8.4   (a) The structure of h-BN is an infinite tessellation of the B-N single bond in a hexagonal lattice similar to graphene (b) These layers stack together to form the bulk material, with weak interactions holding the individual layers in place.

weak van der Waal's interactions, thus leading to lubricating properties as seen in graphene and TMDs. Further similarities with graphene include the distance between layers, which is 3.3 Å in the case of h-BN, and that each boron or nitrogen atom is connected to three neighboring atoms through covalent bonds. However, contrary to the situation in graphene, there is no π-conjugation in h-BN, leading to a negligible free carrier concentration at room temperature and a bandgap of around 6 eV. This opens up the possibility of the use of h-BN as an atomically thin insulator.

Boron nitride is remarkably stable given that both boron and nitrogen atoms are in strong covalent bonds, under no structural stress. The material is mostly immune to thermal decomposition (up to 1000°C), or oxidation in air, as opposed to graphene. It is etched by, or is soluble, in very few solvents, including strong acids. Thus, for device formation and patterning, physical etching mechanisms such as reactive ion etching, or electron beam ablation are used. The strong bonds between boron and nitrogen in the basal plane lead to high mechanical strength in-plane, whereas the atomically thin structure leads to low flexural rigidity normal to the plane. Further, the strong bonding also increases the phonon stability and propagation speed, leading to high thermal conductivity, despite poor electrical conductivity. Boron nitride can be doped with carbon atoms to introduce free carrier concentration at room temperature to increase electrical conductivity.

Hexagonal BN layers can be produced by mechanical and chemical exfoliation from bulk h-BN material as in the case of graphene. Chemical exfoliation is generally carried out in organic solvents such as dichloroethane ($C_2H_4Cl_2$), with the help of sonication. However, these methods lack repeatability and are dependent on the quality of the initial bulk h-BN for purity and size of flakes. The best method for producing large quantities of high-quality h-BN is the CVD process. Typical precursors used for CVD of h-BN include boron trifluoride (BF_3), boron trichloride (BCl_3), borane (BH_3), or diborane (B_2H_6). The typical substrate used for this process is copper foil for ease of availability and good catalytic performance. The use of separate gaseous precursors for boron and nitrogen requires precise control over their flow rates to allow for 1:1 stoichiometry of the deposited BN thin film. However, this can be resolved by the use of precursors that act as a source of both boron and nitrogen atoms, such as borazine ($B_3N_3H_6$, a BN-equivalent of benzene) or ammonia borane (H_3N-BH_3). Further, as in the case of TMDs, the h-BN produced using CVD is not monocrystalline, and in some cases not a single layer as well. The nucleation and growth of the h-BN layer take place simultaneously at multiple locations on the substrate leading to a polycrystalline layer. The grain size in this case depends on the density of nucleation, which depends on the preprocessing conditions for the substrate.

8.2.4 OTHER 2D MATERIALS

Apart from these materials, there are some more that show promise for applications in electronic devices. These include phosphorene, silicene, germanene, MXenes, and so on. Phosphorene is the single atomic layer forming the bulk material black phosphorus, which consists of layers of phosphorene held together by weak van der Waal's interactions. The process of Scotch tape exfoliation can be used to obtain phosphorene from black phosphorous. Phosphorus atoms are hexagonally distributed, however, they are distributed out-of-plane because of the sp^3 hybridization of phosphorus atoms. Three of these orbitals participate in bond formation with the neighboring phosphorus atoms, leaving a lone pair of electrons in the other orbital. Phosphorene has a bandgap of around 1.8 eV that decreases with an increase in the number of layers in a thin film (around 0.3 eV for bulk) and carrier mobilities of up to 1000 cm^2/V-s, which make it a promising candidate for TFT applications. However, phosphorene is unstable in ambient temperature and pressure, reacting with oxygen and moisture in the air to form acids. Thus, encapsulation layers are generally used by researchers to stabilize transistors based on phosphorene. Synthesis of single or few-layer phosphorene is carried out using mechanical or liquid exfoliation of bulk black phosphorus. As of now, there is no known CVD process to produce phosphorene.

Silicene and germanene are proposed to be single-layer materials consisting of honeycomb distribution of silicon and germanium atoms respectively. They are considered to be sp^2 hybridized leading to a structure similar to graphene. Both materials do not exhibit a completely planar structure as seen in graphene but have periodic out-of-plane distortions. Unlike other 2D materials discussed so far, these materials do not occur in bulk material form (stacked layers like graphite), thus, they have to be synthesized using very precisely controlled experiments. This makes research on these materials difficult, however, given the potential promises of having familiar, stable, large-scale 2D material for electronic device fabrication, many research groups across the world are making efforts towards facile synthesis of these materials.

8.3   ONE-DIMENSIONAL MATERIALS

One-dimensional materials are material systems with only one significant dimension – length. These materials consist of long structures only a few atoms across (typically less than few tens of nanometers). The existence of 1D systems has been theoretically postulated for decades, however, in 1991, Sumio Iijima showcased the structure of CNTs proving the existence of a free-standing 1D material system. Following the discovery, several materials have been shown to form stable 1D systems. Over the past three decades, the synthesis and characterization of these materials have been studied extensively, leading to great insight into their structure and material properties. These materials carry great promise from a flexible and stretchable electronics standpoint, because of their ability to conform to any structure. Thin films formed using 1D materials tend to retain their properties even after the application of strain, indicating their potential applicability in stretchable electronics. Prominent among these materials is CNTs, which can be thought of as a rolled-up version of graphene. CNTs have been extensively studied over the past decades and several interesting applications have been showcased for CNT-based materials. Other important 1D materials include silver nanowires, gold nanowires, silicon nanowires, ZnO nanowires, BN nanotubes, and so on. All these materials exhibit different electronic and mechanical properties, thus providing a variety of applications to be covered with the use of 1D materials. Further, their properties are affected by the diameter of the nanostructures, thus providing another method to tune some of the properties. The way 1D materials are classified is a little vague, because completely 1D material systems, which are a long chain of a single atom, are not stable. Thus, 1D materials are said to be those that have long chains that are few atomic distances across, and in some cases have an internal lattice structure. These materials have very high length-to-diameter ratios, up to 10^7 in some cases, and are referred to as quasi 1D materials. The large length and small diameter provide a very high surface-to-volume ratio for 1D material systems, thus leading to enhanced sensitivity to surface activities. Thus, like 2D materials, 1D material systems can also be used for application in sensor systems.

8.3.1   Carbon Nanotubes (CNTs)

Carbon nanotubes are made of tubes of sp^2 hybridized carbon atoms arranged in a hexagonal honeycomb structure. CNTs can be considered as tubes of rolled-up graphene sheets. The diameter and length of the tubes depend on the synthesis process. Most of the synthesis processes, however, provide a mixture of CNTs of various diameters and lengths, which can be separated through post-processing. The number of "walls" in a CNT can vary as well, and in some cases, determine the properties of the final thin film. CNTs are classified as single wall (SWNTs), double-wall (DWNTs), or multi-wall (MWNTs) depending on whether they have single, two, or multiple layers of concentric tubes respectively (Figure 8.5). For example, MWNTs can be thought of as having multiple SWNTs of different diameters and similar lengths, nestled inside each other. As discussed earlier, the sp^2 hybridized carbon lattice forms a planar hexagonal honeycomb lattice through C-C sigma bonds $120°$ apart. This lattice is unstrained because the natural angles of the standalone

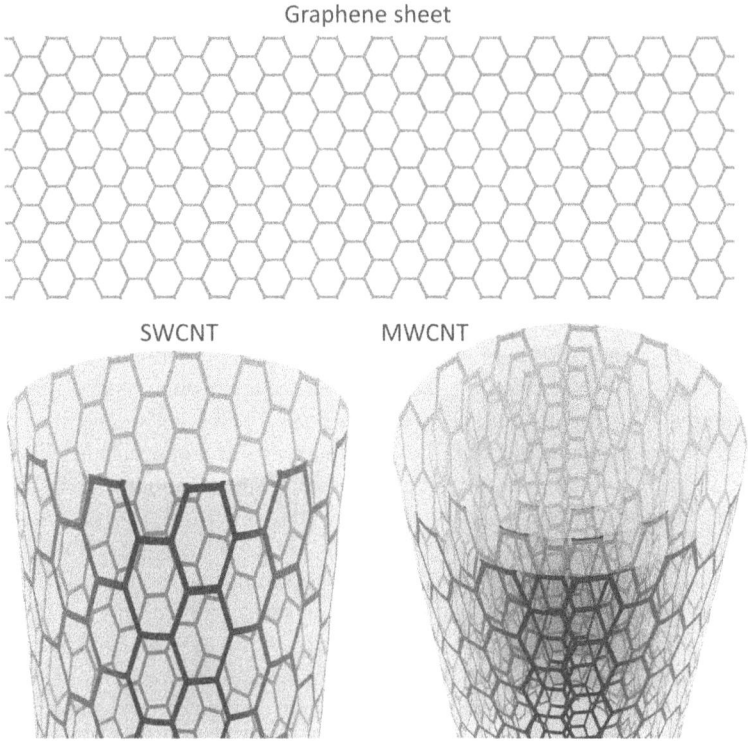

FIGURE 8.5 A single wall carbon nanotube (SWCNT) can be thought of as a graphene sheet folded into a tube. A multi wall CNT is formed when several single walled CNTs assemble concentrically.

bonds are the same as those in the lattice. However, in the case of CNTs, because of the curvature of the graphene sheets into a tube form, there is a strain on the C-C sigma bonds forming the hexagonal lattice. This has two implications. First, there is a significant restructuring of the electronic states, leading to electronic properties different than those of graphene. Second, because the strain on the bonds is related to the radius of curvature, there is a minimum diameter associated with a stable freestanding SWNT.

The properties of CNTs are dependent on their structure. Because a nanotube is considered to be a rolled version of graphene, the structure of the nanotube can be determined based on the seam that was used to roll the structure. This seam is determined by the vector used for forming the circumference of the tube, as shown in Figure 8.6. Consider n and m to be two integers determining the vector on the graphene lattice, with lattice vectors a_1 and a_2. The vector is then given by:

$$C = na_1 + ma_2$$

The length of this vector is the circumference of the tube, which can be used to determine its diameter ($|C|/\pi$). The integers n and m thus define the structure and

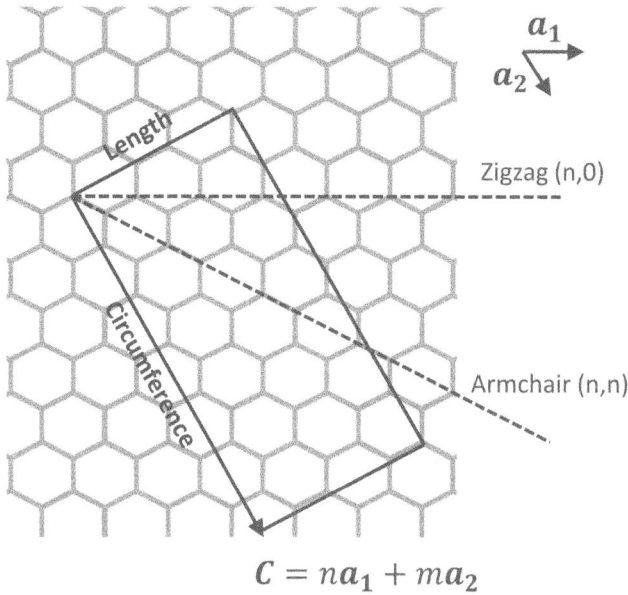

$$C = na_1 + ma_2$$

FIGURE 8.6 The structure of carbon nanotubes can be obtained from the associated graphene sheet and the way "folding" occurs. Based on the basis vectors a_1 and a_2, the chiral number pair (n,m) determines the circumference and the edge configuration of the carbon nanotube.

diameter of the tubes, leading to their physical properties. The pair (n,m) is known as the chiral index, or simply chirality, of the nanotube. To avoid rotational symmetries, the values of n and m are only considered to be positive. For negative values, there are rotations that lead to positive values, preserving the structure and properties of the nanotube. The angle α between the unit vector a_1 and the vector C is also frequently reported and can be indicative of nanotube properties.

$$\alpha = \tan^{-1}\left(\frac{\sqrt{3}m}{m + 2n}\right)$$

There are infinite possibilities for the pair (n,m), leading to infinite types of CNTs. However, only a few types are stable and commonly observed in experiments. Two types of CNTs that are most commonly reported are zigzag and armchair. The zigzag configuration occurs when the vector C is parallel to the unit vector a_1, leading to a situation with the pair $(n,0)$ and $\alpha = 0$. The armchair configuration occurs when the vector C creates an angle of 30° with the unit vector a_1. In this case, the chiral pair is of the form (n,n). The hexagonal honeycomb structure of CNTs is relatively stable, however, there can be defects leading to the formation of pentagons or heptagons. There can be carbon vacancies in the hexagonal lattice or incorporation of impurity atoms. In fact, in some cases, carbon "nanobuds" have been reported attached to the nanotube. These consist of fullerene or Buckyball molecules attached to the side of the CNT. All of these defects alter the local electronic structure of nanotubes leading to changes in thin-film properties. Thus, the control over CNT synthesis, particularly from the point-of-view of nanotube type and quality, is key for their commercial success.

Several processes have been reported for the synthesis of CNTs. The most popular synthesis process is the arc discharge process, used by Iijima in his landmark paper in 1991. The setup for the arc discharge process is shown in Figure 8.7. The process involves creating an electric arc discharge between two carbon (graphite) electrodes. The arc discharge increases the temperature of the electrodes. The high temperature and low pressure maintained in the chamber cause the carbon atoms to vaporize and condense in the form of CNTs and other carbon-containing material, such as graphene flakes, fullerene molecules, and so on. The electrode distance is maintained such that an electric arc of a specific voltage can be struck. The process involves the consumption of electrode material, thus requiring the assembly to be designed such that one of the electrodes can be moved to maintain a constant distance between the two electrodes, eventually requiring the electrodes to be replaced. The process is simple and does not require gaseous precursors that need expensive mass flow controllers and safety infrastructure. A key drawback is that many types of carbon-containing materials are formed along with CNTs, thus requiring some post-process filtration. However, the process has been very well studied over the past few decades which helps fine-tune it to specific requirements. A similar process involving the brute-force vaporization and redeposition of carbon material is the laser ablation method. In this case, a focused laser beam is used to create localized

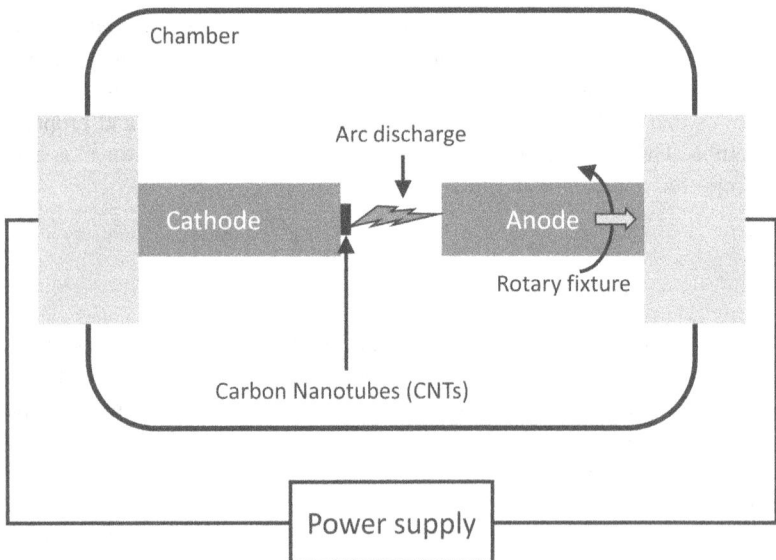

FIGURE 8.7   The arc discharge process is carried out by establishing an electrical arc discharge between two graphite electrodes. The electrodes get consumed during the process, requiring a mechanism to move one of the electrodes to maintain constant distance between them.

high temperatures to vaporize carbon atoms and cause redeposition as CNTs. This process also creates many forms of carbon materials, and thus, requires some post-processing to filter out CNTs.

For a scalable and reliable solution, we turn to the CVD process. The CVD process for CNTs is actually very similar to graphene. In this process, a hydrocarbon is introduced in a heated quartz chamber containing a substrate with a transition metal catalyst. The hydrocarbon breaks down due to the high temperature (pyrolysis), and the carbon atoms dissolve in the transition metal. When the carbon solubility of the transition metal catalyst is reached, the carbon atoms are precipitated in the form of CNTs. The most commonly used precursor gas for CNT growth is methane, however, other gases such as ethane, ethylene, carbon monoxide, ethyl alcohol, acetylene, etc. have also been reported. The most commonly used metal catalysts are iron and cobalt. The CVD process provides much more control over the growth conditions compared to the arc discharge or laser ablation processes. Further, there is significantly lower contamination because of other carbon materials in this process. An interesting variation of this process is the use of metal catalysts in nanoparticle form. This provides control over the diameter of the CNT produced because one metal nanoparticle typically gives rise to one nanotube. However, the process for the fabrication of metal nanoparticles of a given size increases the cost and complexity of the overall process.

Carbon nanotubes are known for some very interesting electrical, mechanical and thermal properties. Because the basic C-C bond structure is similar to that of graphene, it is expected that CNTs show properties similar to graphene. However, the rolling and stitching of graphene sheets impose confinement along the direction of the circumference. This leads to quantization of the allowed electronic states. Further, the resulting electronic structure and allowed electronic states depend on the magnitude of quantization, which depends on the circumference of the nanotube. Thus, the chiral index of the nanotube determines its electronic properties. Thus, some nanotubes behave metallically, while others are semiconducting depending on the chiral index. In the case of semiconducting CNTs, the bandgap also depends on chirality. Given that CNTs of different chiral indices are produced randomly in processes such as the arch discharge, it can be theoretically predicted that of the single-walled CNTs produced, two-thirds will be semiconducting, and one-third will be metallic. This has been observed in experimental results as well. In semiconducting CNTs, the bandgap is 0.4–0.7 eV, depending on the chirality. In the case of metallic CNTs, the free electron concentration for conduction is obtained from delocalized π-electrons. Also, as discussed in graphene, electrons can achieve ballistic transport in defect-free CNTs because of the lack of phonon scattering. Even in the case of semiconducting CNTs, ballistic transport can be observed for the few free carriers available for conduction. The metallic and semiconducting properties of the nanotubes pertain to single-walled CNTs (SWNTs) of a particular chirality. In the case of MWNTs, the individual nanotubes forming the multi-walled system can be metallic or semiconducting. Also, there is very good electronic coupling between nanotubes in a multi-walled system. Thus, even if a single nanotube in an MWNT system is metallic, the entire nanotube behaves metallically. Thus, most of the electronic applications of CNTs utilize their metallic character.

Several other interesting structures have also been theorized such as carbon nano-torus, which consists of CNTs bent into a toroid shape. There has been some experimental evidence of their existence, however, a reliable synthesis method is yet to be developed. Carbon nanotori are mainly studied theoretically for their electrical and optical properties but seem to be promising material systems if they can be produced reliably.

8.3.2 METAL NANOWIRES

Metal nanowires are one-dimensional structures made from metal atoms, such as silver, gold, copper, platinum, and so on. In most cases, the nanowires retain the properties of the bulk metal, thus, silver nanowires have very high electrical conductivity, just like bulk silver. In some cases, the radius of the nanowire can be small enough for carrier confinement effects to be visible. However, in most cases, metal nanowires are seen as a unique structure with the properties of the metal preserved. Thus, the primary application of metal nanowires is for the formation of conductive thin films. Metal nanowire-based thin films form a mesh network structure with individual nanowires overlapping several others to form conductive thin films. These nanowire-based films retain their conductivity even upon application of lateral strain because the nanowires can slide against each other and maintain electrical contact. This is particularly useful for flexible and stretchable electronics applications. For conductive films, silver nanowires (AgNWs) have been extensively studied for several reasons. First, silver is the most conductive metal known to humans. Given that nanowires retain most of the properties of the bulk material, AgNWs are extremely conductive. Second, silver nanowires can be easily synthesized in large quantities and are more resistant to oxidation at room temperature compared to nanowires of other metals such as copper and aluminum. Finally, compared to metals that are more stable, like gold and platinum, silver is more affordable. Thus, AgNW thin films form a middle ground between affordability and stability, while providing excellent conductivity.

Silver nanowires can be synthesized using a process called the polyol process. This process involves the precipitation of metal crystals from chemical baths containing metal salts (Figure 8.8). Nanowire precipitation is obtained by using a capping agent that restricts the growth of precipitating metal crystals in a single direction. For example, AgNWs are synthesized using silver nitrate ($AgNO_3$) as the metal source, and ethylene glycol ($[CH_2OH]_2$) as the solvent, along with polyvinyl-pyrrolidone (PVP) as the capping agent. The capping agent forms a cylindrical film around the silver crystals encouraging growth only in the axial direction. This leads to the formation of high aspect ratio nanowires. This process is very scalable because of the use of a simple chemical bath for nanowire precipitation and the relatively low cost of all the chemicals involved. The process allows for some control over the size of the nanowires produced by changing the relative concentration of the chemicals and the temperature of the bath. In some cases, ionic salts are added to the bath to act as nucleation sites. A similar process can be applied for the synthesis of other metal nanowires such as copper, gold, and so on. However, the existence of a capping layer around the nanowire can increase its contact resistance with other nanowires.

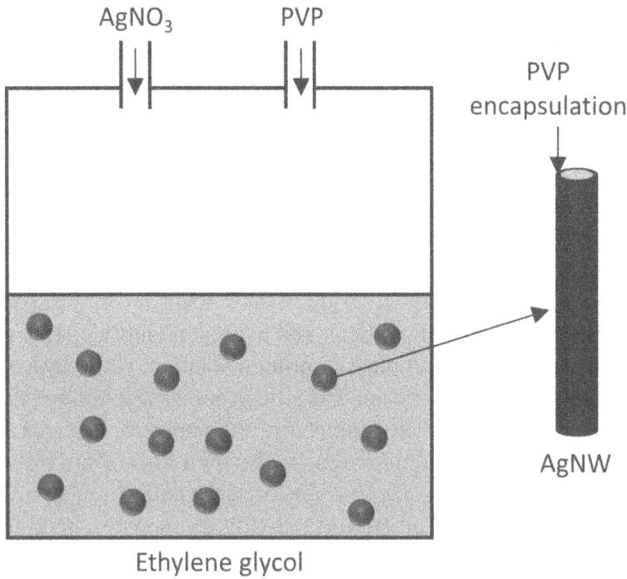

FIGURE 8.8 The polyol process consists of having a metal nanowire precipitate inside a salt solution. An encapsulation layer ensures the growth of metal precipitate in a single direction, resulting in nanowires and nanorods. Example, silver nanowires (AgNW) are formed using silver nitrate in ethylene glycol solution, with PVP as the encapsulation layer.

This can increase the overall sheet resistance of the network, a key performance parameter for conductive thin films. Thus, the capping layer is removed after nanowire synthesis. In the case of silver nanowires, the surrounding PVP layer can be washed away using wet chemical baths after the synthesis.

8.3.3 SEMICONDUCTING NANOWIRES

Semiconducting nanowires are one-dimensional structures formed by semiconducting materials. The most promising material for semiconducting nanowires is, of course, silicon. There is great interest in silicon nanowires because they provide an opportunity to create TFTs on any substrate with the best semiconductor known to humans – single crystal silicon. Thus, we can obtain the performance of state-of-the-art electronic systems with the flexibility that the TFT architectures provide. Thus, a strong interest has been shown in silicon nanowire synthesis, characterization, and device fabrication by the scientific community. While there has been some progress on the synthesis and characterization of silicon nanowires, there is very little progress on the application front. The key problems with the use of silicon nanowires for FET fabrication are the alignment of nanowires on a substrate and the precise positioning of the source, drain, and gate structures. Other challenges include the control over the diameter and length of each silicon nanowire during synthesis. Semiconducting nanowires are most useful if each nanowire can be formed into a FET, as an individual nanowire and not as a mesh network thin film. Thus, while AgNWs and CNTs are progressing as viable conductive thin film options, the progress for SiNWs is limited because of these challenges.

Even so, there has been considerable work done on the synthesis front. Again, CVD is the most promising method for growing silicon nanowires. The process is similar to the CVD of CNTs – a precursor gas containing silicon (silicon tetrachloride, $SiCl_4$) is flown at high temperature over metal droplets, typically gold. These droplets are formed by annealing a substrate deposited with a thin layer of gold. The use of gold for CVD of silicon nanowires is particularly popular because the Au-Si alloy has a eutectic point at 363°C, which reduces the temperature for nanowire growth significantly. Silicon tetrachloride breaks down in the presence of hydrogen gas at high temperatures into silicon and hydrogen chloride. The silicon atoms are then absorbed in the gold droplets forming a solution, while hydrogen chloride escapes the chamber in gaseous form. The silicon soon saturates and precipitates in the form of silicon nanowires. The diameter of the nanowire depends on the diameter of the Au-Si droplet, which depends on the thickness of the initial gold film. However, this is only true for the mean diameter of the nanowire forest. There can be a significant difference in the diameters of individual nanowires, making subsequent FET formation and integration in a large circuit difficult. Another problem with SiNW integration is the control over the crystal orientation of the grown nanowire because many electronic properties of the semiconductor lattice are dependent on crystal orientation.

The process described here is also known as the vapor-liquid-solid (VLS) process because the precursor gases dissolve in liquid droplets followed by solid nanowire precipitation. A variation of this process is the VSS process, wherein the temperature of growth is lower, causing the metal nanoparticle to exist as a solid. Another process for obtaining silicon nanowires is the use of selective epitaxy on silicon wafers. In this process, single crystal silicon wafers are used as a seed for homoepitaxy. However, instead of exposing the complete substrate, a masking layer (such as silicon dioxide) is deposited on the substrate and patterned into holes using lithography. Single crystal silicon is then grown in these cavities using the epitaxial process. This process leads to the formation of silicon nanowires at a known location with a known diameter and crystal orientation, thus making it easier for integrating into electronic systems. The process of selective epitaxy can also be used to fabricate various other nanostructures such as silicon nanotubes (SiNTs), III-V nanowires, and so on.

8.3.4 Other 1D Materials

One-dimensional material structures, both nanotube and nanowire, based on many other materials have been reported in the past. These include oxides of transition metals such as copper and zinc, which have semiconducting properties, and nitrides of silicon and boron, which have insulating properties. In particular, boron nitride nanotubes (BNNTs) have been widely studied on account of them being isostructure with CNT. Like CNTs, BNNTs can be considered as a rolled-up, one-dimensional structure of the two-dimensional BN sheet. In this case, as well, arc discharge, laser ablation, and CVD process can be used for synthesis. However, because BN is insulating, it cannot be directly used as an electrode. Thus, arc discharge is carried out using conductive compounds containing boron, such as yttrium boride (YB_6) or hafnium diboride (HfB_2), in an atmosphere of nitrogen gas. A refractory metal with a high melting point is used as the other electrode to stabilize the arc plasma. The CVD process for the synthesis of BNNTs is very similar to that of CNTs. The precursors for boron can be diborane (B_2H_6) gas, along with nitrogen or ammonia as the nitrogen source. In some cases, a combined boron and nitrogen source, such as borazine ($B_3N_3H_6$), is used. A transition metal catalyst is used for seeding the nanotube. Another interesting process for producing BNNTs is the ball mill process. This process can be used to produce incredibly large quantities of BNNT at relatively low costs. The process relies on creating defect sites in boron particles for nitrogenation to take place. This is done by milling boron particles in the presence of ammonia. The mixture is then annealed at 1000–1200°C in nitrogen ambiance to obtain a high yield of BNNT.

8.4 ZERO-DIMENSIONAL MATERIALS

Zero dimensional materials are nanostructures with no physical dimension being significant. Indeed, they are three-dimensional particles, but they do not have any appreciable length, width, or height, thus they are called 0D particles. They are also referred to as nanoparticles, nanocrystals, or quantum dots (QDs). Like other nanostructured materials, nanoparticles can be formed from materials of various types such as metals, semiconductors, insulators, and in some cases, a combination of these. Typically, particles with all three dimensions less than 100 nm are considered to be nanoparticles. Even at the significantly reduced dimensions, the crystal structure of the bulk material is generally preserved in the particles. However, given quantum confinement effects for charge carriers, the properties of nanoparticles can be significantly different than those of bulk materials. This is particularly true as the dimensions of the nanoparticles reduce, increasing the quantum confinement effect. Apart from this, the surface-to-volume ratio for nanoparticles is extremely high, leading to some very interesting chemical properties. For example, reaction rates can be significantly enhanced because of the availability of a lot of surface area. This leads to the use of nanoparticles primarily as catalysts for many industrial chemical reactions.

8.4.1 METAL NANOPARTICLES

We cannot talk about metal nanoparticles without mentioning the Lycurgus cup from the ancient Roman Empire (circa fourth century AD). The chalice was made using a metal nanoparticle composite in a glass matrix. Its creators used silver, gold, and platinum nanoparticles to create a nanocomposite such that it shines green in reflective light and red in transmitted light. This is because the optical properties of nanoparticles, much like all their properties, depend on their size. While Lycurgus cup is widely regarded as the first application of "nanotechnology" by humans, there have been many other places where colloidal suspension of tiny gold particles has been used for artwork to obtain golden color (for example in ancient Egypt). In modern scientific literature, Michael Faraday was the first to discuss the link between the optical properties of nanoparticles and their size, in his 1857 lecture titled "Experimental Relations of Gold (and Other Metals) to Light". Faraday, at one point, clearly mentions that "… known phenomena appeared to indicate that a mere variation in size of its particles gave rise to a variety of resultant colors". It was in 1908, that G. Mie developed a theory for the scattering of light by nanoparticles by solving Maxwell's equations. This was still in the classical physics realm but provided a quantitative basis to predict the color of a colloidal suspension based on the nanoparticle size. With advancements in quantum mechanics, the theory for the optical properties of nanoparticles has been further developed. For metal nanoparticles comparable or smaller than the wavelength of light used to excite them, the free electrons start oscillating due to the electric field associated with the light wave. The oscillations resonate at a particular frequency that depends on the properties of the metal and the size of the nanoparticles. This is called localized surface plasmon resonance, and causes absorption of light at the resonant frequency, leading to the characteristic size-dependent colors of the nanoparticle colloids.

In flexible electronics, the most common usage of metal nanoparticles is to obtain conductive thin films. This is done by spin-coating or inkjet printing metal nanoparticle solutions on a substrate. These techniques produce a dispersion of nanoparticles, often stacking on top of each other, providing conductivity through the thin film because of the electrical contact between adjacent particles. This is similar to the mesh structure obtained with metal nanowires, however, a key difference is that metal nanowire thin films can sustain conductivity for much larger strains because of the sliding of nanowires with respect to each other without losing electrical contact. In the case of nanoparticles, the thin films provide conductivity for reasonable strains associated with flexing and bending. Nanoparticles of noble metals such as silver, gold, and platinum are usually considered for flexible conductive thin film formation because these are resistant to the formation of native oxide on the surface, which can be particularly problematic given the high surface to volume ratio of nanoparticles. The stability of metal nanoparticles against oxide, nitride formation, or reactivity with other impurities remains a key challenge in the reliability of nanoparticle thin films. Further, metal nanoparticles have been reported to be toxic and carcinogenic given that their tiny size allows them to penetrate the cell membrane. In fact, some metals that react with oxygen exothermically can cause an explosion when exposed to air in nanoparticle form because of the high rate of reaction on account of the high surface-to-volume ratio. Thus, metal nanoparticles need to be synthesized and processed in a controlled environment.

Metal nanoparticles can be synthesized using several techniques that are classified into two basic types – the top-down and bottom-up processes. In top-down processes, macroscopic structures such as bulk material or thin films are broken down to nanoscale size using mechanical, electrical, or thermal force. In bottom-up processes, nanoparticles are "built" or grown from a single or few atom seeds up to the desired size. The ancient gold and silver nanoparticles used in artwork and pottery were likely synthesized using the top-down method of milling. In this process, bulk metal pieces are ground down to nanometer size using mechanical force in a ball mill. The milling time, size of starting metal pieces, and rotation rate determine the size of the resulting nanoparticles. Another top-down method is through the use of thermal energy to break thin films of metal into nanodroplets. Other top-down methods include the use of plasma, laser, or arc discharge to eject a plume of material from a solid target, followed by condensation of the ablated particle vapor. These methods have the advantage of being low-cost and scalable, however, there is very limited control over the size of the resulting nanoparticles, which is critical given that nanoparticle properties are heavily dependent on size. Thus, some form of post-processing is required to filter out nanoparticles of a particular size range. For example, a wide distribution of nanoparticle sizes is obtained through ball milling, which can be segregated through simple techniques such as air classifiers (elutriation).

The most commonly used bottom-up process is precipitation. This can be done by creating super-saturated solutions of metal salts by mixing two or more solutions together, or by precipitating metal using a reducing agent. For example, the gold colloids discussed by Michael Faraday in his landmark lecture were created by reducing aqueous chloroauric acid ($HAuCl_4$) solution. Another example is the reduction of the aqueous solution of chloroplatinic acid (H_2PtCl_6) using sodium borohydride ($NaBH_4$). The size of the nanoparticles produced can be controlled by adjusting the concentration of reagents, bath temperature, pH, and mixing speed. However, colloidal suspensions are generally unstable, leading to the coagulation of nanoparticles to form micro- or macroscopic precipitates. Thus, some stabilizers are required to prevent the coagulation of nanoparticles. This process is extremely scalable and can be used to produce industrial quantities of nanoparticles with reasonable repeatability. The process also affords control over nanoparticle size and produces uniform pseudo-spherical particles. Further, the colloidal suspension thus produced can be used as nanoparticle ink for printing, spraying, or spin coating any substrate to form a metal nanoparticle thin film. Another bottom-up process for nanoparticle synthesis is through thermal decomposition of a precursor gas containing the metal ion (pyrolysis). In this process, aerosol droplets of a liquid solution containing a metal salt are passed through a high-temperature furnace. The solvent evaporates and the salt decomposes reducing the aerosol into solid nanoparticles. The aerosol droplets are typically produced using ultrasonication (ultrasonic spray pyrolysis).

Interestingly, nanoparticles of certain metals such as silver can be synthesized using biological sources. These so-called green synthesis methods involve the use of cellular chemistry to break down metal salts and precipitate nanoparticles of very specific sizes. For example, the bacterium *Pseudomonas stutzeri* can produce silver nanoparticles extracellularly in the presence of aqueous Ag^+ ions in the periplasmic space. Similarly, the fungus *Fusarium oxysporum* has been reported to produce silver nanoparticles. The size of the nanoparticles depends on the species of the organism used and the initial concentration of silver ions in the solution.

8.4.2 SEMICONDUCTING NANOPARTICLES

Semiconducting nanoparticles, also known as QDs, are nanoparticles made of semiconducting elements or compounds. This is a very important class of nanomaterials because of the unique electronic and optical properties bestowed on the semiconductor lattice because of the small size of the particle. The key to understanding semiconductor nanoparticles is the underlying physics of confinement of charge carriers inside the nanoparticle. Most nanoparticles only contain a few hundred atoms forming the nanocrystalline structure. Because of the limited number of atoms (nuclei), the solution to Schrodinger's equation for electrons is very different from that in bulk crystal. We are no longer able to apply the Bloch theorem typically used for infinite lattices – even microscopic lattices have several billion atoms. The boundary conditions at the end of the nanoparticle along with the limited number of nuclei result in an electronic structure that is very different from that of the bulk semiconductor. This results in very interesting electronic and optical properties that are dependent on the shape and size of the QDs. For example, the electronic states of QDs resemble those in individual atoms (not as part of any lattice), i.e., the electronic states are well-defined and tightly bound instead of being a "band". A key property of QDs is the dependence of bandgap on size. This is a direct consequence of the variation in dispersion of electronic states with the number of atoms present in the lattice. Thus, changing the size of a QD can change the bandgap, leading to a change in the wavelength of light it can absorb/emit. This means QDs for a particular size appear a particular color in reflected light. This property is extensively used in commercial applications of QD thin films, as light-emitting particles (quantum dot LEDs or QLEDs) and as light-absorbing particles (quantum dot solar cells).

Just like metal nanoparticles, QDs can be synthesized using top-down and bottom-up processes. In the top-down approach, physical processes such as laser ablation, arc discharge, plasma discharge, sputter, etc. have been used. For example, irradiating a silicon target with a laser beam of sufficient intensity can lead to the creation of silicon quantum dots (SiQDs) of high purity and crystallinity. However, there is a large variation in the size of the resulting particles. The bottom-up approach includes chemical processes such as colloid formation through reduction or supersaturation or decomposition of precursor materials with heat. For example, SiQDs can be obtained by reducing silicon tetrachloride ($SiCl_4$). These techniques are similar to those discussed in the case of metal nanoparticles and have similar advantages and disadvantages. However, there are several binary semiconducting nanoparticles that have important commercial applications. These include chalcogenides, such as cadmium sulfide (CdS), lead sulfide (PbS), cadmium telluride (CdTe); III-Vs, such as indium arsenide (InAs), indium phosphide (InP); and oxides, such as zinc oxide (ZnO). The processes for the synthesis of these binary systems need to take into account the stoichiometry and the crystal structure of the resulting semiconductor particle. A commonly used process is the colloidal synthesis, wherein precursors for both cation and anion are mixed in a chemical bath to create the desired material. For example, for the development of metal chalcogenide nanoparticles, metal precursors such as cadmium oxide (CdO), zinc oxide (ZnO), lead oxide (PbO), indium chloride ($InCl_3$) are used along with elemental powders of the desired chalcogenide (sulfur, selenium, tellurium). For large-scale synthesis, continuous production approaches are typically used. In this case, precursors for both components of the material are mixed in the same solvent and an aerosol of the solution is created. These aerosol droplets are then heated to cause a reaction between the two precursors within each droplet. The solvent eventually evaporates because of the high temperature leaving dry QDs to be collected. For example, for the synthesis of cadmium selenide (CdSe) QDs, cadmium oxide (CdO) is used as a precursor for cadmium; trioctylphosphine selenide (TOPS, $[C_8H_{17}]_3PSe$), which is obtained by dissolving elemental selenium in trioctylphosphine ($[C_8H_{17}]_3P$), is used as a precursor for selenium; and toluene is used as a solvent along with some surfactants to keep the solution stable. There are several advantages of using this process. First, it can produce very large quantities of semiconducting QDs with a continuous throughput. Second, because the reaction takes place inside each droplet separately, the size of the resulting nanoparticle is dependent on the size of the droplet, which can be precisely controlled using high-quality nebulizers. Thus, the size distribution of the resulting nanoparticles can be very tightly controlled. Third, because the reactions are occurring separately, complex ternary and quaternary QDs can be synthesized with reliable stoichiometry.

Indeed, the future of electronics, particularly that of flexible electronics, seems to be in utilizing the myriad nanostructured materials that have been developed so far and will be developed in the future. In this chapter, we have seen the boundless opportunities provided by the world of nanostructured materials, their unique properties, and several processes for their synthesis. However, if their properties are to be fully utilized, it is important to create a path from synthesis to integration into state-of-the-art electronic systems. In the subsequent chapters, we will look at some of the processes developed to take advantage of these specialized materials for potential commercial applications.

EXERCISES

8.1. Why do electrons behave as if they are massless while traveling through graphene?

8.2. Describe, in detail, the process for obtaining MoS_2 from elemental molybdenum and sulfur.

8.3. We have fabricated zigzag and armchair CNTs with chiral number $n = 10$. What is the diameter of the nanotubes?

8.4. What is one of the bottom-up ways of creating metal nanoparticle inks? Give an example for gold NP solution.

8.5. Why does the bandgap of semiconducting nanoparticles (quantum dots) depend on their dimensions?

Part III

Integration Strategies

9 Substrates, Transfer, and Bonding

9.1 INTRODUCTION

All the different materials and their synthesis processes have one thing in common, that there is nothing in common. There are variations in substrates, temperatures, pressures, deposition conditions to obtain good quality thin films, bulk, or powders of various kinds. This, in addition to the various applications these materials have to be used for, creates a large number of possibilities. To achieve integration of the complete system on a single platform, there is a need to transfer materials from one substrate to another. The processes involved in transferring, handling, and bonding these materials need to be compatible with all the materials in question. The process parameters have to be such that there is no damage to functional materials. The materials used for fabricating the substrate and as the bonding agent need to be in accordance with the final application. In particular, the total flexibility required by the use case needs to be accounted for while determining the appropriate integration strategy. This is also applicable to the materials and processes used for creating the encapsulation or packaging for the electronic devices. In this chapter, we will navigate the complex labyrinth of the substrates that are commonly used for flexible electronics applications, their properties, and the processes by which functional thin films can be transferred and bonded to these substrates. Further, we will also look at some of the emerging techniques to package flexible electronic devices. It is important to note that the materials and processes discussed in this chapter are the current state-of-the-art, not necessarily the best or most optimal integration strategies for flexible electronics. Indeed, there are many challenges associated with the integration of heterogeneous material systems into a completely functional electronic device, and many research groups around the world are focused on solving them.

9.2 FLEXIBLE SUBSTRATES

There are several characteristics of an ideal substrate for flexible electronic devices:

1. First and foremost, the substrate material should be flexible. This means the flexural rigidity for a given configuration and flexing axis should be as low as possible. This boils down to having a low Young's modulus for a substrate of given dimensions, or the ability to create ultrathin versions of the substrate, given a material.

DOI: 10.1201/9781003010715-12

2. The substrate material should have a high elastic strain limit to account for large strains in the top and bottom layers associated with the bending motion. This will ensure an adequate minimum bending radius for various applications.

3. The restoring force generated inside the substrate, upon the application of stress, should be high to allow for the substrate to retain its original shape upon removal of stress. Thus, the substrate material should be elastic up to the minimum bending radius required.

4. Ideal substrates should be able to withstand high temperatures, particularly if the integration strategy involves direct deposition of materials on these substrates. Some bonding materials require high anneal temperatures to cure, which is only possible if the substrate is able to withstand them. Further, post-process annealing is performed for some integration strategies to enhance device performance. For example, annealing silver nanowire (AgNW) based thin films can lead to better contact between adjacent nanowires and an increase in the conductivity of the thin film. These anneals can only be performed with substrates compatible with high temperatures.

5. From a usability point of view, ideal substrates should be lightweight and easy to handle independently. For example, many materials can be made flexible by reducing their thickness, however, in many cases, very thin substrates cannot be independently handled without crumbling.

6. Substrates should be easy to manufacture and low-cost to allow for scaling up of device manufacturing. For example, substrates that can be manufactured in bulk using large chemical baths or through the roll-to-roll process are ideal.

7. The substrates, particularly in the case of wearable electronics, should be non-toxic and biocompatible.

8. To avoid the buildup of stress because of temperature cycling over its lifetime, it is ideal to have a substrate with a low coefficient of thermal expansion (CTE), or with CTE matching that of other materials in the integration.

9. The substrate should ideally have high thermal conductivity so that heat generated because of processing within the flexible electronic device can be dissipated effectively.

10. The substrate should be electrically insulating to make sure different parts of the integrated system are electrically isolated from each other. However, depending upon the application, there may be some merit in having an electrically conductive substrate that can provide an easily accessible common terminal to interconnect various parts of the circuit, or that can shield the circuit from electromagnetic noise.

11. Ideal substrates should have low water and oxygen permeation to ensure reliable operation of electronics under any deployment condition. In addition, the substrate material should be resistant to chemical reactions to avoid issues such as rust and decay.

12. An ideal substrate should have very low surface roughness, thus, the material and process used for substrate preparation should allow for very smooth surfaces to be fabricated.

13. It is ideal to have a transparent substrate for applications involving optoelec-
tronics such as displays, solar cells, and so on.

These properties have been summarized in Figure 9.1. All of these qualities
define an ideal substrate as well as a superstrate, and the material used for creating
the complete packaging for flexible electronic systems. However, it is impossible to
obtain a material having all of these ideal properties because many of these properties are conflicting, i.e., having one desirable property implies the loss of another
from a basic physics perspective. For example, it is stated that an ideal substrate
should be flexible, which requires the Young's modulus of a material to be low, given
the dimensions and axis of flexure are fixed. However, low Young's modulus leads to
a low spring constant for a substrate of given dimensions, leading to a low restoring
force. Thus, there needs to be a balance between the requirement for high restoring force and low flexural rigidity. Similarly, the requirement of high thermal conductivity and electrical insolation results in a very narrow set of materials. Highly
crystalline materials such as diamond demonstrate high thermal conductivity due
to effective phonon transport through the lattice, while being electrically insulating because of the lack of free carrier concentration at room temperature owing
to the tight binding of electrons in the lattice. However, such materials are neither

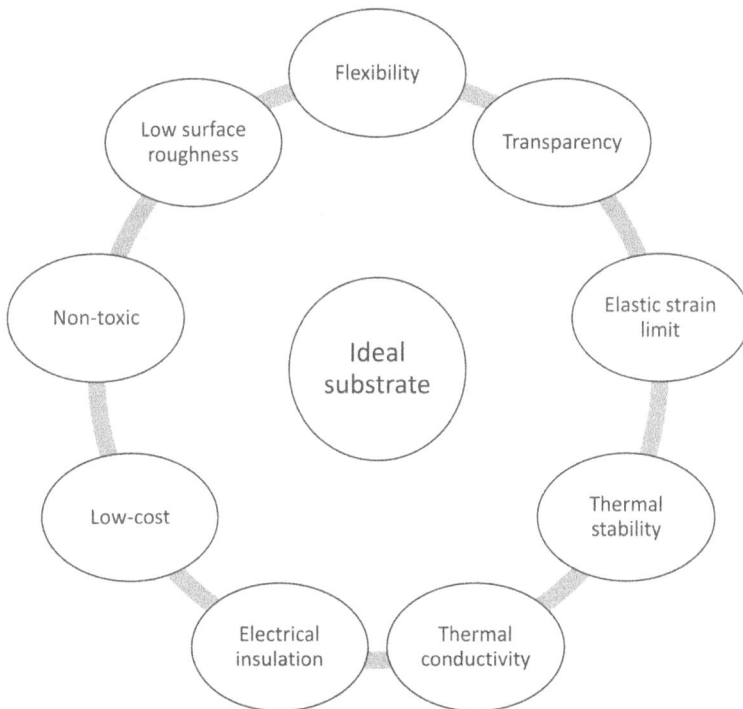

FIGURE 9.1 Properties associated with an ideal substrate material for flexible electronics
applications.

flexible, because the tightly bound lattice leads to a high Young's modulus, nor easy to fabricate, because highly crystalline materials require sophisticated fabrication techniques such as the epitaxial process. Hence, the search for ideal substrates for flexible electronic applications leads to compromises based on the application at hand, functional materials being used, the conditions of deployment, and so on. Indeed, it is interesting to note that most of the properties described here are fulfilled by the most commonly used substrate in the electronic industry – the single crystalline silicon substrate. Silicon is mechanically, thermally, and chemically stable; has low CTE; has high thermal conductivity; has low electrical conductivity (in the intrinsic state); is easy to manufacture and handle; can have atomically smooth surface finish; and can withstand high temperatures. However, it does not qualify the first and foremost property required from a flexible electronic substrate – flexibility. While silicon can be made flexible by reducing its thickness, it leads to other problems related to ease-of-handling. Thus, flexible electronic applications primarily use three types of substrates – metal foils, glass, and polymer sheets.

9.2.1 METAL FOIL SUBSTRATES

Metal foil substrates are one of the first substrates to be used for flexible electronics research. Thin steel foils were used to work with organic electronic materials and metal oxide-based thin-film transistors (TFTs) in the 1990s. Steel foils possess many properties listed for an ideal flexible substrate. They are resistant to chemicals, mechanically stable, have high thermal conductance, have low CTE, and can withstand very high temperatures (up to 1000°C). Further, their high electrical conductance can be used to create an easily accessible common ground terminal or for shielding from electromagnetic noise. In particular, the use of steel foil substrates for the fabrication of organic electronic material was popular because of their ability to act as a barrier for oxygen and moisture. Steel foils can also be produced in bulk through a roll-to-roll production process, significantly reducing their cost. Most of these properties are common among all metal foils, making them desirable for flexible electronics applications. Commonly used metals as substrates in foil form include steel, copper, aluminum, Kovar (nickel-cobalt-iron alloy), and so on. The selection of a particular metal as a substrate depends on the application being considered. Figure 9.2 shows a flexible copper foil that can be used as a substrate for flexible electronics applications.

There are several advantages to the use of metal foils as flexible electronic substrates, however, they offer some key disadvantages leading to their reduced popularity. One such problem is the apparent plasticity, or the lack of restoration forces, in a metal foil substrate. We have all experienced the plasticity of metal foils when using aluminum foils to pack food items. This behavior is unique to metals because of their peculiar bonding structure where electrons are shared among a sea of positively charged kernels. Thus, when displaced due to the application of a large strain, metal atoms do not experience restoration forces that covalently or ionically bonded atoms typically face upon being displaced from their equilibrium position. The resulting apparent plasticity can be both detrimental and advantageous for the use of metal foils for flexible electronic applications. In applications where it is

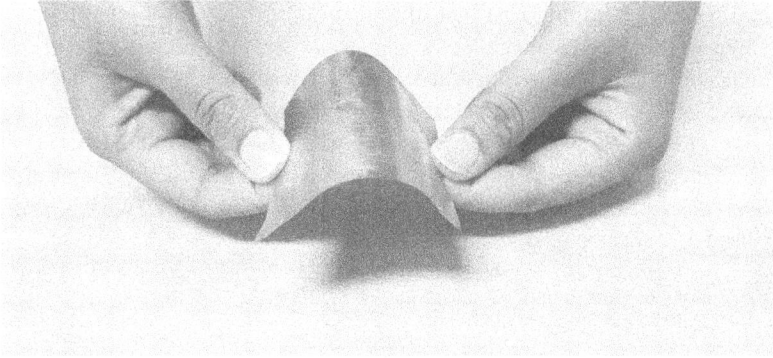

FIGURE 9.2 Any metal foil can be considered for use as a flexible substrate if it is fabricated with sufficient thickness for a given bending radius.

expected that the flexible system should conform to a surface, the plasticity of metal foils can be useful. However, in cases where it is important that the system maintains its original shape, this can be a disadvantage. Another major problem with the use of metal foils for substrates is their surface roughness. The process for creating thin foils of metals involves the use of giant rollers to cold press thick sheets of metals until they reach the required thickness. Generally, cold rolling is used for creating flexible sheets, because hot rolling very thin sheets are impractical from a handling perspective. If the rollers used for cold rolling contain microscopic defects or ridges, they can be transferred to the substrate during the rolling process. Typical RMS surface roughness of metal foils is in the order of a few microns. Thus, the defects on the surface can be large compared to the stack being formed for electronic materials (typical TFT stack height is 1 μm). This can lead to shorts in the device resulting in lower yield and reliability issues. To obtain a better surface finish, the process needs to be carried out with specialized rollers with high-quality surfaces. The rollers should be kept free of dust particles or other contaminants during the process to obtain an ultra-smooth metal foil surface. These constraints increase the cost of production of the foils for applications in electronics, making their commercial use impractical.

To reduce the cost of manufacturing, metal foils are sometimes post-processed to obtain a high-quality surface finish. For example, a cold-rolled steel foil can be subjected to chemical mechanical polishing (CMP) process to obtain a better surface finish before deposition or transfer of electronic materials. The CMP process for metal foils is typically carried out with silica or alumina slurry, leading to a post-process surface roughness of a few nanometers. Another commonly used process for increasing metal foil surface quality is the use of spin-on passivation layers. For example, a layer of spin-on-glass (SOG), which is a colloidal solution of silica particles, is spin-coated on a metal foil substrate and annealed to form a thin film. The thin film is significantly more planar than the underlying substrate, providing a smooth surface for the fabrication of electronic circuits. These layers also provide an insulating layer to separate one electronic device from another. If the substrate is to be used as a common terminal, the passivation layer can be patterned using lithography and etched to create metal vias. However, the trade-off with the use of passivation layers for planarization and electrical insulation is the flexibility of the substrate. Thick passivation layers lead to better planarization and electrical insulation; however, they are susceptible to cracking when the flexible substrate is bent. On the other hand, thin layers that can withstand bending strain do not offer reasonable surface planarization. Apart from SOG, benzocyclobutene (C_8H_8, BCB) is also commonly used as a surface passivation layer.

9.2.2 FLEXIBLE GLASS SUBSTRATES

Glass slides and substrates are commonly used in many scientific investigations. Glass, like any other brittle material, can be made flexible, by reducing its thickness. The thickness at which glass can be called flexible depends on the required bending radius, but most flexible glass substrates are less than 200 µm in thickness. Glass-based materials have been used for various applications for thousands of years, thus providing a very deep knowledge base related to the properties of various kinds of glass. The earliest uses of glass can be traced back to the ancient Egyptian and Mesopotamian civilizations. It is an amorphous solid with silicon dioxide or silica (SiO_2) as its primary constituent. In crystalline form, silica is referred to as quartz, which is also a commonly used material for applications ranging from crystal oscillators and scientific instrumentation.

Glass possesses several properties associated with the ideal flexible substrate. Glass sheets can be made flexible by reducing the thickness of the sheet, and in recent years, a roll-to-roll process for flexible glass sheets has been established. Further, glass sheets are easy to produce because of the existence of a highly scaled infrastructure from the extraction of ore to purification and processing. Glass sheets can also be very low-cost because the two base elements used in glass, silicon, and oxygen, are the two most abundant elements in the Earth's crust by weight. Much like steel, the mechanical, chemical, and optical properties of glass depend on the additives and the fabrication process used. Commonly used types of glass include soda-lime glass, which incorporates additives like sodium carbonate (soda), calcium carbonate (lime), among other materials; and borosilicate glass, containing boron trioxide. Among these types, soda-lime glass is the most produced and is commonly used for windowpanes, glass jars, bottles, and so on. It is popular because the additives lower the glass transition temperature making it easier to work with and provide shape. Glass substrates are optically transparent, proving to be an ideal substrate for optoelectronic applications. Glass is already a popular substrate choice for fabrication of displays, touch screens, photovoltaics, and so on (Figure 9.3). This is because apart from being transparent, glass is also resistant to many chemical reagents and can withstand fairly high temperatures (500–700°C for soda-lime glass). This makes it easy to transition these integration strategies onto a flexible glass substrate. Flexible glass substrates can also be made transparent to selected wavelength ranges by changing their chemical composition and processing method (changing the "color" of the glass). This provides more opportunities for use in specialized applications. Glass substrates can have very smooth surfaces (RMS roughness around 1 nm),

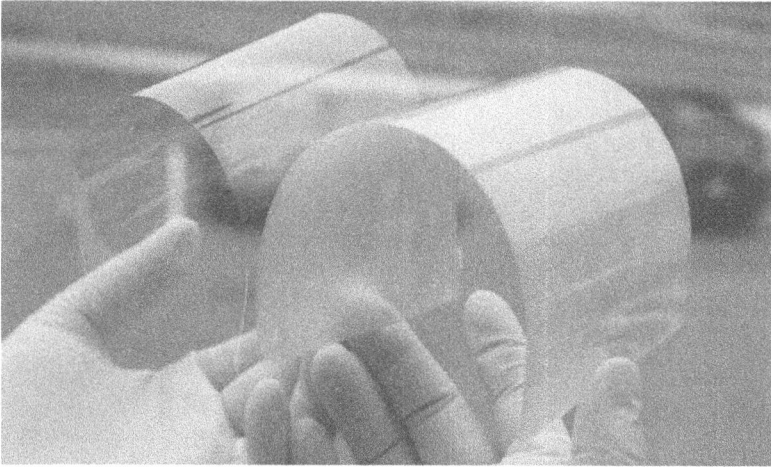

FIGURE 9.3 Digital images of flexible glass (left, thickness = 100 μm) and long TOWFs-coated flexible glass (right, thickness = 127 μm). (Reprinted from [4] under the Creative Common CC BY License.)

depending on the process of fabrication. For example, Corning's propriety fabrication process can provide flexible glass substrates of thickness 100 μm with average surface roughness < 0.5 nm (Willow® glass). This reduces the post-fabrication planarization step that is required for most metal substrates.

Glass substrates tend to have higher surface energy compared to their metal foil and polymer counterparts. Surface energy is the property of a surface that determines how well another material is able to deposit and adhere to it. High surface energy substrates are desirable for the material being deposited to "wet" the surface and eventually form a stable thin film. A classic example of a low-energy surface is the surface of a lotus leaf that makes the water droplets bead up instead of allowing the water to form a thin layer on top. Possessing high surface energy allows materials to form a thin film on the glass substrate. This can be an important property because most integration strategies for electronic devices rely on deposition and adherence of materials on the substrate. Glass sheets also have very low water/oxygen permeability. This is important for applications involving oxygen or moisture sensitive materials such as organic semiconductors. Glass containers are routinely used to store moisture or air-sensitive materials (often in conjunction with rubber O-rings for reversible sealing). In fact, glass sheets are so adept at providing a hermetic barrier, that the limiting factor for a glass encapsulation to act as a barrier depends on the edge sealing. In most cases, silicone sealants can be used to provide a hermetic seal for glass packaging.

The above-mentioned properties make flexible glass an ideal substrate for flexible electronics application. However, there are some limitations to the use of glass. The biggest problem with flexible glass substrates is their flexural rigidity. Although lowering the thickness of the sheet increases flexibility, the rigidity of a glass substrate is much higher compared to those of polymer sheets of the same thickness.

This requires glass substrates to be substantially thinner compared to polymer sheets, for the same flexibility, thus, reducing their ability to be handled independently. For example, a 50-μm glass substrate can be flexible but is prone to shattering if handled incorrectly because of the low thickness of the underlying material. On the other hand, while metal foils also have high rigidity, and consequently require to be much thinner to provide a desired flexibility; they are not vulnerable to catastrophic failure upon handling because of their malleable nature. For example, the thickness of the aluminum foil used for wrapping food items in the kitchen ranges from 20 to 25 μm. It is greatly flexible and does not shatter upon bending. However, there have been many advances in recent years in the development of flexible glass substrates leading to innovative uses particularly in flexible display applications. A commercially available example is the use of a flexible protective glass superstrate on the Galaxy Fold Z2 by Samsung.

9.2.3 POLYMER SUBSTRATES

Polymers are fast becoming the substrate material of choice when it comes to flexible electronics applications. Because of the vastness of synthetic organic chemistry, there are almost unending polymers to choose from. Further, a change in a particular ligand or moiety can lead to specific changes in the properties of the final polymer. Thus, there is a scope to tune the properties of the polymers according to the requirement of the final application. While metal foil and glass sheet properties can also be changed by changing additives and processing methods, the range of properties achievable using polymers cannot be matched. For example, we have polymers that have reversible stretchability (elastic limit) from a few percent to 10s of times (1000s of percent) their original length. There is similar diversity in chemical resistance, optical transparency, hermetic barrier properties, surface energy, and all the other properties that are important for consideration in flexible substrates. Typically, polymer thin films can be fabricated using spin-coating and curing. This process provides for a very smooth surface finish with negligible surface roughness. However, the major disadvantage of polymers is their thermal stability. Most polymer substrates have a glass transition temperature of less than 200°C, which means they cannot be used for direct deposition of thin films using CVD or other high-temperature processes.

The most commonly used plastic substrates are poly(ethylene terephthalate) (PET) and poly(ethylene naphthalate) (PEN). PET is the polymer of ethylene terephthalate ($C_{12}H_{14}O_6$), which is an ester of ethylene glycol ($[CH_2OH]_2$) and terephthalic acid ($C_6H_4(COOH)_2$), while PEN is a polymer of ethylene naphthalate ($C_{14}H_{10}O_4$), which is an ester of ethylene glycol ($[CH_2OH]_2$) and naphthalenedicarboxylic acid ($C_{10}H_6(COOH)_2$). Their structures are as shown in Figure 9.4. Both PET and PEN belong to the polyester family. They provide good chemical and mechanical stability. They are highly flexible and can be made transparent for specific wavelength ranges. Similar to glass and metal, they have a long history of industrial usage which means the infrastructure to produce large quantities of these materials is already present. However, their key disadvantage is their thermal stability. Both materials deform severely when exposed to temperatures above 200°C and cannot be relied

Polyethylene terephthalate (PET)

Polyethylene naphthalate (PEN)

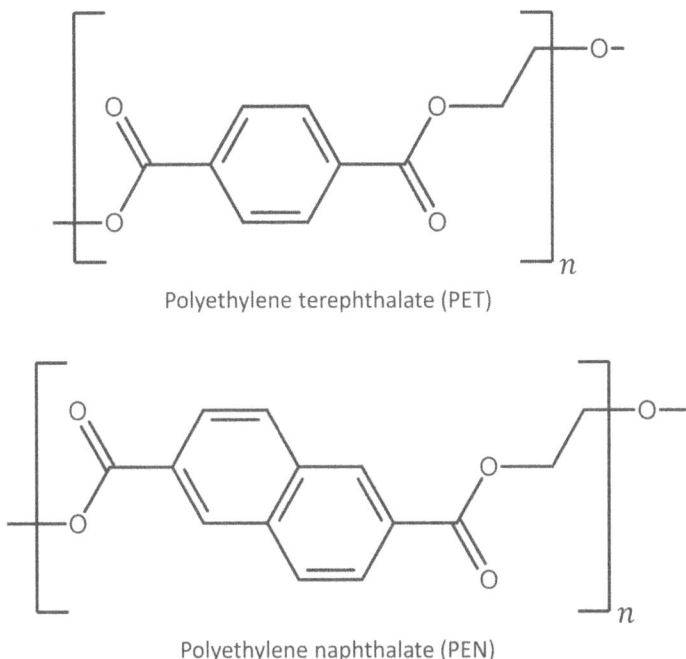

FIGURE 9.4 PET and PEN are polymers of the polyester family, being ethylene glycol esters of the terephthalic and 2,6-naphthalene dicarboxylic acids respectively.

on as substrate materials at moderately high temperatures. PEN has a slightly better response to high temperatures because the presence of naphthalene rings stabilizes the polymer.

Polyimide is another polymer commonly considered as an ideal substrate for many flexible electronic applications. Polyimide is a polymer of the imide monomer, which is a functional group consisting of two acyl moieties attached to a nitrogen atom ($R_1CO-NR_2-COR_3$). Imides are generally formed from dicarboxylic acids through a reaction with ammonia. Polyimides are polymers with the central nitrogen molecule attached to an organic group with another attached to the acyl group ($[-R_1-CO-NR_2-CO-]_n$). Polyimides share some common properties derived from their base structure and composition; however, they can be tuned based on the functional groups R_1 and R_2. The most important property of polyimides is their relative stability at moderate temperatures compared to other polymer materials. Kapton (DuPont de Nemours, Inc.) is a commercially available polyimide material, which is stable up to 400°C (Figure 9.5). Thus, Kapton sheets can be used as a substrate for carrying out many fabrication processing steps directly. Kapton is also resistant to many chemicals routinely used in the semiconductor fabrication industry. However, it absorbs light in the visible blue region, thus giving it a yellow tinge. This can limit its applications in the flexible display/photovoltaics space. Another problem with the use of polyimide as a substrate is its limited elastic strain, which is reported to be less than 1% for Kapton. This is very low compared to most polymer substrates and can be a problem in the

Kapton, poly (4,4'-oxydiphenylene-pyromellitimide)

Polysiloxane

Poly-dimethylsiloxane

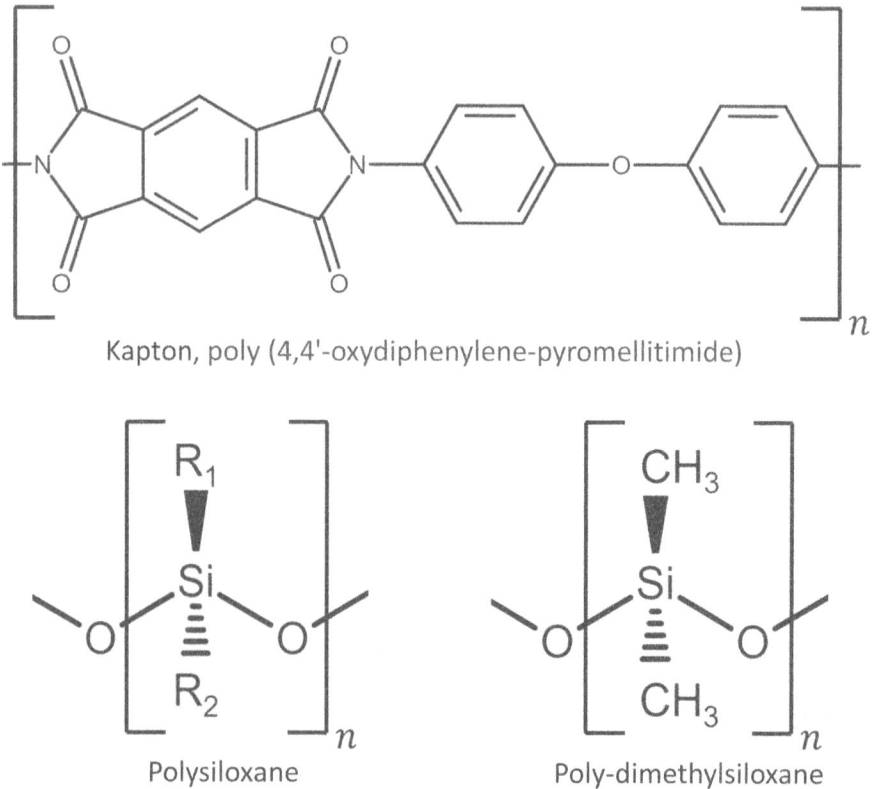

FIGURE 9.5 Polyimides are polymers of the imide monomers characterized by the presence of the -N-CO- structure. Kapton, poly (4,4'-oxydiphenylene-pyromellitimide), is a commercially available polyimide from DuPont. Polysiloxanes, or silicones, are polymers of the siloxane monomers characterized by the -O-Si-O- structure.

use of Kapton for stretchable electronics applications, or applications requiring very large strains. This does not limit the flexibility because very thin sheets of Kapton can be fabricated and handled reliably. Polyimide is also commonly used as a substrate for fabricating flexible printed circuit boards (fPCBs).

Silicones are polymers consisting of a siloxane group attached to other organic moieties. The siloxane group consists of an oxygen atom attached to two silicon atoms that are attached to three other groups. Silicone polymers typically include a -Si-O-Si- chain with the Si atoms being attached to other functional groups ($[-SiR_1R_2-O-]_n$). The simplest of the siloxane polymers is polydimethylsiloxane (PDMS), in which the silicon atom is attached to two methyl groups. Thus, the PDMS chain consists of a repetition of the structure $[-Si(CH_3)_2-O-]_n$. Polymers from the silicone family are typically transparent, rubber-like materials with very low oxygen/moisture permeability. Thus, they are commonly used as sealants and adhesives for waterproofing applications. Apart from these, they possess very high reversible stretchability (elastic strain limit); they are resistant to chemical reagents;

TABLE 9.1

A Qualitative Comparison of the Three Broad Classes of Flexible Substrates

Property	Metal Substrate	Glass Substrate	Polymer Substrate
Thermal stability	Very high	High	Very low
Flexural rigidity(at a given thickness)	High	Very high	Very low
Surface roughness	Very high	Low	Low
Oxygen/moisture permeation	Low	Very low	Low
Transparency	None	Very High	High
Elastic strain limit	Low	Very low	Very high

and can be formed into thin films that can be handled independently. Commercially, PDMS is available in the form of Sylgard 184 (Dow Inc.), which is a two-component kit consisting of a base and curing agent to be mixed in the ratio of 10:1 by weight. The thermal curing process can take place at room temperature upon mixing the reagents but is greatly accelerated with temperature. For example, PDMS thin films using Sylgard 184 kit can take 24–48 hours to completely cure at room temperature, whereas only 10–15 mins at 150°C. Once cured, Sylgard 184 is a transparent, soft, smooth solid that is resistant to many chemical reagents used in semiconductor fabrication. It has a very high elastic strain limit, with reported values being in excess of 100% for Sylgard 184. This makes it an ideal substrate for stretchable electronics applications. However, the surface energy of PDMS is such that it is not wet easily causing problems with stiction with other thin films. Further, PDMS does not respond well to temperatures above 150°C.

Table 9.1 provides a qualitative comparison between metal foils, glass sheets, and polymer sheets for the properties listed for ideal flexible substrates.

9.3 TRANSFER PRINTING

Because of the various limitations of substrate materials discussed in the previous sections, many of them are not usable for direct deposition, growth, lithography, etching, or processing of functional electronic thin films such as silicon, III-Vs, organics, metal oxides, and so on. Thus, there is a need to develop a process to transfer some of the components of a flexible electronic system to a destination substrate, once a specific set of steps have been completed on the host substrate. These components are invariably in the form of a patterned thin film, or a set of patterned thin films, to be detached from the host substrate, where the processing has been conducted, and transferred to a destination substrate, for further processing or device completion. The process of transferring a set of devices precisely from one location on a substrate to another substrate is referred to as transfer printing. In the simplest example of transfer printing, the host and the destination substrates are pressed together to facilitate the transfer of devices from the former to the latter. In this case, the most important condition for the transfer process to be successful is that the adhesion of the device stack to the host substrate should be lower than that with the destination substrate. This can be made possible using several techniques. The adhesion of a set of thin films on a substrate depends on the surface energy of the substrate, the properties of the thin film, the deposition process, and the post-processing involved. All of these can be controlled to obtain a device that is loosely adhered to the host substrate. The host substrate can be chosen such that the surface energy of the substrate is low. This allows for thin films deposited or grown on the substrate to easily detach from the host substrate. However, using a substrate with low surface energy can lead to problems with deposition and growth of the thin films to begin with, thus, careful selection of deposition processes is required to obtain high-quality thin films on such substrates.

In many cases, devices are detached from the host substrate before the transfer process by etching away some part of the support structure under the device layers. It is common practice to use a "sacrificial" layer to connect the device stack with the host substrate. This layer can then be etched partially or completely to release the devices before being transferred to the destination substrate. For example, in the device-last approach for making devices using flexible silicon, a silicon-on-insulator (SOI) substrate is used. In this case, the oxide layer can be treated as a sacrificial layer and etched away using HF to release the silicon layer on top (see Chapter 4 for details). The composition of the sacrificial layer and the etch chemistry depends on the device layer and the host substrate. They are chosen such that the etch chemistry has a high etch rate for the sacrificial layer and is also highly selective to the device and the host substrate. Further, the etch chemistry is chosen such that there is control over the rate of etching. In practical situations, the thickness of the sacrificial layer is very small (of the order of a few hundred nanometers) compared to the distance of the etch (tens of microns to hundreds of microns). Thus, the undercut etch process can be mass transport limited, because of the small cross-sectional area available for the flow of reactants and reaction products compared to the volume of the material to the etched. This low surface-to-volume situation can reduce the etch rate, as the undercut progresses, changing the dynamics of the etch. Thus, it is important to model the etch and plan the etch parameters carefully to obtain a successful undercut. Further, in most practical cases, it is not advisable to completely eliminate the sacrificial layer because doing so can release the device layer from the host substrate even before the transfer process. Thus, a small section of the sacrificial layer is preserved to act as a loose connection between the device layer and the host substrate. This creates additional constraints on the process window for the etch process because the etch rate should be uniform across the substrate. If not, then some devices may be completely released from the substrate, while others are still firmly attached, subsequently leading to a poor transfer yield.

The direct transfer of devices from one substrate to another also has one interesting consequence. Because the top of the host substrate (with the devices) is pressed against the top of the destination substrate, the devices transferred to the destination substrate are inverted. To avoid this problem, transfer printing is generally done using an intermediate substrate, called a "stamp", in two steps – pickup and printing (Figure 9.6). The stamp is a substrate used to host the devices during the transfer

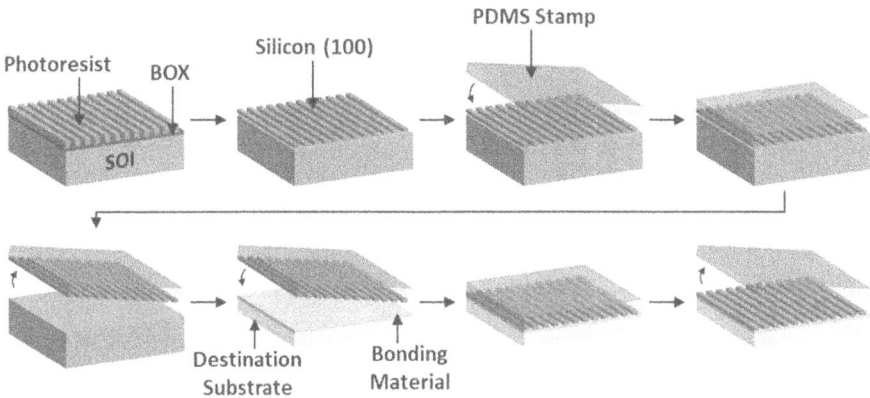

FIGURE 9.6 The transfer printing process involves the use of a "stamp" as an intermediate substrate for picking up loosely held structures from the host substrate and placing them on the destination substrate. (Adapted with permission from [2], © John Wiley and Sons, 2016.)

process. In the pickup step, the stamp is pressed against the host substrate to transfer the functional devices onto the stamp. In the printing step, the stamp is pressed against the destination substrate to transfer the devices onto the desired location on the destination substrate. As discussed earlier, it is important that in both steps, the adhesion between the host substrate and devices be lower than that between the devices and the destination substrate. In the case of the use of a stamp, the process can still consist of using a sacrificial layer to reduce the adhesion between the host substrate and the devices. This facilitates the pickup step. However, in the case of the printing step, the devices are positioned on the stamp that already has a reasonable adhesion with the device layer. Hence, some additional processing needs to be done to release the devices from the stamp and onto the destination substrate.

9.3.1 Viscoelastic Stamp

The transfer of devices from one substrate to another can be facilitated by tuning the mechanics of the process. If a polymer-based stamp is used, because of the viscoelastic properties of the stamp, the adhesion of the stamp can be modulated based on the speed of retraction. Thus, the "peeling" rate, i.e., the rate at which the stamp is retracted from the surface of the host/destination substrate can determine the adhesion of the device layers on it. From theoretical models and experimental results, it has been found that the critical energy release rate is dependent on the peeling velocity. The higher the peeling velocity, the more the adhesion on the viscoelastic stamp surface. This provides a convenient method for modulating the adhesion of the device components on the surface of the stamp simply by adjusting the peeling velocity. The peeling velocity is kept high during the pickup step of the process, to allow for high adhesion between the device layers and the stamp. In the printing step, the peeling velocity is kept low to reduce the adhesion between the devices and the stamp to a value lower than that between the devices and the destination substrate. The mechanical tuning of adhesion allows for a simple process for transfer printing, however, there are some limitations to this technique. The model does not work if the host/destination substrate is also viscoelastic. Further, the pragmatic rates at which peeling can be obtained are limited, reducing the process window for the use of this method. Another method relying on mechanical retraction to increase surface adhesion uses the frictional forces acting on the device layers. In this method, the stamp is retracted by applying a shear force, i.e., the stamp is retracted in the direction orthogonal to the destination substrate surface. The frictional forces between the device layer and the destination substrate oppose this motion causing the device layer to delaminate from the stamp. However, this may put additional pressure on the device layer resulting in loss of performance.

The most commonly used material to create viscoelastic stamps for transfer printing is PDMS. There are several reasons for this. First, stamps of particular sizes can be created easily by creating molds for curing liquid PDMS. Second, the surface properties of stamps can play an important role in transfer printing, particularly when no adhesives are used. Highly smooth PDMS stamps can be created by using polished silicon substrates for molding the top surface of the stamp. This is typically done by placing a silicon substrate piece of the desired size into the mold to form the top surface of the stamp. Third, the surface of the PDMS stamp can be activated for better adhesion to certain surfaces or device layers using oxygen plasma exposure. This provides more options for the integration strategy. Finally, PDMS is easy to handle, non-toxic, and resistant to many organic solvents. A critical disadvantage of the use of PDMS, or any other viscoelastic material, as a stamp for transfer printing is the possibility of minor changes in the relative position of the functional devices during the transfer printing process because of the elastic nature of the stamp.

9.3.2 THERMAL RELEASE STAMP

Another very commonly used method for transfer printing relies on the use of thermal release tape (TRT), which is a unique adhesive tape that has temperature-dependent adhesion. The TRT generally has high adhesion at room temperature but significantly reduced adhesion above a particular transition temperature. This property of the TRT is used to release the device layer from its surface in the printing step. This is usually combined with mechanical techniques such as control of peeling velocity and application of shear force for enhanced adhesion control. The use of TRT increases the range of adhesion that can be obtained – from very high for the pickup step at room temperature and fast peeling, to very low for printing step at high temperature and slow peeling. The major disadvantage of this process is the use of high temperatures for release assistance. This can be detrimental to the device layer or the destination substrate, putting additional constraints on the integration strategy. The use of adhesive tape (TRT) can also cause residues to remain on the device layer after the release process, which may require additional cleaning steps to completely remove. This can increase the chances of contamination and further complicate the integration process.

9.3.3 SOLVENT RELEASE STAMP

In some cases, the stamp or the adhesive used for transfer printing can be dissolved away in a solvent. The choice of solvent to be used for dissolution depends on the integration strategy, the devices to be transferred, and the destination substrate, while the choice of the stamp depends on the choice of the solvent. The most commonly used solvent for solvent-assisted transfer printing is acetone. It provides a wide variety of adhesives and tapes that can be dissolved in it. Acetone baths can also remove organic contamination from the device layer and destination substrate and are routinely used as a cleaning bath post-transfer printing in any case. Hence, it is a common solvent choice for completing the transfer printing process. Given the dissolution of the adhesive results in complete removal of the tape as well as any adhesive residue, this process provides a very clean transfer of the functional devices. Further, the process also facilitates a much higher transfer yield compared to mechanical or thermal methods because of the complete release of the devices in the printing step. However, a key disadvantage of this process is the limitation that all the materials, i.e., the functional device layer stack and the destination substrate used, should be resistant to the solvent used. In the case of acetone, this can be a tough constraint to meet, particularly if the integration involves the use of organic functional layers. To avoid this problem, in some cases, water-soluble tapes are used for transfer printing. The use of water as a solvent can provide more options for functional material selection, however, it can have its own effect on the devices, such as hydroxylation and subsequent oxide formation on the surface of the device layers. The solvent-based process does provide a promise of higher transfer yield and cleaner transfer to destination substrate if the constraints of solvent interaction with device layers and destination substrates can be resolved.

9.4 ADHESIVE BONDING

Adhesives are commonly used to bond the device layers to the stamp or the destination substrate. An ideal bonding adhesive should be able to provide variable adhesion, low adhesion during application, and high adhesion upon curing. The process of curing, in this context, creates a solid thin film bonding the device layers with the destination substrate. These bonding materials remain on the destination substrate and become a part of the final system. Hence, upon curing, they need to possess most of the properties of ideal substrates. They have to be flexible, low-cost, non-toxic, easy to handle, and stable under deployment conditions. In addition, ideal bonding materials should have good thermal conductivity to transfer heat generated in the functional devices to the substrate. Bonding materials that can be cured are usually based on polymers that can be cross-linked along several points in the chain to form large 3D structures. Adhesives of this kind are cured by providing energy for the polymers to form cross-links, form larger molecules and solidify. The energy input can be in the form of heat (thermally cured polymers) or light (UV cured polymers). In some cases, both UV irradiation and annealing are used to obtain curing. For example, one of the most commonly used adhesives for bonding is the SU-8 photoresist family. The SU-8 monomer consists of eight epoxide functional

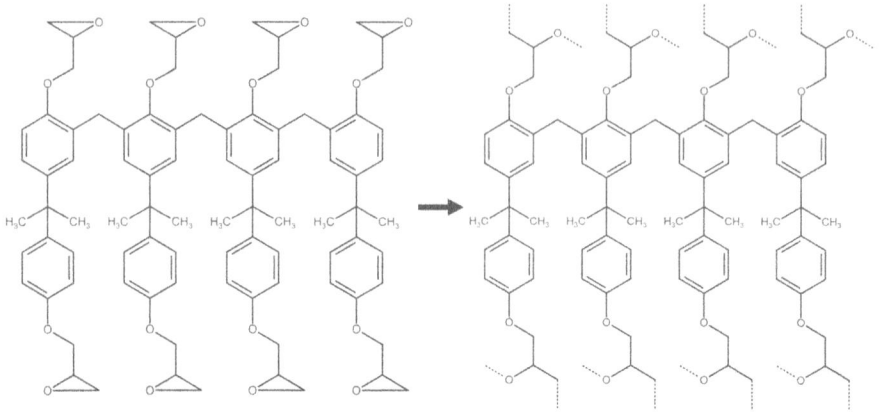

FIGURE 9.7 The SU-8 monomer gets its name from the eight epoxy groups present in the molecule. When cured, these provide eight points of bond formation with other monomers leading to a complex and gigantic 3D molecule.

groups (Figure 9.7). Epoxide functional groups are cyclical ethers with two carbon atoms bonded together and attached to an oxygen atom. One of the commercially available forms of SU-8 is the SU-8-2000 series from Microchem, which was initially launched as a photoresist for high-resolution lithography. It is now routinely used as a structural or adhesion layer for integration strategies involving flexible electronics, MEMS, microfluidics, and so on. It is cured by exposure to UV followed by annealing. Because of the presence of multiple cross-linking sites on every molecule, the resulting cross-linked structure consists of giant 3D molecules with very high molecular weights. This provides mechanical and chemical stability to the cured thin film, making it ideal for use as a bonding material. Cured SU-8 thin films can withstand temperatures up to 200°C, and are insoluble in most organic and inorganic solvents.

Apart from SU-8, many other epoxy-based materials can be used as adhesives. In fact, epoxy adhesives are commonly used in many industries such as carpentry, construction, interior designing, and so on. Molecules with epoxide functional groups readily react with themselves or other organic moieties such as phenols, acids, thiols, and so on to form large cross-linked molecules. Thus, many epoxy-based adhesives are available as a two-part kit, the epoxy resin, and the co-reactant. When they are mixed, the resulting cross-linking hardens the epoxy leading to adhesion. Other polymer adhesives include photoresists, which are already in use in most semiconductor fabrication facilities for lithography applications, and silicones, which can provide highly flexible thin films after curing. Photoresists can also be patterned using UV lithography post-curing and solvent evaporation to aid with integration options.

The process of transfer and bonding the functional devices to their final position forms an important part of any integration strategy for flexible electronic applications. This is particularly so because of the great diversity of materials used for each application and the need for all these material layers to form a single system. Even with all the advancements in substrate materials, transfer techniques, and bonding methods, the heterogeneous integration of various functional devices from different substrates onto a single platform remains a great hindrance for the realization of flexible electronic systems. Indeed, looking at the variety of material devices that need to interact reliably to form an electronic system, it is fortuitous that the integration of state-of-the-art CMOS electronics can be done on a single semiconducting substrate – silicon.

EXERCISES

9.1. List the ideal properties associated with a substrate for flexible electronics applications. Give two examples of sets of properties that are in conflict with each other, if present in the same material.

9.2. What are the key advantages/disadvantages of metal foil substrates compared to flexible glass substrates?

9.3. Name the key classes of polymers used for flexible substrates, along with the properties of each.

9.4. Describe the transfer printing process, along with the kinds of stamps that can be used for it.

10 Barriers, Insulators, and Packaging

10.1 INTRODUCTION

The fabrication of ideal destination substrates and the transfer of functional devices to them constitutes only one part of an integration strategy for flexible electronic applications. Once complete, it is important to protect the devices during deployment using barrier coatings, insulator layers, and packaging materials. The function of each of these materials is different and together they provide a reliable protective cover for the functional device layers from the damaging conditions of the outside world. This is particularly important for flexible electronic systems because of the possibility of heterogeneous material integration. The challenge with such an integration is that some materials are sensitive to temperature, some to moisture, some to oxygen, and some to mechanical abrasion. Thus, the packaging strategy should be such that it can provide shelter to the devices within from all these external interferences. At the same time, there are certain applications where exposure to the environment is part of the deployment scenario, for example, in the case of environmental sensors, such as gas sensors, humidity sensors, and so on. Such an application requires the sensing materials to be directly in contact with the external environment. Further, in some cases a partial exposure may be necessary, for example, in the case of flexible displays, while the light-emitting materials can be packaged, it is important to use high transmissivity (transparent) materials for packaging. Similarly, in the case of flexible solar cells, it is important to make sure all the solar irradiation incident on the device is allowed to interact with the functional thin film for electricity generation, without being inhibited by the packaging material.

In the case of state-of-the-art silicon CMOS electronics, the packaging process is used to reduce exposure of the bare silicon chip to the external environment, as well as to allow ease of handling and connection. For example, the dual in-line package, developed in the 1960s, is a commonly used through-hole packaging strategy for silicon ICs. It provides for a safe package to ensconce the bare silicon chip, and it allows the IC to be connected to other devices on a printed circuit board (PCB). The packaging also allows for easy removal and replacement of the ICs in the case of damage. This is also true for most of the modern packaging methods, such as BGA, QFN, SOIC, TSOP, and so on. Thus, it is important to design the packaging such that it conceals the functional devices while exposing a connection schema to facilitate the creation of large circuits. This dual purpose of packaging needs to be remembered while designing an integration strategy for packaging flexible electronic systems (Figure 10.1).

DOI: 10.1201/9781003010715-13

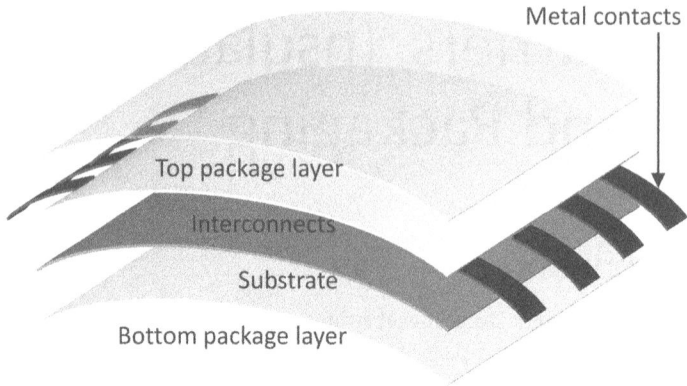

FIGURE 10.1 A typical packaged flexible electronic system will have at least a top and bottom layer of packaging material, along with some metal contacts to connect the system with external circuitry.

10.2 BARRIER THIN FILMS

Barrier materials are an important part of a packaging strategy to make sure external elements such as moisture and oxygen are not able to penetrate through the package and interact with the functional devices. This is primarily important for functional devices based on organic electronics because these thin films are very sensitive to air and moisture. However, in the case of inorganic functional devices as well, permeation and subsequent interaction of oxygen/moisture with functional thin films can cause oxidation, rust, development of trap states, and eventually, functionality issues. Further, ideal packaging materials are expected to protect functional devices from many other compounds present in air such as carbon dioxide, nitrogen, dust particles, and so on. Hence, it is essential to consider the barrier properties of materials used for packaging flexible electronic systems. In everyday use, packaging materials are commonly used to protect food items from contamination. In this case, as well, exposure to moisture and oxygen can lead to bacterial activity and subsequent rotting of food. The conventional choice of materials for packaging food items consists of metal containers (aluminum cans) and ceramic jars (earthen pots). These materials provide very high barrier properties on account of their high density. Barrier properties are quantified in terms of the vapor transfer rate which is the mass or volume of permeation through a unit area of the material surface in one day. The rate is generally calculated for standard deployment conditions and is commonly expressed in g/m^2-day or ml/m^2-day. Ideally, lower transfer rates are desirable to restrict the penetration of undesirable materials into the system.

10.2.1 PERMEATION MECHANISMS

Before considering the materials and integration strategies that can provide a high barrier against key environmental elements, it is important to understand the mechanisms behind their permeation through a thin film. The main mechanism of permeation through a thin film is the diffusion of a permeate from a high concentration region (external world) to a low concentration region (inside the system). This process can take place through the pinhole defects or cracks in the material or a slow process of dissolution or absorption of the gas into the barrier film on one side, movement across the barrier film, and evaporation from the other. Thus, the process of permeation in a medium with no pinholes or cracks involves both solubility of the gas in the material and diffusivity through the material. In case any of these parameters is low, the permeation of a specific gas through the medium is significantly reduced. For example, the diffusivity of a permeate depends on the shape, size, and reactivity of its molecules, and the lattice structure of the barrier material. Thus, dense metallic films have very low material permeability. This is also true for ceramic lattices such as silicon dioxide or aluminum oxide. In particular, high-quality, defect-free crystals of silicon dioxide (quartz) and aluminum oxide (sapphire) are almost impenetrable for oxygen at room temperature. In their amorphous or polycrystalline form, silica and alumina can have some oxygen permeability because of the presence of grain boundaries, defects, and pinholes. Hence, along with the material, the deposition method used, and the post-processing performed on the thin film also influence the barrier properties of the final film.

Consider the barrier film shown in Figure 10.2. The permeation of a gas through a barrier film can be modeled using Fick's law of diffusion and Henry's law of dissolution. Fick's law of diffusion provides the flux of material through a medium as:

$$J = -D\frac{dC}{dx}$$

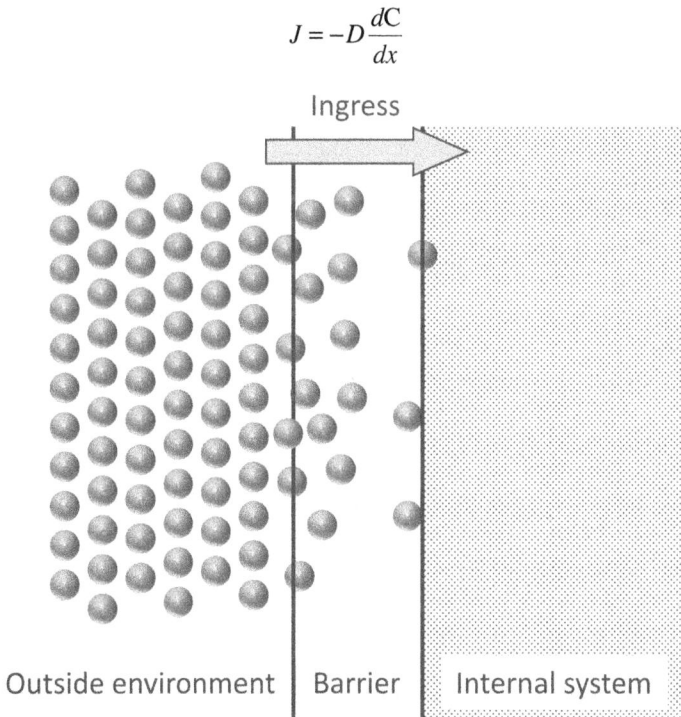

FIGURE 10.2   The ingress of vapor through a barrier film into the internal system. The concentration of the species in the outside environment is very high, causing adsorption, absorption, and diffusion of the molecules toward the internal system.

where J is the material flux, D is the diffusion constant, dC is the change in concentration from one end of the barrier film to another, and dx is the thickness of the film. The concentration of the permeate can be expressed in terms of its solubility coefficient (S) and partial pressure (p). Thus, we obtain:

$$J = -DS\frac{dp}{dx}$$

The constant DS is represented by P, called the material permeability of the thin film. Because both solubility coefficient and diffusion constant are dependent on environmental factors such as temperature, pressure, and concentration of the permeate, it can be concluded from this theory that these factors also influence the material permeability of barrier thin films. Thus, it is important to test barrier films under appropriate test conditions pertaining to the environmental parameters of the deployment scenario.

10.2.2 BARRIER MATERIALS

From the mechanics of permeation, it is clear that a good barrier film should either have a low diffusivity for the permeate or a low solubility. Of course, materials that have both the parameters low for a particular permeate can become excellent barriers for that species. Materials with low diffusivity tend to be those that can form highly dense lattices with very low defect density. Metals and ceramics both form dense lattices, but pinholes and defects cause gas ingress into the material through the grain boundaries in amorphous and polycrystalline materials. The defect density in a thin film depends on the fabrication process. Indeed, metal foils and cans are very commonly used for packaging food items. The same materials can be used for housing flexible electronic devices providing a barrier against oxygen and moisture permeation. However, like ideal substrates, it is ideal to have electrically insulating packaging for electronic systems to avoid unintended short circuits. Further, metal foils, depending on the metal used, can be more expensive to produce compared to other materials, such as plastics – one of the reasons why plastic packaging is now dominating the food industry. Another key problem with metal foils is their plasticity at low thicknesses. This lack of the ability to restore their original shape can be beneficial for some applications but can also lead to reliability issues with some flexible electronic systems. In the case of ceramics, the brittleness of the underlying material leads to the requirement of much lower thickness resulting in a fragile thin film. Further, the diffusivity through the material increases significantly with the reduction in thickness because the probability of having a through pinhole defect is much higher. Thus, polymer materials are becoming a popular choice for flexible electronics packaging, with low cost and ease of manufacturing. Further, polymer thin films, unlike metal foils, can be transparent, which is essential for use in display and solar cell applications. Most polymer thin films also show highly elastic behavior for large strains. However, the major disadvantage of the use of polymers for packaging is their high material permeability for moisture and oxygen.

A commonly used way to utilize the benefits of both ceramics/metal and polymer thin films is the use of multilayered barrier materials. This strategy is commonly used in the food packaging industry, in the form of metal-coated plastic sheets, for example, commonly used for packaging potato chips. Thus, a thin film of ceramic or metal on a plastic sheet can provide the best of both material systems. In principle, the deposition of a thin film of high-density ceramic such as silica or alumina on a polymer substrate should be able to provide a highly impenetrable material that is transparent and easy to handle. However, in practice, this technique can lead to high permeability if pinhole defects and cracks develop on the ceramic thin film in the duration of use, particularly upon repeated flexing. These pinholes lead to the direct contact of the underlying polymer layer with the environment leading to ingress as well as lateral diffusion across the polymer thin film. Thus, multiple small pinholes can be more damaging than a single large defect. To avoid this problem, it is common practice to use multiple alternating layers of ceramic and polymer materials to form a reliable barrier sheet (Figure 10.3). The use of multiple layers drastically reduces the probability of having a pinhole in all the ceramic layers at the same position. As a result, the ingress molecule needs to travel laterally across the polymer layer to the next pinhole defect, leading to a much longer meandering diffusion path. Thus, the permeability of these films is much lower than the polymer sheets or ceramic sheets of the same thickness. Further, the polymer films provide structural support, flexibility, and ease of handling to the thin film. Because the polymer films are usually formed using spin-coating or by curing liquid condensates, they effectively provide smoothening to the barrier layer. This leads to the deposition of better-quality ceramic films on top, increasing the barrier properties

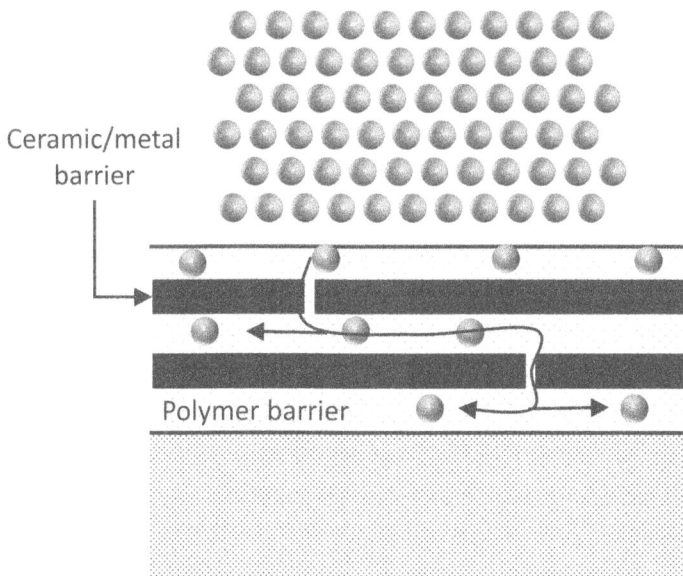

FIGURE 10.3 Multiple barrier thin films are combined to obtain the best properties of individual layers. For example, a ceramic barrier can be used with a polymer barrier – polymer layer provides a backing to the brittle ceramic layer, while the dense ceramic layer (with possible pinholes) increases the diffusion path for the ingress species, reducing the influx.

of the entire structure. These multilayer ceramic/polymer structures have been successfully commercialized and are used in commercially available flexible organic LED (OLED) displays.

10.3 FLEXIBLE DIELECTRICS

It is well-known that advancements in silicon IC technology were led by the dimensional scaling of transistors. The reduction in the size of transistors led to a proportional reduction in all other dimensions associated with the integration, including the gate dielectric thickness. While this meant higher capacitance per unit area for the gate, the reduced thickness of the dielectric led to an increase in leakage current resulting in much higher static power dissipation. This problem was solved by replacing silicon dioxide as the gate dielectric material with materials with high relative permittivity (κ) as the dielectric layers. These high-κ materials allowed for thicker gate dielectric layers, for the same capacitance per unit area as provided by the much-thinner silicon dioxide thin film, thus providing higher resistance to leakage current. This subtle change in the integration strategy helped ward off the effects of the end of the scaling revolution and helped continue with performance improvement in silicon CMOS. In the case of flexible electronic systems, it is imperative that all the material layers used for device fabrication be flexible. Hence, any integration strategy involving the use of transistors (MOSFETs or TFTs) needs to consider material systems that can be used as dielectrics or insulators.

10.3.1 INORGANIC DIELECTRICS

For our flexible electronic integration, we can, of course, make use of the dielectric materials that have been used in the conventional silicon CMOS industry for decades. Silicon dioxide or silica had been used in the CMOS industry before the introduction of high-κ dielectric materials that are mostly oxides or oxynitrides of transition metals. Commonly studied inorganic high-κ dielectric materials include hafnium oxide, hafnium silicate, aluminum oxide, zirconium oxide, and so on. They exhibit excellent properties for consideration in CMOS integration strategy because they provide high capacitance per unit area (on account of large relative permittivity), high resistance per unit area, good interface properties with the semiconductor, and very low defect density. However, these properties are a result of careful control of the deposition process and several steps of post-processing. In many cases, the material is deposited using expensive techniques such as atomic layer deposition or other high vacuum deposition processes. These processes are also carried out at a high temperature and can require multiple high-temperature post-deposition anneals to provide defect-free thin film and a high-quality semiconductor/dielectric interface. These materials can be used for conventional silicon electronics because silicon substrate is able to withstand high-temperature processing. Thus, the use of conventional high-κ materials as dielectrics in flexible electronic systems is only restricted to integration strategies involving fabrication of the complete device structure before being transferred to a flexible substrate, or in case the flexible substrate is such that it is capable of withstanding high-temperature processing (such as metal foils).

In some cases, low-temperature processes, such as sputtering or evaporation, have been used to deposit thin films of high-κ materials onto polymer substrates. Other possibilities include methods involving nanoparticles such as the sol-gel process. However, thin films deposited using these processes tend to be highly defective leading to a possibility of device failure. The defects are also more pronounced once the films are subjected to bending strain. Even with conventional high-temperature deposition processes, the compatibility of these high-κ dielectric materials with many of the semiconductors routinely used for flexible electronics fabrication, such as MoS_2, CNT, graphene, pentacene, ZnO, etc., and the quality of the semiconductor/dielectric interface formed, is still being investigated. Hence, apart from the continued study of conventional high-κ materials on polymer substrates, it is important to consider some organic and polymer dielectric thin films for integration into flexible electronic systems.

10.3.2 ORGANIC DIELECTRICS

Organic dielectric thin films have been studied as a substitute for conventional inorganic thin films for use in flexible electronic devices, particularly those based on organic semiconductors. Any thin film to be used as a gate dielectric of a transistor structure should have high capacitance per unit area (high-κ) and low conductivity. Further, it should have high stability to avoid the formation of defects because of the application of mechanical or electrical strain. It should also be easy to deposit using simple, low-temperature techniques such as spin-coating. Finally, it should be able to form a defect-free interface with the semiconductor under consideration. Given all these constraints, there are only a hand full of polymer thin films that are considered to be viable alternatives to inorganic materials. These include polyimide (PI), polymethyl methacrylate (PMMA), polystyrene (PS), and SU-8 among others. Polyimide forms a highly stable thin film with desirable electrical, mechanical, and chemical properties. The dielectric constant of polyimides is between 3 and 4, depending on the organic moieties in the chain. This is similar to the dielectric constant of silicon oxide (3.9); thus, polyimides are not considered high-κ materials. However, the ability to form stable, high quality, and defect-free thin films on polymer substrates provide a reason to consider them for flexible electronic applications. One of the polyimides commercially available in liquid form is the PI-26xx series from HD MicroSystems. These can be spin-coated to obtain a dielectric layer of any required thickness, particularly with the use of thinners to control the viscosity of the initial solution. While these are cured at relatively high temperatures, the PI-25xx series can be cured at around 200°C for better compatibility with some flexible substrates.

Polyvinylidene fluoride (PDVF) is commonly used in the electronics industry as wire wrap, known for its stability and insulating properties. It is a polymer of vinylidene fluoride or 1,1-difluoroethene (CH_2CF_2). Because of the presence of the highly electronegative fluorine atoms, PVDF has a high vacuum dipole moment, leading to a high dielectric constant ($\kappa > 10$). Thus, PVDF is considered to be a prime candidate as an organic high-κ dielectric material. The total dipole moment and hence the dielectric properties of the thin film depend on the deposition process, and subsequent crystallization process. The crystallization with the highest dipole moment

PVDF   PVDF-TrFE

PVDF-TrFE-CTFE

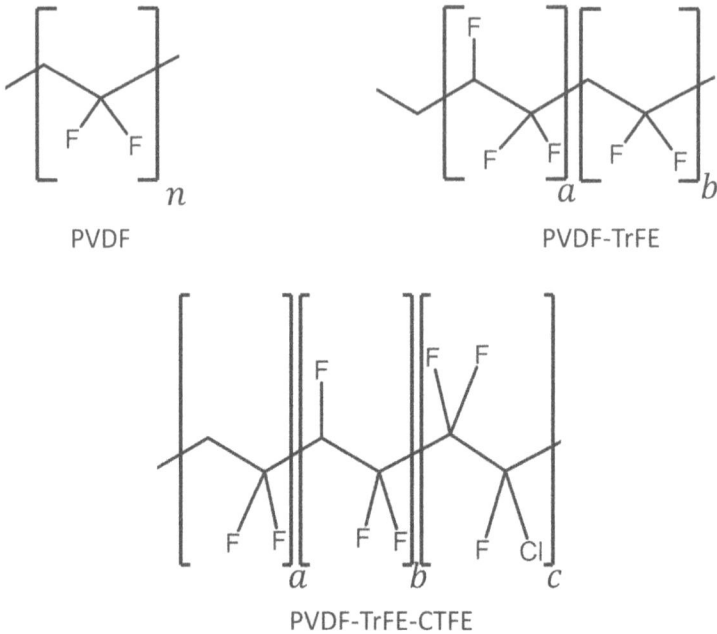

FIGURE 10.4 PVDF and its copolymers.

(called the β-phase) is the all-trans configuration with the alignment of all dipole chains in the same direction. This phase can be induced spontaneously by copolymerization of PVDF with trifluoroethylene (CHF=CF$_2$, TrFE). The copolymer, P(VDF-TrFE), exhibits a high dipole moment and high room temperature dielectric constant. However, P(VDF-TrFE) also exhibits ferroelectric properties that produce remnant polarization when an electric field is applied to the thin film. This leads to a hysteresis that may not be desirable for gate dielectric applications. The hysteresis can be reduced by blending P(VDF-TrFE) with paraelectric polymers, however, the process complicates the integration strategy. In some cases, yet another monomer is added to the copolymer chain to improve dielectric properties while reducing ferroelectric ones. For example, addition of chlorotrifluoroethylene (CFCl=CF$_2$, CTFE) or chlorofluoroethylene (CFCl=CH$_2$, CFE) can break the long-range ordered ferroelectric domains in P(VDF-TrFE) into smaller "nano-domains" (Figure 10.4). These domains align into an all-trans state upon application of an electric field, leading to very high dielectric constants ($\kappa > 50$), with much lower remnant polarization. The use of PVDF and its copolymers provides an opportunity to achieve the capacitance per unit area specifications with a thicker dielectric, thus leading to a lower probability for pinholes and other film defects.

10.3.3 HYBRID DIELECTRICS

Given that both organic and inorganic thin films have desirable properties as dielectric materials, it makes sense to combine them to obtain the benefits of both material classes. Inorganic materials provide a high dielectric constant and low leakage current but have to be processed at high temperatures to obtain a high-quality film. On the other hand, polymers are easily processed at room temperature into defect-free thin films but do not provide a high dielectric constant with low hysteresis. This problem can be solved by using hybrid dielectric materials that combine both inorganic and organic materials into one high-quality, high dielectric constant thin film that can be solution-processed at room temperature. The simplest way of obtaining such a hybrid structure is to deposit multilayer thin films with alternating films of inorganic and organic materials. The inorganic thin films, which are typically sol-gel processed, provide a high dielectric constant, while the organic thin films cover the pinhole defects and grain boundaries in the inorganic thin film to provide a highly insulating structure. Although the presence of organic part reduces the dielectric constant and the presence of defects in the inorganic film reduces the insulation, the overall thin film provides a reasonable compromise between the two properties along with low-temperature processability. Another way of creating a hybrid inorganic/organic dielectric film is the use of inorganic nanoparticles incorporated into an organic polymer matrix. Nanoparticles of high-κ inorganic dielectric materials can be incorporated into organic dielectrics such as polyimide, PMMA, P(VDF-TrFE), etc., to enhance their dielectric properties. The concentration of the nanoparticles in the polymer matrix also provides a parameter to tune the film properties and adjust competing properties like dielectric constant and resistivity at optimum levels depending on the application.

10.4 PACKAGING STRATEGIES

Packaging strategies form the backbone of any integration. They are the culmination of the integration and finalization of the electronic system to be deployed for an application and include the external layers, the substrate, the electronic system, the contacts, and so on (Figure 10.5). We have been taking the example of silicon chips and the impact electronics have had on the world. It is important to note that silicon CMOS design and fabrication are only one part of the puzzle. The chips need to be packaged in such a way that they can be connected with other electronic chips to form a complete circuit capable of carrying out tasks. The PCBs can only be used to connect *packaged* silicon ICs with other packaged electronic components. In the conventional electronic system, the silicon wafers are diced into individual chips after all the steps related to device fabrication and interconnect formation are complete. The chips are picked up (peeled off from the adhesive tape) and placed onto the package structure using highly precise instruments. The metal pads on the chip, that connect to the internal circuitry, are connected to those on the package. These interconnections provide the path for the internal circuitry to be connected to external components. In some modern versions, there are more than one silicon chip ensconced in the same package providing additional functionality and reducing the number of external components needed to make the system work, the so-called system-in-package (SiP) integration. Apart from these, the packaging in conventional electronic systems performs one more very vital task – heat dissipation. The heat generated by the silicon chip when it processes information needs to be dissipated

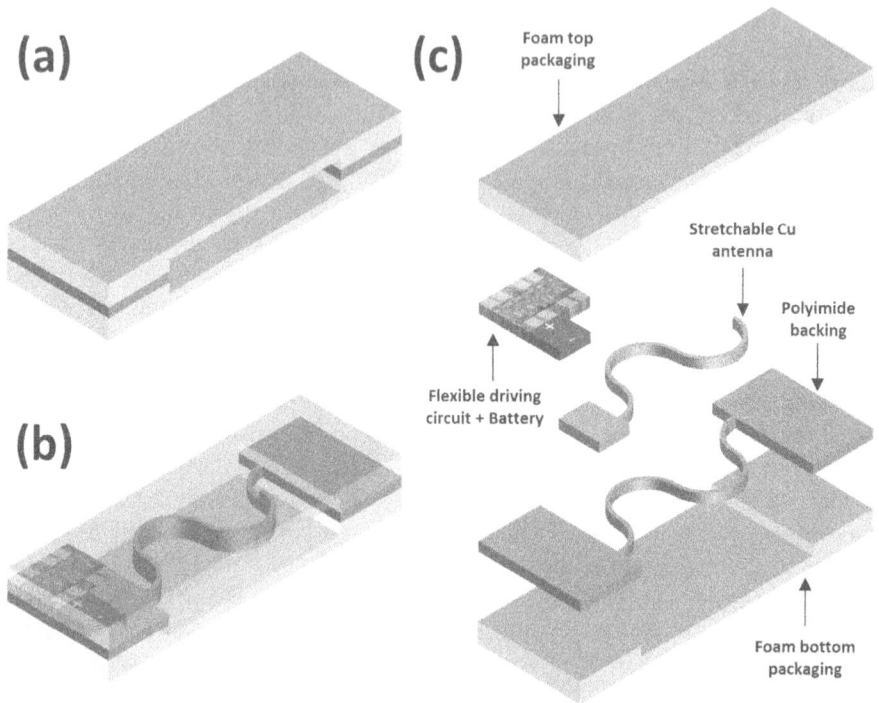

FIGURE 10.5 The packaging of a flexible electronic system will require unique design of materials and processes to accommodate the various material systems each module is based on.

efficiently to the outside world. Without this feature, the trapped heat would raise the temperature of the processor, altering the electronic and design characteristics of the transistors and permanently damaging the chip. Thermal considerations are particularly important for state-of-the-art silicon processors because they are able to integrate more transistors per unit area, which leads to higher power density.

As a result, to realize the true potential of flexible electronic systems, it is important to have a robust packaging strategy that provides for insulation from atmospheric ingress (barrier layers) and electrical noise (dielectric layers), while providing thermally conductive channels for heat dissipation. Further, packaging should provide a medium for making connections between the internal circuitry and the external components. Thus, the packaging strategy needs to be an integration of the barrier layers, dielectric layers, and conductive vias to achieve the desired goal. This, along with the basic requirement that all components, materials, and thin films used in the packaging strategy have to be flexible, at least up to the level required by the end-user application, makes the packaging strategy in flexible electronics very challenging. This is why, despite being such a vital part of the complete system, packaging barely finds any mention in the flexible electronics literature. However, the same material systems that have been discussed as destination substrates, barriers, and dielectrics need to be integrated into a package. For example, the packaging strategies used in commercially available flexible organic light-emitting diode (OLED) panels utilize transparent polymer thin films as a superstrate to allow for light emission. However, these films tend to scratch easily owing to the softness of the material. Thus, recently, some flexible OLED displays have been capped with a flexible glass film to improve the hardness of the package and reduce vulnerability to scratches. These display packages typically have an interconnection bus attached to the flexible printed circuit board (fPCB) controlling the transistor backplane. The interconnects can be attached to an external device or another fPCB as part of a larger integration. The same is true for many versions of flexible solar cells that are also commercially available. These are capped with transparent polymer/glass thin films to allow for the transmission of sunlight to the photovoltaic devices. The interconnections are much simpler, with only two-wire connections in most cases, that are connected directly to the photovoltaic array at the module level. The complexity of packaging strategy increases significantly with the requirement of a larger number of output pins, as is the case with memory devices, or with the requirement of heat conduction, as is the case with high-performance processing devices.

At the end of the day, the integration strategies related to transfer, bonding, and packaging define the outer appearance, aesthetic, and usability of an electronic device. To the average user, these can sometimes matter more than the performance or efficiency of the core system inside the package. There is still a long way to go before we will be able to walk into an electronics store and purchase a flexible version of the IC555 for integration into our hobby project, but that is the path we need to take to unlock the true potential of flexible electronic systems.

EXERCISES

10.1. What is the dominant permeation mechanism for a vapor through a barrier thin film?

10.2. What are the advantages/disadvantages of ceramic barrier coatings compare to polymer ones? Explain how the use of multiple barrier coatings combining ceramic and polymer will help in creating a better barrier film.

10.3. Say we need to have a capacitance per unit area of 1 fF/μm^2 for our circuits. We have the option of using silicon dioxide ($\kappa = 4$) and aluminum oxide ($\kappa = 10$) as dielectric films. What will be the thicknesses of these films? If the resistivity of the films is the same, what is the change in leakage current going from SiO$_2$ to Al$_2$O$_3$ dielectric?

10.4. We are planning to select a barrier thin film for flexible electronic applications. Suppose the requirement maximum vapor transmission rate is given as 1 mg/m^2-day. The maximum absolute humidity of the deployment location is going to be 30 g/m^3. If we are constrained to use a film of thickness 10 μm, what should be the diffusion coefficient of water vapor through it?

10.5. What are some of the organic alternatives for high-κ dielectric thin films for flexible electronic devices?

11 Flexible Printed Circuit Boards

11.1 INTRODUCTION

In the previous chapters, we have discussed some integration strategies to bring the devices based on flexible materials and substrates together into a package. While this in itself is an important exercise and researchers around the world are working toward it, the ultimate goal of the study of flexible electronics is to create a completely flexible electronic system, resembling the ones we currently see around us, such as smartphones, computers, watches, and so on. All these systems are made possible by the use of printed circuit boards (PCBs) to support the individual components that are usually separately created using their own process integrations. For example, to create any computing machine, we need to bring together a processing unit, say an arithmetic logic unit, and a memory to store information about the things being computed and the instructions for the said computation. In conventional electronic systems, these are brought together through a PCB to which both the processor chip and memory chip are connected, and which interconnects them through metallic interconnects. These are connected to other components such as display, sensor systems, actuators, transceivers, etc., through the PCB as well. The PCB then becomes the backbone of the system, or the "motherboard", providing interconnection between different components for data transmission as well as power supply. It is important to note that the use of PCB provides a unique opportunity in terms of the integration strategies used for fabricating individual components, in that they can be different. For example, the integration strategy used for the fabrication of memory chips is different than that used for processor chips, which is different from that used for sensor chips and so on. All we need to make the complete system work is that each component is compatible only in the way the input/output is defined and processed. In conventional systems, this comes down to the voltage levels defining the digital bits and the standard communication protocols used to communicate between chips. This is a critical advantage of using PCBs to integrate electronic systems because this helps in optimizing individual integration strategies to fabricate processor, memory, display, sensors, and so on.

We can draw an analogy for flexible electronic systems from the conventional systems based on a PCB supporting and interconnecting individual components. The fact that we can use different integration strategies for different parts of the systems is even more important in flexible electronics, because of the variety of materials, processes, and integrations possible. The fabrication processes can be optimized at the individual component level and they can be interconnected using a PCB backing. However, for the final system to retain flexibility, the PCB used for supporting and interconnecting flexible components should also be flexible. Further, in conventional

DOI: 10.1201/9781003010715-14

systems, the metal interconnects on a PCB can be in multiple layers depending on the complexity of the system. In the case of flexible PCBs (fPCBs) as well, it will be important to have multiple metal layers, while maintaining physical flexibility, to obtain a reasonable flexible alternative to present electronic systems. There is also a middle ground – the use of fPCB with rigid conventional silicon components. This approach, called flexible hybrid electronics (FHE), is already in use in many commercial applications.

11.2 FLEXIBLE PCB DESIGN

In the first chapter, we discussed the origins of electronic systems through a patent by Albert Parker Hanson in 1904 [1]. The patent was essentially for an early version of a fPCB. It included a flexible substrate with multiple levels of copper foil interconnects and vias designed to provide interconnections to different levels (Figure 11.1). The modern fPCBs have the same basic principles, although the materials and fabrication processes have become far more sophisticated. These fPCBs can be used to interconnect rigid conventional electronic components like surface-mount silicon chips, resistors, capacitors, LEDs, and so on. This use of fPCBs with rigid components is a "hybrid" integration strategy combining the best of both worlds. Flexible PCBs, like conventional PCBs, consist of various layers of materials such as base substrate, metal layer, adhesive, and cover layer. Apart from these basic layers that repeat in the case of multilayer PCBs, there are several other materials such as encapsulation layers, surface finishes, stiffeners, and so on.

11.2.1 FLEXIBLE PCB MATERIALS

The base material of a PCB should be electrically insulating and mechanically stable, and still have reasonable flexibility required by the application. The material should also be compatible with the metal layer, adhesive layer, cover layer materials, and the materials used for fabrication processes such as patterning, etching, dicing, and so on. The typical substrates used for fPCB manufacturing include polyimide, polyether ether ketone (PEEK), or polyester (Mylar, PET, PEN) thin films. The most commonly used substrate material is polyimide because of its electrical, thermal, chemical, and mechanical stability. Polyimide thin films cost significantly more than polyesters, however, their thermal and chemical stability compared to polyester thin films gives them a distinct advantage as a base for fPCBs. Thermal stability is of prime importance for PCB manufacturing because the soldering of components is performed at elevated temperatures. Polyesters are used if the integration has specific design requirements, for example, if transparency is required as part of the system design.

For the conductive film, the metal thin film that is most commonly used is copper, for the same reasons it has been used in conventional PCBs. Copper is a highly conductive metal that provides a low-cost, chemically resistant thin film. While silver can provide better conductivity, its cost is prohibitive for large scale use; and while aluminum can provide a low-cost thin film, it is not as conductive and can be easily corroded by many chemical reagents; and while gold and platinum can provide the

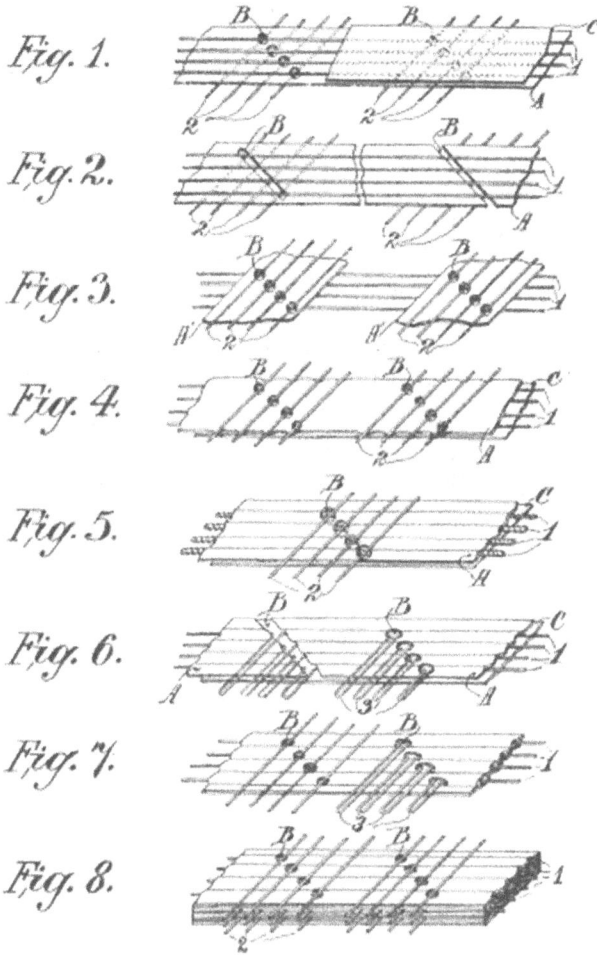

FIGURE 11.1   A figure from the patent filed by Albert Hanson in 1904 with the USPTO, titled "Electric cable". Patent granted in 1905. US782391A.

best chemical stability, their costs are hugely prohibitive. Some metallic alloys such as stainless steel, beryllium copper, nickel-chromium have some of these desirable properties, however, because electroplating is a critical process step in PCB manufacturing, a single element such as copper is more practically feasible. Further, it is not harmful or toxic for humans to touch. Thus, in high-volume commercial applications, copper provides an ideal compromise between low-cost, stability, processability, and conductivity. Other metals are used sparingly in very specific applications. For example, in systems that have weight as a key constraint, aluminum can be used, whereas nickel-chromium alloy (Nichrome) can be used when resistance to corrosion is required. Copper, like most metals, forms a native oxide on its surface when it comes in contact with air and moisture, however, the process is self-limiting because the native oxide layer itself acts as a diffusion barrier for oxygen to interact with underlying bulk copper. This can be a nuisance because it increases the electrical contact resistance of exposed copper electrodes, however, the self-limiting nature of the reaction prevents copper from "rusting". Thus, a copper/polyimide bilayer is the most common choice for fPCB manufacturing.

Adhesives play an important role in the PCB fabrication process (both rigid and flexible). They are used to adhere conductive metal foils to base substrates, to join multiple layers of the structure, or to attach a cover layer to the assembly. The adhesives are chosen such that they are compatible with all the materials being used in production, i.e., base substrate, metal, cover layer, patterning chemicals, etc., and have the same mechanical properties after curing as that of the base substrate. Adhesives are typically based on some form of organic polymers such as epoxies, acrylics, polyimides, and so on, and vary widely depending on the integration strategy. In some cases, the adhesive between the base substrate and the metal layer can be completely avoided if the metal thin film is directly deposited on the substrate using techniques such as sputtering. The thickness achievable with this process is low (<1 micron) compared to that required for PCB manufacturing (>10 micron). Thus, the sputtered metal thin film is typically used as a seed to grow metal selectively on the base substrate using low-stress processes such as electroplating. Cover layer or coverlay is a layer used to provide protection to the conductive paths on the PCB from moisture and oxygen ingress (barrier layer), while allowing access to contact pads for soldering or component assembly. The cover layer material is generally the same as that of the base substrate to improve mutual compatibility, for example, polyimide for polyimide base, polyester for polyester base, and so on. The contact holes in the cover layer can be pre-formed using mechanical drilling or laser patterning or can be patterned using photolithography after deposition on the top surface of the PCB. The structure of a bilayer fPCB is shown in Figure 11.2.

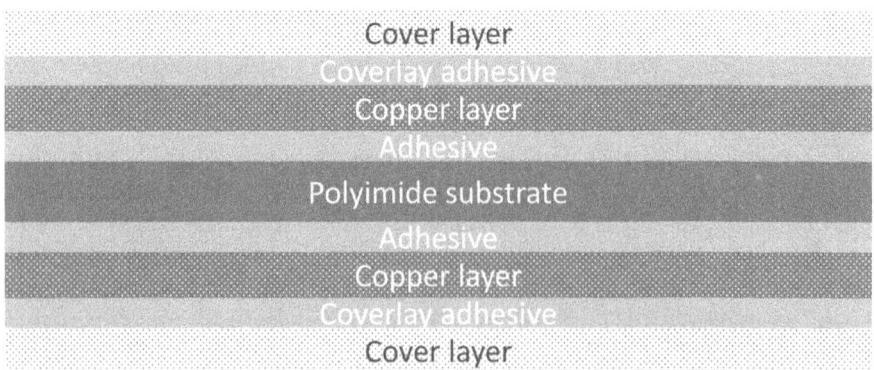

FIGURE 11.2 A double layer flexible PCB with polyimide substrate, copper metal layers (top and bottom), and cover layer encapsulation. The thickness of individual layers is between 15 and 25 μm, leading to the complete stack being flexible (thickness 200–300 μm).

11.2.2 Flexible PCB Fabrication Process

Flexible PCBs, like conventional ones, are fabricated by patterning the metal layer (copper) attached to the base substrate (polyimide). The copper layer is typically patterned using a mask created from the PCB design. The mask pattern is transferred onto a photoresist using photolithography, followed by copper etching using a wet chemical bath. In the case of multilayer PCBs, a covering film is added to the structure before the next copper/polyimide bilayer is overlaid. The interlayer vias are processed by drilling holes in the overlaid covering film followed by copper electroplating to fill the holes with copper to form conductive vias. The process of photolithography, etching, overlaying, drilling, and electroplating is repeated until the circuit is completed. This is the same process used for the fabrication of conventional PCBs, the difference being the use of flexible copper/polyimide bilayers instead of rigid copper/FR4 boards. The use of flexible base substrates provides some unique advantages from a manufacturing process perspective. The processing of fPCBs does not need to be sheet-by-sheet, as is the case for their rigid counterparts. The key advantage of the use of flexible materials is that they can be processed in a roll-to-roll fashion. The flexible base substrate such as polyimide or polyester can be supplied in the form of a roll that is continuously unwound as it is processed. This can increase the efficiency of the process and provide a much higher throughput (finished product per unit time) compared to the sheet approach. The processes such as adhesive lamination, metal foil attachment, cover layer bonding, can all be performed with individual webs drawn from the respective rolls. In the case of processes such as masking and photolithography, a static web can be used which is subsequently transported to the etching bath before being stored as a roll again.

11.3 FLEXIBLE HYBRID ELECTRONIC (FHE) SYSTEMS

The use of fPCBs, even with conventional surface mount components, provides several advantages. The use of fPCBs with rigid, conventional components leads to "semi-flexible" or FHE systems. The advantage of this strategy lies in the use of components that are already commercially available to create a flexible system. This also provides an opportunity to use devices based on state-of-the-art, scaled silicon CMOS technology. Further, the fabrication and patterning of fPCBs are done using well-established and scalable processes. This leads to high-performance electronic systems that have the ability to reversibly bend, flex and twist. This provides designers more options for creating innovative system designs and use cases, such as flexible display panels, solar cells, foldable smartphones, rollable TVs, and so on. Even in cases where the final product is not subject to bending cycles by the user, the use of fPCBs provides the designer the ability to create free-form structures in 3-dimensions. In fact, this is the most common use of fPCB technology in commercially available electronic systems today. Another advantage of the use of fPCBs is the lower weight and thickness of the overall assembly compared to that obtained with conventional PCBs. This is of great significance for modern applications with wearable and healthcare monitoring systems becoming the norm. The use of fPCBs in these systems allows more components to be packed into the same thickness, resulting in higher performance and functionality. Finally, the lower thickness also allows for the heat generated in the circuit components to flow out more efficiently, resulting in better thermal performance of the overall system.

However, this raises the question that if the use of fPCBs is so advantageous, and if the technology to manufacture and assemble fPCBs is widely available, then why not replace all the PCBs with flexible ones? There is a one-word answer to that question: cost. The use of flexible layers increases the manufacturing complexity of the process. For example, the use of roll-to-roll processing requires that the tension in the entire web, sometimes a few feet in width, be uniform throughout the process. This requires high-end manufacturing tools leading to an increase in costs. Further, in some designs, the holes in the cover layer need to be overlaid on the conductive pads with an alignment error window of a few microns. Such an alignment accuracy is difficult to obtain and maintain throughout the R2R process. For manufacturing a large number of PCBs, it is common practice to create a single large area PCB, say using R2R processing, that is then diced into the final size. The use of R2R processing and the subsequent handling of fPCBs after dicing becomes a challenge requiring sophisticated manufacturing equipment, leading to higher costs, especially in the case of multilayer fPCBs. Further, the handling challenges remain even after the fabrication of the PCB is complete. The soldering of rigid components on a fPCB and the subsequent mounting of the PCB onto the final system requires specialized tools and expert operators. This adds to the cost of the system fabricated using fPCBs. It should be noted that attaching rigid components to fPCB reduces its flexibility locally. Thus, if a large number of rigid components are attached, the flexibility may reduce to a point where it is indistinguishable from a rigid PCB. Thus, the density of components on a fPCB cannot be as high as that on a rigid PCB. This is true even in cases where the fPCB is only used for free-form design and is not subjected to bending cycles. In the case of applications that require reversible bending cycles, the density of the components needs to be reduced further. Indeed, in practice, the part of the fPCB at the axis of rotation only contains interconnects to other areas populated with components. This is also done to protect the components from the strain caused by bending. Bending strain can be detrimental to the adhesion of the components on the fPCB as well as to the reliability of the soldering of the components to the interconnects. This increases the quality of adhesion and soldering required for reliable operation of the system in the long run, adding to the cost of manufacturing. Finally, the flexibility of multilayer PCBs reduces with the increase in the number of layers, because the stacking of flexible substrates, metal thin films, covering layers, and adhesion layers creates a much thicker final PCB, increasing its flexural rigidity.

11.3.1 Small Silicon Chips

The FHE strategy can only provide limited flexibility along some axis, but not complete conformality to a surface. This is particularly true if large silicon chips are used as rigid components, which completely eliminates the local flexibility of the structure. However, the scaling of transistors over the last four decades has provided an opportunity to fabricate very small silicon chips (only a few millimeters square) with remarkable processing and storage capabilities. Hence, it is possible to design an integration strategy where these small silicon chips are interconnected using a fPCB, such that there is sufficient distance between them to retain system-level flexibility. This is particularly true for electronic systems that can be broken down into individual components. For example, in the case of internet of things (IoT) applications, the electronic system includes a sensor system, a transceiver, a memory, and some processing capabilities. All of these systems can be integrated on a fPCB in the form of small chips, with minimal external components, such that the flexibility of the system is optimized. Even in the case of systems requiring large processing or memory capacity, individual discrete components with small footprints can be interconnected on a fPCB to obtain the desired performance. For example, a certain memory capacity can be obtained by interconnecting many smaller memory chips in parallel on the same data bus and intelligent addressing design. This "distributed" system can then be more flexible than that based on a single or few large area chips. However, such an integration will occupy a much larger area and will require high quality, precise and reliable soldering and adhesion of all the small footprint components. Thus, designers of such systems may have to consider the trade-off between flexibility and system footprint. Another key problem with the use of small chips is that the power dissipated in the system is concentrated at these points. Compared to an integration with large area chips with built-in heat sink and thermal management, a system based on small discrete chips may result in much higher power density and temperature for the same power dissipation (Figure 11.3). Such a system will then require much better thermal design and implementation, adding to the total cost. However, this approach does provide a way to obtain flexible, high-performance electronic systems with state-of-the-art silicon components and well-established fabrication processes.

11.3.2 RIGID-FLEX INTEGRATION

The integration of the complete electronic system on a fPCB requires sophisticated equipment and processes in place, leading to an increase in the overall cost of production. Whereas flexibility in the complete system can be beneficial for some applications, many end users only require flexibility along a particular axis of motion. For example, in a foldable smartphone, it is important to have flexibility along the central axis of the screen, while the remaining device can be completely rigid. This realization led designers to develop an integration strategy that combines the fPCB and conventional rigid PCB in the same system. The final PCB thus includes parts assembled on a fPCB and those on a rigid PCB connected together in a "Rigid-Flex" PCB. The resulting system is flexible along some key axes providing the user an illusion of using a flexible electronic system. The integration can also be used to produce electronic systems distributed in 3-dimensional space owing to the flexible interconnectivity provided by the fPCB and the mechanical support provided by the rigid PCB. Rigid-flex PCBs can be fabricated through a uniform integrated process for the entire system, or the rigid and flex parts can be fabricated separately and connected together during system assembly. The latter process has been popular over the years, with the former one now becoming more common owing to better integration

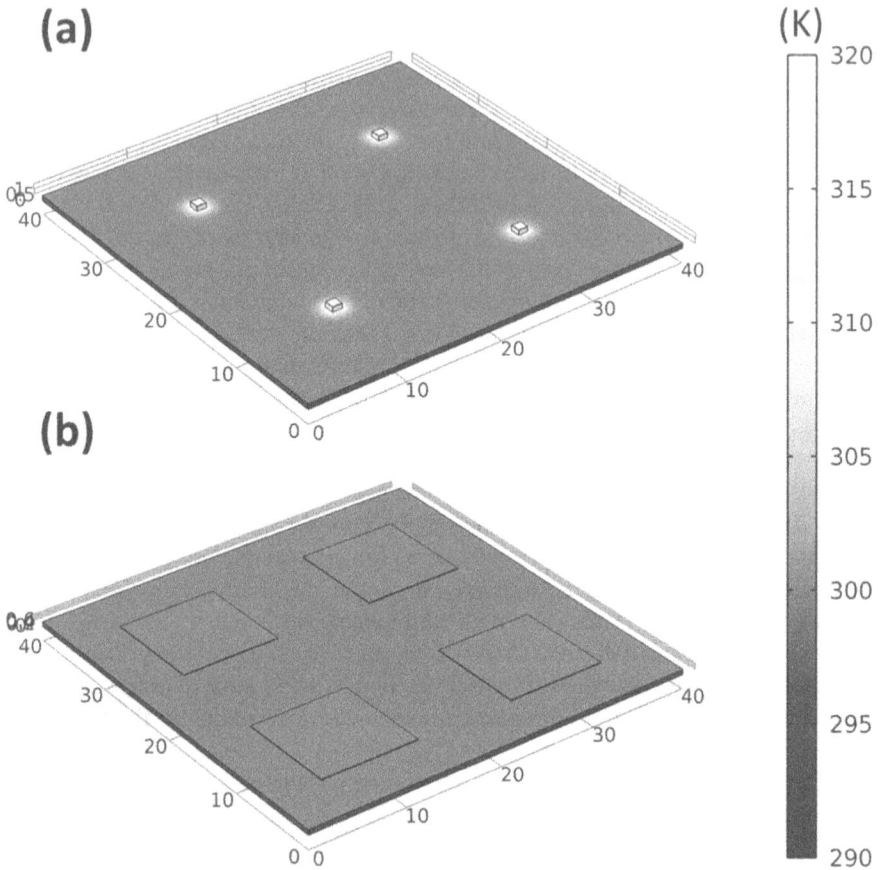

FIGURE 11.3   Heat dissipation in a hybrid integration with small silicon chips can lead to hotspots because of the dissipation of heat in small areas. In the case of fully flexible systems, the large area and lower thickness facilitate faster dissipation of heat. (Adapted with permission from [2], © John Wiley and Sons, 2016.)

efficiency. Indeed, currently, the most common commercial use of fPCBs is as a bendable interconnect between several pieces of rigid PCBs.

In case the flexible and rigid parts of a PCB are manufactured separately, the processes followed for fabricating these individual parts are similar to their regular processes. The circuit is designed in such a way that different modules of it are on different rigid PCB parts, while an interconnection port is provided for interconnecting them. The fPCB part can then have these interconnection buses between the different parts of the system. Such an integration strategy provides several advantages. It provides an opportunity to use separate tooling for the rigid and flexible parts of the PCB leading to cost-optimized processes for each. This results in a reduction of the overall cost of production because flexible parts are typically smaller compared to rigid parts. Further, because flexible parts are only being used as an interconnect bus, they do not require assembly, or a high-quality requirement for

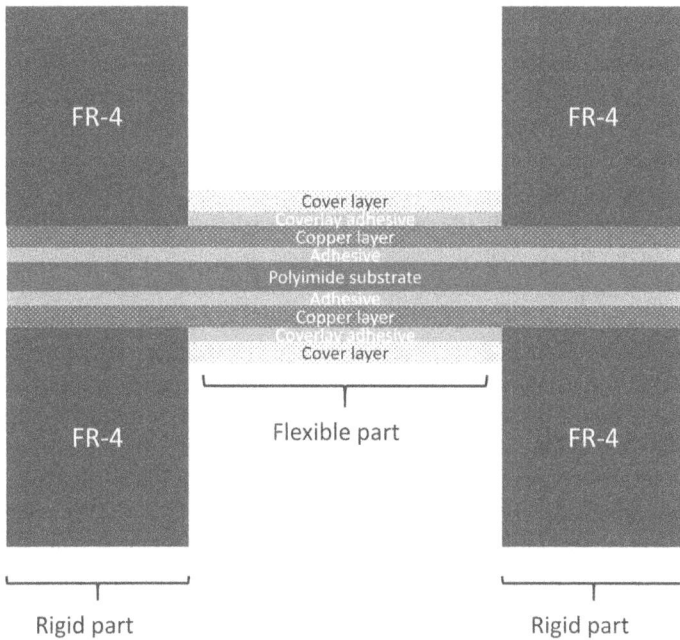

FIGURE 11.4   The cross-section of a typical rigid-flex assembly where the flexible part is used to connect two rigid PCBs that have the assembled electronic components.

adhesion and soldering, thereby further driving down costs. Thus, this integration strategy for rigid-flex assembly is very popular and can be seen in many commercial systems (Figure 11.4). In the case of a combined integration strategy for fabrication, the process typically starts with a single layer of the flexible or rigid part. The process for overlay covering, drilling and plating is followed in a similar way as for individual rigid or fPCB fabrication. The difference is that the rigid and flexible layers are already patterned into the final design structure and combined into a single PCB structure at the final stages of processing. This strategy allows the designers to optimize the PCB area, given that there is no need for separate connection ports on either the flexible or rigid parts. The tail end of the process, which includes surface finishing, milling, forming, etc., is undertaken by the rigid-flex assembly as a single unit leading to seamless and more reliable interconnections. However, because this part of the process is common for the complete structure, the tools used for it need to be able to handle a (partly) flexible PCB. Thus, some parts of the process require sophisticated tooling, increasing the overall manufacturing costs.

The benefits of rigid-flex integration have resulted in the use of this strategy for many commercial applications, not only related to flexible or wearable electronic systems. We find rigid-flex PCBs in the dashboards of automobiles, inside computers, printers, laptops, and most of the electronic systems around us. The flexible part is mostly used for interconnection, thus, having no electronic components attached to it. It has, in many cases, replaced bulky cabling for interconnection inside modern electronic systems.

11.4 FULLY FLEXIBLE SYSTEMS

A hybrid flexible system consists of fPCBs, either in part (rigid-flex integration) or as a whole, along with rigid electronic components. The use of this integration strategy has provided us lighter and slimmer devices, that in some cases are also dynamically flexible. However, this is only one step forward on the road to obtaining truly flexible electronic systems. Such a system would require the use of materials discussed in the previous chapters to form the internal device structure on flexible substrates, and the use of barrier films, coatings, and packaging strategy to obtain completely flexible electronic components. These components will then be integrated into a working electronic system on a fPCB. Such a system would have complete flexibility along any axis of bending, given a minimum bending specification. At present, there are very few demonstrations of such a complete system on a flexible platform. These are mostly from research groups demonstrating simple flexible circuits. For example, a flexible solid-state lithium-ion battery connected to a flexible light-emitting diode, or an array of sensors connected to a simple readout circuitry. These are typically packaged inside a flexible polymer encapsulation such as PDMS. The fabrication, integration, and demonstration of these systems prove that completely flexible stand-alone systems are possible. However, there are several components that come together in an electronic system for it to be truly stand-alone. These include a processor, memory, transceiver, sensors, actuators, antenna, and other peripheral components. A system cannot be completely flexible until all these components are flexible, and they are integrated on a flexible platform. Over the past two decades, research focus has shifted from materials and individual devices to integrated systems, leading the way for further development. The way forward then is to create more and more complex forms of flexible systems so that eventually we are able to develop fully flexible versions of the modern-day electronic systems.

EXERCISES

11.1. What are the considerations for materials for the substrate and metal layers of flexible PCBs? Which materials are commonly used for substrate and metal layer?

11.2. What are flexible hybrid electronic systems? What are their advantages and disadvantages?

11.3. What are the advantages of rigid-flex integration? Provide an example of a commercial product employing rigid-flex assembly.

11.4. Why can we not use small silicon chips integrated on flexible PCBs to create a fully flexible system?

12 Printed Electronics

12.1 INTRODUCTION

To conclude the section on integration, we will discuss one of the most promising technological developments in electronics manufacturing – printing. The growing popularity and use of printing technologies for electronics manufacturing have led to the development of a new class of electronic devices and systems – printed electronics. The technology relies on the use of traditional printing techniques, typically used for graphical design and book publication, for "printing" electronic systems. These techniques include screen printing, inkjet printing, slot-die techniques, spray coating, engraving, and so on. There are several benefits of using these techniques for manufacturing electronics. First, they offer a low-cost process for obtaining patterned electronically functional thin films on a flexible substrate compared to traditional photolithography. Second, there is less wastage of precious precursor materials compared to other deposition techniques such as chemical vapor deposition, sputtering, evaporation, and so on. Third, many of the printing processes are compatible with roll-to-roll (R2R) fabrication technique leading to very high throughput. A major disadvantage of the use of printing techniques is that the line resolution, deposition thickness control, and alignment accuracy are lower compared to traditional semiconductor manufacturing techniques. Thus, printing techniques are typically used for preparing the flexible substrate with large area structures at a high rate and low cost, before using other techniques to determine finer details. The choice of the printing method, of course, depends on the substrate to be used, the application at hand, and the resolution and alignment accuracy it requires. Apart from these considerations, the printing technique and the tools used need to be adjusted for the material to be deposited and the rheological properties of its "ink" and the materials already present on the substrate.

12.2 INK FORMULATION

The formulation of the ink plays a critical role in determining the success of thin-film deposition for any printing technique. The ink formulation typically consists of particles of the material, dissolved in a suitable solvent. The material, the method of fabrication of the material particles, and the choice of solvent are critical for the formulation of usable ink. Further, in most cases, a stabilizer or surfactant is added to the formulation to prevent the conglomeration of material particles. Parameters related to ink rheology, such as viscosity and surface tension, and those related to material particles, such as size and tendency to conglomerate, are important in determining the "printability" of the ink. These parameters need to be considered against the choice of printing technique to determine the optimum ink formulation. For example, the viscosity of the ink used for inkjet printing is lower than that used for screen printing.

DOI: 10.1201/9781003010715-15

The most common application of printing techniques for flexible electronics is for obtaining conductive traces on flexible substrates. These traces are formed using conductive inks consisting of conductive nanomaterials dispersed in a solvent. Commonly used conductive nanomaterials include metal nanoparticles, carbon nanotubes (CNTs), graphene flakes, reduced graphene oxide (rGO) flakes, metal nanowires, and so on. In the case of CNT and other carbon-based nanomaterials, water or alcohol (typically isopropyl alcohol) or a mixture thereof can be used as a solvent. This reduces the cost of ink formation significantly. However, it is important to add surfactants to improve the dispersion of carbon particles in water and prevent their conglomeration. Typically, a molecule containing an organic tail and an inorganic head, such as sodium dodecylbenzene sulfonate (SDBS), is used as a surfactant. The presence of surfactants in the ink can increase the contact resistance between adjacent nanomaterials and decrease the conductivity of the final thin film. Hence, some studies use volatile surfactants that evaporate with the solvent leaving behind a more pristine nanoparticle thin film. Metal nanoparticles of various metals such as silver, gold, copper, have been extensively studied for the formulation of conductive inks, with silver being the most popular choice. This is because silver provides a highly conductive and chemically stable thin film at a relatively low cost (compared to gold). There are several processes for obtaining silver nanoparticles, such as polyol, pyrolysis, milling, and so on (discussed in Chapter 8). The average size of the nanoparticles is important for determining the printability of the ink, and the properties of the final thin film. Particles of less size result in better ink fluidity and better packing density of the film, resulting in better conductivity. Typical average particle sizes used in commercially available silver inks range from 10 nm to 100 nm diameter. In some cases, smaller silver nanoparticles (10–20 nm) are added to inks based on larger copper particles (50–60 nm), so that they can occupy the pores between two copper particles. This leads to better packing density and film conductivity. There are several organic conductive molecules gaining popularity for thin film formation as well as printed electronics. For example, polyaniline, polypyrrole, Poly-3,4-ethylenedioxythiophene (PEDOT), and so on, have been studied as conductive organic compounds (see Chapter 5). The key problem with the use of organic compounds is their insolubility in common solvents to form a stable ink. The solubility of PEDOT in water is increased by adding a sulfonated polystyrene monomer to create the copolymer, poly(3,4-ethylenedioxythiophene) polystyrene sulfonate (PEDOT:PSS). PEDOT:PSS is soluble in water as a colloidal dispersion and can be used for printed electronics to obtain a conductive thin film (Figure 12.1).

Printing techniques have also been used to create insulating and semiconducting thin films. The ink formulation, in this case, includes the use of insulating or semiconducting nanoparticles dispersed in an appropriate medium. For example, zinc oxide nanoparticle inks are commercially available for printing semiconducting thin films. Novel nanomaterials such as molybdenum disulfide (MoS_2) are also available in printable ink form, as a dispersion of monolayer flakes in a solvent. Dielectric or insulating inks consist of organic polymers or ceramic nanoparticles dispersed in a solvent.

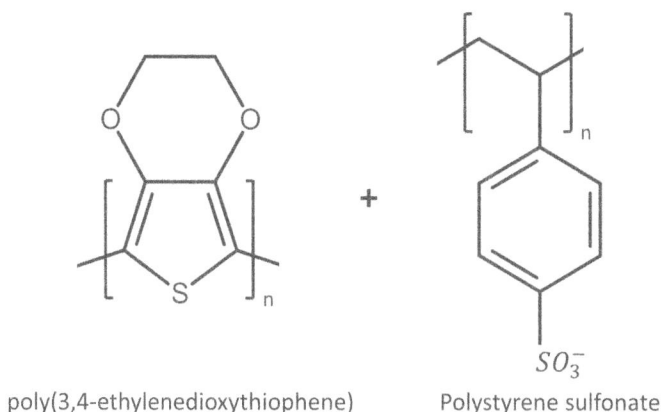

poly(3,4-ethylenedioxythiophene)   Polystyrene sulfonate

FIGURE 12.1 PEDOT:PSS is a mixture of two separate polymers – poly(3,4-ethylene-dioxythiophene) (PEDOT), and polystyrene sulfonate (PSS).

12.3 INKJET PRINTING

Inkjet printing is the most popular and deeply researched printing technique for the fabrication of large-area flexible electronics. In principle, the technique is relatively simple – a solvent-based ink is squeezed out of a chamber through a nozzle, one drop at a time, and deposited on a substrate. The solvent then evaporates, leaving a thin film of the desired material on the substrate. However, there is a significant amount of engineering and optimization that goes into each of the steps described above, from ink formulation to substrate wetting and thin film formation. The mechanism used for squeezing the ink droplet out of the chamber can be based on piezoelectricity or thermal expansion.

Regardless of the details of the technique, there are several advantages of using inkjet printing. First, the process can be conducted at room temperature with minimal anneal temperatures required for solvent evaporation and film densification. This increases the number of flexible substrates that can be used with inkjet printing. Second, the process can be controlled precisely using the mechanism used for droplet formation and stage movement. Because every droplet requires an actuation cycle, the formation of each droplet can be digitally controlled using the control electronics. In some implementations, the size of the droplet can be controlled based on the actuation strength, giving another control dimension. Also, the positioning of the droplet on the substrate can be controlled using stage mechanics. Thus, inkjet printing provides a way to obtain highly precise patterns on a substrate. Further, because these controls are electronically programmable, inkjet printing provides a method to obtain dynamically controllable designs, i.e., every substrate can have a different design based on a pre-coded algorithm. For example, the process can be programmed to take into account defects on a substrate (say through an optical image) and make adjustments to the design, or the process can be repeated until a certain thickness of the thin film is obtained. Further, inkjet printing allows for designs to be made parametric such that bespoke designs can be created based on customer specifications. Third, the design is digitally stored, thus very low costs are incurred to make changes, compared to other techniques that may require redevelopment of physical masks or molds. Fourth, it is a non-contact method, both in terms of physical force and chemical exposure. The substrate is only exposed to a small force of the fall of the droplets, and the substrate only comes in contact with the solvent and the film-forming material. In contrast, the conventional lithography process exposes the substrate to several chemicals such as adhesion promoter, photoresist, developer, PR remover, and so on. This is particularly important if the substrate or a thin film already present on the substrate, is sensitive to mechanical or chemical perturbation. The non-contact nature of the process also helps in obtaining patterns on uneven, irregular, or rough surfaces. Finally, inkjet printing is a very scalable method, providing an opportunity to use multiple printing heads at the same time, or sequentially. This technique is commonly used in printing color images on white paper, wherein droplets of cyan, magenta, yellow and black inks are used in different ratio combinations to produce different color pixels (CMYK color model). Such a technique can be modified for fabricating complex multilayer electronic structures over a large area, or for creating designs with different materials side-by-side on a single substrate.

While there are several advantages that make the use of inkjet printing popular for large-area flexible electronics, it is also important to look at some of the key disadvantages of using the inkjet printing process. First, the minimum feature size achievable using inkjet printing is much larger compared to that using photolithography. While the former can provide minimum line widths of a few microns, the latter can produce feature sizes of a few nanometers with state-of-the-art lithography equipment. Hence, inkjet printing is not usable for extremely scaled designs that are seen in modern silicon CMOS processing. Second, inkjet printing is a sequential deposition and patterning process, i.e., each droplet follows the previous one, thus not allowing for a large area to be deposited at once. In contrast, deposition and patterning in semiconductor manufacturing are processes that take place for the complete wafer simultaneously. While inkjet printing technique can be used in conjunction with the fast-moving R2R process, it is still slower compared to other printing techniques such as screen printing, spray coating, etc., where a large area can be printed at once. Notwithstanding these drawbacks, in this section, we look at the details of the inkjet printing technique and the various parameters that determine the final film properties.

12.3.1 JET FORMATION

The formation of droplets is a complex dynamic process that requires careful design of the chamber and the actuation mechanism, and strict adherence to ink parameters. The basic principle is to push the liquid through a small orifice, called the nozzle, from where the droplets emerge and fall onto the substrate. The process of printing can be carried out in two distinct modes: continuous inkjet (CIJ) or drop-on-demand (DoD). In the CIJ method, there is a continuous stream of droplets emerging from the nozzle. The substrate is attached to a movable stage to control the design

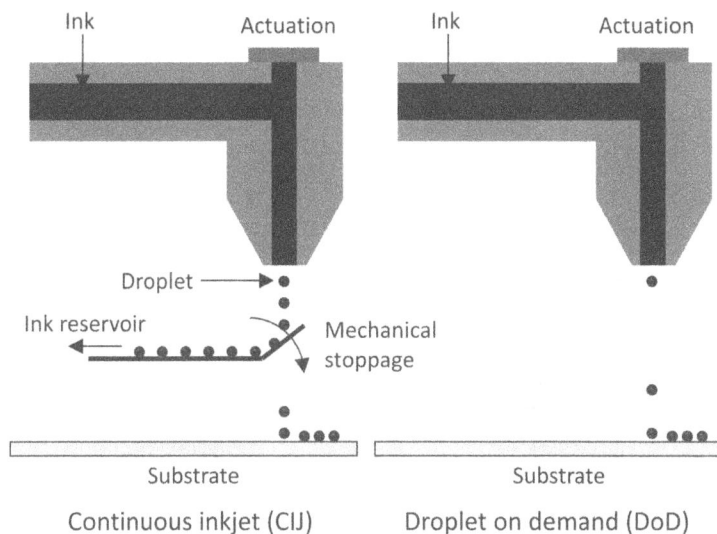

FIGURE 12.2 In the continuous inkjet (CIJ) method, the droplet stream is continuously generated. It is diverted to an ink reservoir in case it is not required on the substrate. In the droplet on demand (DoD) method, the droplets are only produced when they are required on the substrate.

formed and its location on the substrate. In some designs, there is a provision to stop the stream of droplets from reaching the substrate as shown in Figure 12.2. In the case of DoD, the droplets are only formed when the deposition is required on the substrate. In both cases, the actuation can be provided by creating a pressure wave inside the chamber using a piezoelectric membrane, thermal expansion, or acoustic actuation. In the case of piezoelectric actuation, a piezoelectric membrane, such as lead zirconium titanate (PZT), is electrically actuated causing it to change shape. This phenomenon is used to create the pressure wave inside the chamber that pushes the ink droplet out of the nozzle. In thermal actuation, an electric heater is actuated to produce thermal expansion and temporary vaporization of the ink leading to the pressure wave (Figure 12.3).

The key difference between CIJ and DoD, i.e., the ability to switch the droplet stream on/off, provides for the advantages and disadvantages of the technologies. With CIJ, because the droplet stream is continuous, there is consistency in the size of droplets obtained, leading to a simpler design of the actuation mechanism and nozzle. Further, there is no danger of the ink drying out and clogging the nozzle head, which is ever-present in the DoD method. On the other hand, the DoD method provides more freedom and precision in the design to be printed, without the wasteful diversion of the ink stream. For this reason, most of the laboratory setups employ the DoD methodology, while industrial ones use CIJ. DoD is particularly important in the case of printing systems involving multiple heads and materials because the duty cycle of each nozzle is very low. Hence, it is not ideal to have a continuous stream of droplets from all the nozzles at all times. However, prolonged inactivity of

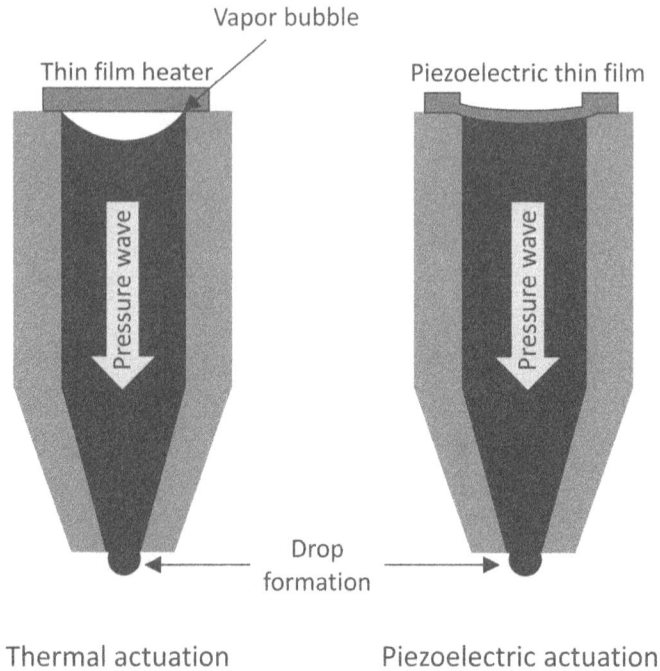

FIGURE 12.3   In the case of thermal actuation, a thin-film heater is used to create a local-ized hotspot that leads to the momentary evaporation of a layer of solvent, causing a pressure wave to travel through the chamber and eject a droplet. In the case of piezoelectric actuation, the deformation of the thin film creates the pressure wave in the chamber.

a nozzle head can lead to changes in the fluid at the opening of the nozzle, which may result in sedimentation or clogging. This can affect the size of the droplet when the nozzle is fired next, or in the worst case, lead to a missed droplet altogether. This is avoided in some systems by creating nozzle headcovers that are shut during inactiv-ity, or by firing the nozzle into a separate bin a few times before firing for substrate printing. In any case, the design of the nozzle and the parameters of the ink formula-tion need to be considered together in order to determine an optimum inkjet printer system design.

The two most important parameters for ink formulation are surface tension and viscosity. Surface tension originates from the fact that molecules at the surface of a liquid are attracted to other liquid molecules in the bulk more than the molecules of the atmosphere. The molecules at the surface are attracted inwards and laterally towards each other, causing the liquid surface to behave like a thin membrane under tension (hence the name surface tension). The magnitude of this tension depends on the difference between the attractive force between the liquid molecules and that between the liquid and surrounding molecules. Thus, liquids with higher cohesive forces, such as water, have higher surface tension. Mathematically, surface tension is defined as the energy required to increase the surface area of the liquid by one unit

(SI units – newton per meter, N/m). Because the increase in surface area of a liquid requires energy, a free volume of liquid (a drop), under no external influence, always assumes the shape with the least surface area per unit volume, i.e., a sphere. On the other hand, viscosity is the resistance of a liquid body to deformation arising out of the internal frictional forces acting between adjacent layers of liquid because of relative motion between them. Thus, a liquid with high viscosity will resist change in shape on account of the friction between the moving molecules. Mathematically, viscosity is defined as the shear stress required to generate a particular relative velocity gradient inside the liquid. If the velocity gradient created by the shear stress (τ), is represented by dv / dx, then viscosity is given by, $\mu = \tau / (dv / dx)$. The SI unit for viscosity is pascal second (Pa-s), whereas centipoise (cP) is a more commonly used unit from the CGS system (1000 cP = 1 Pa-s).

Now, let us look at the creation of a jet from a nozzle and its conversion into a falling droplet. When energy is imparted to the liquid ink in the chamber, creating a pressure wave, a certain amount of ink is displaced out of the nozzle (in the form of a cylinder for a circular nozzle head). The amount of ink displaced depends on the amount of energy provided because the formation of the cylindrical jet involves the creation of a new liquid surface area. This cylinder quickly breaks away from the main fluid body because of the lack of energy to create more surface area and because of the Plateau-Rayleigh instability. The stream of jet then spontaneously forms into a single droplet, or into multiple droplets, depending on the interplay between viscous, internal, and surface tension forces (Figure 12.4). In some cases, the energy supplied may be insufficient to eject enough liquid to form a droplet, leading to failed firing of the nozzle. Ideally, it is required that every firing of the nozzle creates a distinct single droplet every time, without any satellite droplets. The conditions for this situation are summarized using the Ohnesorge number (Oh). Ohnesorge number is a dimensionless number defined as:

$$Oh = \frac{\mu}{\sqrt{\rho \sigma d}}$$

Where μ, ρ, and σ are the viscosity, density, and surface tension of the liquid, and d is the nozzle or drop diameter. The Ohnesorge number describes the dominance of the viscous forces over those of surface tension and inertia (through liquid density and size). For a liquid to eject out of the nozzle, it should have low viscosity compared to surface tension and inertia, thus requiring a low Ohnesorge number. However, if the viscosity is too low compared to surface tension and inertia, the droplet can eject at a high velocity cause it to break into multiple droplets. In literature, the inverse of Oh, denoted by Z ($=1/Oh$), is commonly used as a dimensionless constant to describe the properties of the ink. A value of Z between 1 and 10 is desirable for the reliable formation of a single droplet on every nozzle firing. For values lower than one, the viscous forces dominate and dissipate the pressure wave within the chamber leading to failure in droplet ejection, while for values higher than ten, the ejected droplet leads to the formation of multiple satellite droplets.

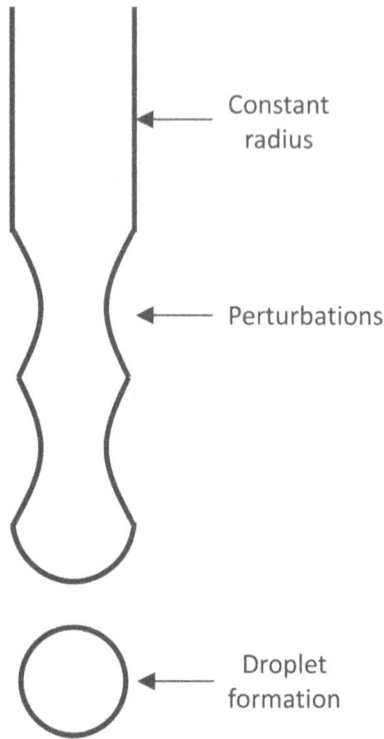

FIGURE 12.4 A continuous stream of jet spontaneously forms into a single droplet, or multiple droplets because of the Plateau-Rayleigh instability, depending on the interplay between viscous, internal, and surface tension forces.

It is clear that the rheological properties of the ink can play a critical role in the success of droplet formation. However, the choice of the solvent, which determines these properties, cannot be dictated by these considerations alone. The feasibility of nanomaterial dispersion, evaporation temperature, chemical stability, toxicity, and cost are the other factors that are commonly considered while selecting a solvent for ink formulation. Hence, it is important to have control mechanisms for parameters such as surface tension and viscosity even when the solvent in question is fixed. For this reason, some additives are added to the ink formulation to change these parameters. For example, the addition of 10^{-3} M SDBS in water leads to a change in its surface tension from 0.072 N/m to 0.040 N/m without appreciable change in its viscosity. On the other hand, water and alcohols such as ethylene glycol can be mixed in certain ratios to obtain a solvent of a specific viscosity. Finally, given the final composition of the ink, the properties of surface tension and viscosity can also be changed by changing the temperature of the chamber. Both viscosity and surface tension decrease with temperature, but at different rates, hence, an optimal temperature for droplet formation can be arrived at. Thus, the formulation of the ink and the design of the printhead are done with all these parameters and considerations in place.

12.4 OTHER 2D PRINTING TECHNIQUES

While inkjet printing is the most popular technique under research for printing large-area electronics manufacturing, other printing techniques such as spray coating, screen printing, flexographic printing, and electrohydrodynamic printing, are being studied to offer viable alternatives for specific applications.

In screen printing, a porous mesh screen containing the ink is pressed against the substrate to transfer the ink onto the substrate. There is a stencil or mask separating the substrate and the screen for determining the pattern of the ink. It is an ancient method, first appearing in ancient China, at the end of the first millennium. The ink is typically applied to the screen using a blading process, while the reverse stroke is used to press the screen against the substrate to obtain the pattern (Figure 12.5). Given the stark difference in process, the ink formulations and the properties of the inks used for screen printing can be very different from those in inkjet printing. In particular, it is important to have a higher viscosity ink to facilitate its application on the screen. Further, the transfusion through the screen requires the particle size to be strictly regulated, and very little coagulation of nanoparticles can be allowed before application on the screen. The key advantage of this process is the speed at which a large area of the substrate can be printed, greatly increasing the process throughput compared to inkjet printing. The disadvantage of this process is the use of a physical mask which does not allow for dynamic variation in the pattern design, as in the case of inkjet printing. Further, the process involves the physical contact of the screen and the substrate, which may not be appropriate for delicate or uneven substrates.

Spray coating is also a process that has been in use in industrial applications for many decades, particularly in the surface coating industry, such as painting automobile parts. The process involves atomization of the ink to create a stream of very

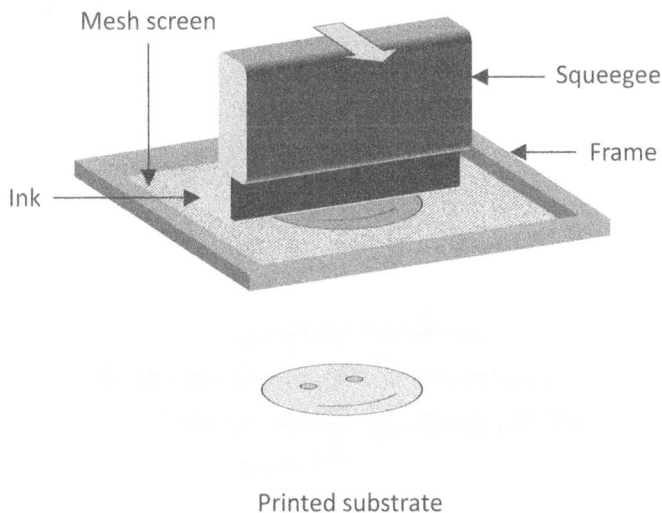

FIGURE 12.5 Screen printing involves the use of a mesh screen to transfer ink onto a substrate. The ink is typically applied using a squeegee to fill the mesh apertures with ink before pressing the screen against the substrate for pattern transfer.

fine droplets (a few microns in diameter). These droplets are then carried up to the substrate using a gas stream, where they impinge and deposit. The solvent evaporates (like inkjet printing) leaving a thin film of the required material on the substrate. The process is typically followed by a low-temperature anneal for sintering and film densification. To obtain a pattern, a stencil mask can be used to restrict the droplets from reaching the substrate. The coverage of the substrate area depends on the shape of the nozzle, and hence the aerosol stream, and the distance of the nozzle from the substrate. The process typically uses the movement of the spray head or the substrate to cover a large area and can involve repeated deposition over an area to obtain a reasonably thick film. The spray nozzle produces atomization of the ink by applying very high energy (or pressure difference) across the nozzle head creating a high surface area of the liquid (a large number of small droplets). One of the ways in which this energy is provided is through high frequency (ultrasonic) vibration of a piezoelectric crystal (the typical frequency of vibration is 20–200 kHz). The advantage of spray coating is that it is a compromise between the ability to digitally control the design (better than screen printing) and throughput (better than inkjet printing). The process, like inkjet printing, is completely non-contact.

Flexographic printing is based on the R2R process used for printing on paper, for example, in the printing of currency notes. The process involves the use of a printing plate with pattern structures raised from the surface, similar to a rubber stamp. The printing plate is attached to a roller that gets wetted in the ink from an ink bath and deposits the ink onto the substrate through physical contact. The pattern is obtained because the substrate only makes contact with the raised structures on the printing plate. This process is an extremely high throughput process, capable of printing a large area substrate (large width), and is perfectly suited for R2R manufacturing process (Figure 12.6). However, the key disadvantages are the

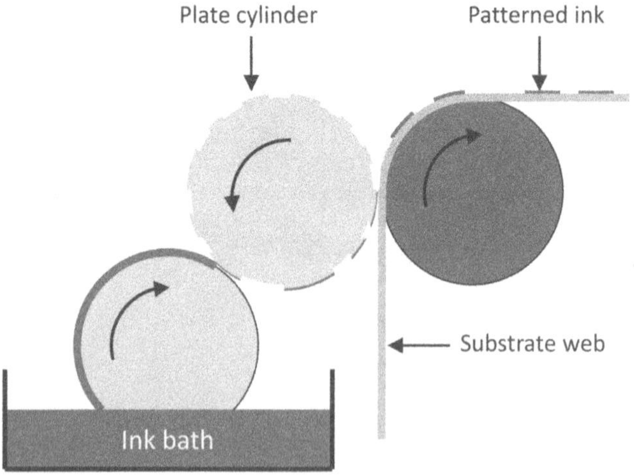

FIGURE 12.6 Flexographic printing uses a pre-patterned plate with structures raised from the surface. Ink is applied to the plate, followed by pressing against the substrate to obtain the patterned ink on the substrate. It is typically deployed using the R2R production method for higher throughput.

high setup cost involved, particularly for the manufacturing of the printing plate with high-resolution structures. This is particularly problematic if a change in the pattern may be required. Also, the process involves physical contact with the substrate web, reducing the choice of substrate usable. Finally, because there is no mask placement (like in screen printing or spray coating), or the possibility of real-time correction (like in inkjet printing), this process provides a very poor alignment accuracy for a pre-patterned substrate.

Electrohydrodynamic (EHD) jet printing or e-jet printing makes use of the electrostatic force to eject ink droplets from the nozzle, as opposed to thermal, piezoelectric, or acoustic energy used in inkjet printing. An electric field is created between the nozzle head and the substrate, by applying a DC potential difference between them. This causes the mobile ions in the liquid to accumulate at the surface, leading to tangential stress on the surface. At a sufficiently high field, this stress overcomes the surface tension to form a new liquid surface, leading to the ejection of the droplet. The process, like inkjet printing, is digitally controllable and can provide droplets on demand. However, this only works with substrates and inks that are able to withstand high electric fields (of the order of MV/m).

12.5   THREE-DIMENSIONAL (3D) PRINTED ELECTRONICS

3D printing is a revolutionary technology designed to produce three-dimensional designs by depositing or curing material layers sequentially. The process has become very popular for quick prototyping and hobbyists all over the world use 3D printers to create everything from toys, tools, action figures, and even consumer products. The reasons for their popularity include ease of use, the ability to create a prototype design for demonstration or experimentation, and the ability to print complex single-part structures that are not achievable using conventional approaches such as injection molding. In commercial space, 3D printing is gaining acceptance because of the ability to create parametric designs or create bespoke solutions. The use of 3D printing for mechanical prototyping is becoming more widespread with time, however, we are also seeing a surge in the use of 3D printing for electronic prototyping. This includes printing of designs for enclosures and support of electronic systems and PCBs. Further, recently, the 3D printing process has been proposed as a process to integrate flexible electronic devices and functional materials layers into conventional mechanical designs.

There are several different ways in which 3D designs can be printed using different kinds of materials. All of these methods are clubbed together in the term 3D printing techniques. The most popular commercial techniques widely used for 3D printing are stereolithography (SLA) and fused filament fabrication (FFF). SLA relies on the curing (or solidification) of a liquid polymer upon exposure to a UV laser. The process of curing is initiated because of the formation of crosslinks between polymer chains using the energy provided by UV light. The UV laser is controlled by a computerized mechanism that processes the 3D design to determine the points that need to be solidified in each layer (two dimensions). The laser is then used to expose the surface of the liquid polymer bath, solidifying it at the points required by the design. After the exposure of each layer, the stage moves vertically (third dimension) to allow for a new layer of liquid to be solidified into the next layer. Thus, the complete design is "printed" layer-by-layer in three dimensions. The SLA process produces prototypes with very high-quality surface finish and aesthetics; however, the cost of the equipment and consumables is high compared to other 3D printing techniques. Further, for a given printer, the design can be printed using a single material. In the FFF process, a solid wire-like filament of the material is extruded through a heated nozzle. The heat from the nozzle causes the polymer filament to melt and fuse with the rest of the design. The extrusion, melting, and re-solidification of the filament are carried out at each point in the design, thus requiring the design to be converted into a 3D grid. Further, the amount of material ejected at each point is fixed, limiting the minimum feature size and resolution achievable using this process, compared to the SLA technique. However, the major advantage of FFF compared to SLA is the ability to include multiple materials in the same design. This is particularly important from the printed electronics point-of-view because different materials such as conductive materials, insulating materials, light-emitting materials, etc., can be combined into the same design. FFF also allows for the use of conventional thermoplastics, such as polylactic acid (PLA) and acrylonitrile butadiene styrene (ABS) increasing the acceptability of 3D printing for commercial applications, particularly in the automobile industry. Apart from SLA and FFF, several other processes have been developed for 3D printing, such as digital light processing (DLP), selective laser melting (SLM), laminated object manufacturing (LOM), and so on. In the context of flexible electronics, the process used for 3D printing depends on the materials and applications involved.

Polymers are a class of materials most widely used in 3D printing, because of the tunability of their mechanical properties using heat (FFF) or light (SLA). Some of these polymers can be used for fabricating components of flexible electronic systems, such as substrates, encapsulations, barrier layers, packaging, and so on. Further, these polymers can also be used as matrices for nanomaterials such as CNTs, graphene flakes, metal nanowires, etc., to form electronically functional composites. These composites, when printed alongside other materials in a layer-by-layer fashion can provide some unique opportunities for electronic system design. However, to obtain the desired properties in the deposited structure, uniform dispersion of these nanomaterials into the polymer matrix, with no agglomeration over time, is essential. Apart from polymer composites, materials with different electronic conductivities can also be 3D printed into any desired shape. This allows for the printing of electronic components such as resistors, capacitors, inductors, metal lines, sensors, antennas, and so on, and their interconnections. Thus, a large number of circuit designs are possible, particularly because 3D printing allows for components to be placed in the vertical dimension as well, as opposed to the conventional PCB design, where only planar designs are possible. This feature, along with the fact that the designs are digitally stored, provides system designers the freedom to develop unique architectures and free-form shapes for the same circuit. An example of the power of 3D printing for electronic system design was demonstrated by Kong *et al.*, in 2014 [5]. This study demonstrated an array of quantum dot LEDs based on CdSe/ZnS quantum dots. The design of the system started with a layer of silver paste and PEDOT:PSS printed inside a cavity to form metallic contacts. The active material

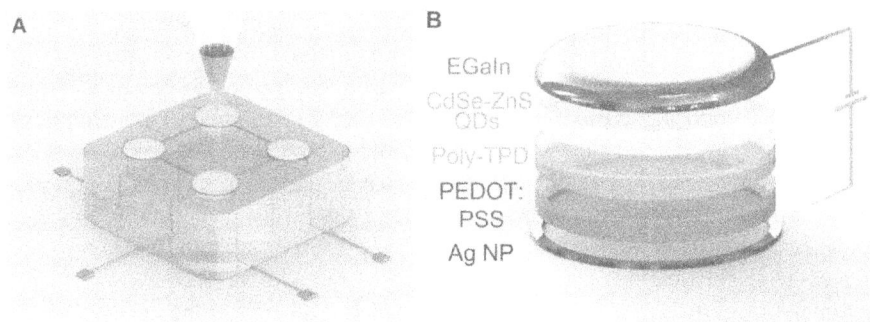

FIGURE 12.7   The layer-by-layer schematic of various components in the 3D printed quantum dot LED. (Adapted with permission from [5]. Copyright 2014 American Chemical Society.)

was dispensed on this contact layer in the form of a CdSe/ZnS solution, followed by a liquid metal (EGaIn) to form top contact (Figure 12.7). Such a system design can be printed on any surface, including a curvilinear one, and can be formed into an array of QLED pixels. The integration of such a system along with the capability of producing passive components, metallic interconnects, and encapsulation layers along all three spatial dimensions provide tremendous opportunities for the integration of free-form electronics in the future.

12.6   NANOIMPRINT LITHOGRAPHY

Nanoimprint lithography (NIL) represents a class of printing processes used to obtain high-resolution patterns. There are many variations to the process, however, the underlying technology remains the same. The process involves the use of a pre-patterned mold to press upon a viscous resist creating variations in its thickness. The resist can then be used to pattern the film stack underneath. The mold can be created using high-resolution lithography, or other patterning techniques such as 3D printing, such that there are topological variations in its surface profile. The process typically begins by spin coating a thin film of photoresist on the substrate. The mold is then pressed against the substrate to imprint the pattern on the surface of the resist. At this stage, the resist is required to be above its glass transition temperature so that it can be easily deformed and cast into the required pattern. Following this, the resist is solidified either by cooling the thermoplastic polymers to a temperature below their glass transition temperature, or by curing the uncrosslinked polymers using heat treatment or UV irradiation. The mold is then removed from the substrate and the patterned resist is used to transfer the pattern onto the underlying thin film. In some cases, certain thin films, typically polymer nanocomposites, can be directly patterned into the required shapes using a mold imprint.

The most important advantage of the NIL process is the ability to create high-resolution patterns using a simple process. Line resolution of a few tens of nanometers is easily achievable using NIL, which would conventionally require high-end photolithography equipment that accounts for a significant portion of the cost of

CMOS manufacturing because of the use of precision control systems and optical lenses to obtain nanometer-scale resolution. Additionally, the photoresists used in the conventional lithography systems are precisely engineered to obtain a certain thickness and optical sensitivity at a particular wavelength, whereas, the simplicity of the NIL process permits the use of any polymer or composite with tunable fluidity as the resist material. Further, the NIL process can be used to transfer large-area patterns onto thin films in a single step. The process thus provides the ability to create high-resolution patterns at a much faster pace compared to photolithography. The process is also compatible with flexible substrates and with the R2R manufacturing technique (Figure 12.8). A key process associated with NIL is the creation of the mold. The mold surface needs to be patterned into high-resolution topographies, typically using conventional photolithography and etching processes. Silicon is a commonly used material for mold creation because of the compatibility of silicon wafers with conventional bulk micromachining techniques. Other mold materials include metals like nickel or soft materials like PDMS. In the latter case, the mold can be used to create patterns on curvilinear surfaces as well. The process of creation of the mold can be long, cumbersome, and expensive given the resolution and accuracy required. However, once the mold has been created, it can be repeatedly

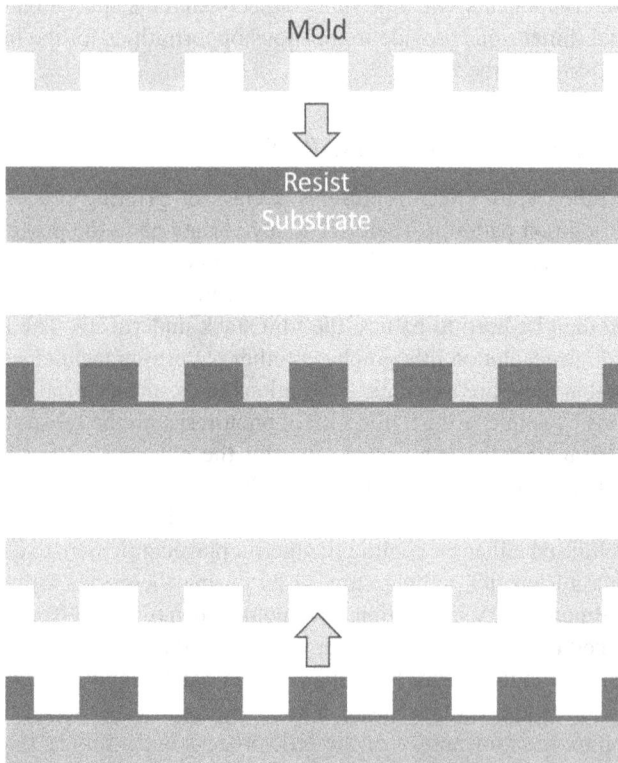

FIGURE 12.8   Nanoimprint lithography involves the use of a mold to pattern a resist into a predetermined 3D pattern with very high resolution.

used many times, thousands of times for some materials, leading to a reduced cost over the mold lifetime.

The NIL process is a promising approach for high throughput patterns on flexible substrates, however, it does have its downsides. One of the major problems, compared to the conventional photolithography process, is the accuracy of aligning the mold pattern with patterns already present on the substrate, referred to as overlay accuracy. The overlay accuracy in conventional lithography is obtained using mask registration by moving the mask or wafer stage to obtain alignment before exposure is carried out. This process is tricky with a mold because of the need for transparent (or through etched) registration windows. Further, the topological patterns on the mold would require higher clearance from the wafer leading to less accurate registration. Another problem with the use of molds for pattern transfer is damage and wear over successive uses. In NIL, the pattern transferred onto the substrate is completely dependent on the quality of the mold, and if a mold incurs nanoscopic damage to the pattern during use, all the subsequent patterns will have that damage. Hence, it is very important to use a mold material that preserves its pattern, at a nanoscopic scale, even when successively impressed under substantial pressure against a substrate over time. Finally, the selection of the mold and resist material is constrained due to the need for a defect-free separation of the mold after pattern transfer. This is particularly difficult because the nature of the resist needs to go from fluid to solid while being in contact with the mold, and the mold is pressed against the resist with a substantial force for pattern transfer. The mold then needs to separate cleanly from the resist surface, without taking any resist debris with it. This typically requires the use of an antiadhesion coating on the mold surface, increasing the cost and complexity of the process. For example, a silicon mold can be coated with a monolayer of perfluorodecyl trichlorosilane ($C_{10}H_4F_{17}$-$SiCl_3$), commonly known as FDTS, to reduce the surface energy and discourage adhesion with resists. Recognizing the potential of the NIL process for electronics manufacturing, some semiconductor equipment manufacturers, such as Canon Inc., have introduced a line of commercial NIL production equipment.

12.6.1 Self-Aligned Imprint Lithography (SAIL)

SAIL is an improvement on the NIL process, such that it uses a single patterning step to pattern multiple thin films in a self-aligned manner. The trick is to use the three-dimensional nature of NIL pattern transfer to obtain multiple steps of thin-film patterning. The process begins with the deposition of all the thin films required for the device design on the substrate. For example, for a thin-film transistor (TFT) array, the gate metal, dielectric, channel, and source/drain layers are all deposited sequentially. A resist is then deposited on the substrate and patterned into a 3D structure such that the patterns to be transferred onto different thin films, exist on different height levels. The entire substrate is then subjected to sequential anisotropic etching for the underlying layers while using the resist as the mask. After etching a specific layer, the resist itself is anisotropically etched to remove the pattern with the lowest height, resulting in a change in the mask design for etching the next thin-film layer. The process continues until all the thin films are etched to form the device. It is

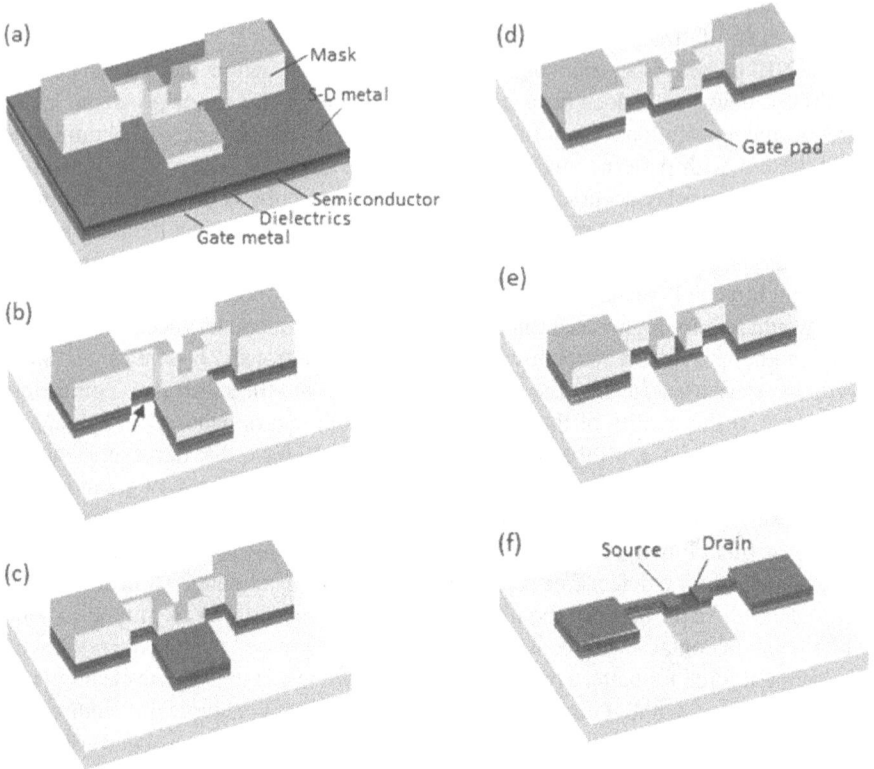

FIGURE 12.9 Schematic illustration of fabricating bottom-gated TFT with 3D SAIL. (a) Mask creation by 3D SAIL and plasma etching; (b) etching stack entirely; (c) remove bottom level of mask; (d) etching until the gate pad; (e) remove next bottom level of mask; and (f) etching top metal to define the channel. (Reprinted from [6] under the Creative Commons Attribution 3.0 license.)

schematically illustrated in Figure 12.9. The SAIL process shares all the advantages and disadvantages of the parent NIL process. It can produce high-resolution, large-area designs with high throughput R2R process once a robust mold is prepared. Further, it solves the overlay accuracy problem, because the pattern for different layers is transferred in a single step leading to self-aligned designs.

The introduction of printing techniques such as inkjet printing, spraying, 3D printing, and NIL for bulk electronics manufacturing has been a boon for the flexible electronics industry because flexible substrates can now be directly subjected to the deposition of patterned thin films without the need for high-end manufacturing equipment. The use of direct printing techniques also reduces the overhead associated with transferring thin-film devices onto flexible substrates. Further, with imprint techniques such as NIL and SAIL the resolution of the devices produced using printing has become comparable to that achieved using standard optical lithography techniques. It is only a matter of time before these processes are widely adopted for commercial electronic device manufacturing.

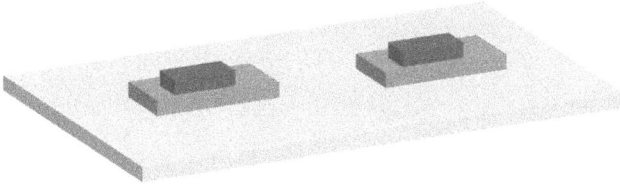

FIGURE 12.10 Example of a patterened multi-layer thin-film structure.

EXERCISES

12.1. Describe the advantages/disadvantages of CIJ and DoD methods for inkjet printing.

12.2. We plan to create an ink based on water as the solvent. What is the nozzle diameter required to obtain an Ohnesorge number of 0.2 at 30°C? (consider the properties of pure water).

12.3. Describe the screen printing and flexographic printing techniques for obtaining patterns on flexible substrates in detail.

12.4. Given the pattern in Figure 12.10, design a mold that can be used to achieve it using the SAIL process.

12.5. Enumerate the advantages and disadvantages of the nanoimprint lithography (NIL) process.

Part IV

Applications

13 Flexible Processors

13.1 INTRODUCTION

In the previous sections, we have looked at the materials and processes that can be used to fabricate flexible electronic devices and systems. The integration strategies discussed in the last section provide an insight into the way forward for large area, complex electronic systems that can resemble or even outperform some of the state-of-the-art rigid electronic systems currently with us. It is important to develop a system core that enables us to control, interact, and communicate with such large and complex systems. These are the central processing units (CPUs) that form the core of most of the electronic systems around us. Hence, we start with the section on applications of flexible electronics by looking at ways in which we can create processing units using flexible materials and processes. We will look at the conventional processing systems and the basic principles that govern their operation so that these can be replicated with flexible materials. The processing unit, and indeed most of the conventional electronics relies on Boolean algebra, which provides a mathematical framework for calculations based on fixed states of variables. In normal algebra, variables can take any value from an infinite set such as integers, real numbers, complex numbers, and so on. However, in the case of Boolean algebra, variables are only allowed to take values from a fixed set of states. In the case of digital systems, these states are limited to only two states – 0 and 1, and these are implemented in digital circuits using the absence or presence of charge (voltage) or using variations in switch behavior, such as on/off. The switches are themselves implemented using transistors, which is where semiconductors such as silicon play an important role. At the outset, it might look absurd that little pieces of silicon inside a chip can make a "decision", or can perform complex tasks such as controlling displays or powering Siri, however, it should be realized that such everyday tasks are performed by processing units routinely consisting of billions of transistors interconnected to form complex circuits. It is, in a sense, the power of interconnection of billions of simple things into one entity, not unlike the human brain that, through the mere interconnection of simple neuron cells, can achieve something as remarkable as consciousness. This is where it is important to investigate, in some detail, the key reason behind the spectacular success of the silicon CMOS industry – dimensional scaling. The scaling of transistors to smaller sizes allows for more device density, i.e., the number of transistors per unit area of silicon. This leads to processors that can perform more functionality, allowing developers to encode more and more complex algorithms over time. Finally, the last piece of the puzzle is the speed at which these transistors operate. An algorithm can be complex, however, if it is slow, it may not be as impressive or useful. Thus, a computer algorithm that is able to predict the weather is impressive, however, it is not useful if it takes two days to predict tomorrow's weather. The speed of the processors depends on how quickly the transistors

DOI: 10.1201/9781003010715-17

are able to switch an output node between the two states of Boolean algebra, i.e., from zero to one, and back. This depends on the current carrying capacity, or the on-current, of the transistor. Thus, to create the electronic magic that we are all used to, we require transistors that can carry large amounts of current and can be packed into a very small area. This is an important point to keep in mind while designing processors using flexible materials and processes.

13.2 CONVENTIONAL ELECTRONICS

13.2.1 DIMENSIONAL SCALING

The problem of speed and scaling needs to be quantitatively understood before we can design flexible processors. Conventional transistor design relies on the metal-oxide-semiconductor (MOS) structure which consists of a semiconducting channel forming a capacitor with a conductive gate electrode. The voltage applied at the gate terminal, with respect to the substrate, dictates the charge concentration in the semi-conducting channel leading to switching on or off of the transistor. Analysis based on this design gives the on-current for a transistor as:

$$I_d = \left(\frac{W}{L}\right) \mu \, C_{ox} \left(V_{dd} - V_t\right)^2$$

where I_d is the saturation drain current, or the on-current of the device, μ is the mobility of the charge carriers in the semiconductor, V_{dd} is the voltage applied, V_t is the threshold voltage, C_{ox} is the gate capacitance per unit area, W and L are the width and length of the transistor channel. This current needs to charge the output node, which in most CMOS designs, is the gate capacitor of another transistor. The time taken to charge this capacitor to the voltage V_{dd}, with current I_d is given by:

$$T = \frac{C_g V_{dd}}{I_d}$$

where C_g is the capacitance of the gate. It is modeled by multiplying the gate capacitance per area (C_{ox}) with the width and length of the transistor. It is clear from the equation that for the switching time to be less, the drain current needs to be more. The switching speed is a vital parameter in processor design because it governs the speed at which information is processed. Apart from these, another important parameter is the power consumed during processor operation. We know that the energy required for charging a capacitor through a resistor is CV^2, half of which is stored in the capacitor, while the other half is dissipated in the resistor. Now, if the transistors are operated continuously, it is reasonable to assume that this energy will be consumed in switching time T. Thus, the power dissipated by the transistor is given by:

$$P = \frac{CV^2}{T}$$

Now, let us consider the parameters that govern the drain current, the switching speed, and the power dissipation. From the above equation, we have the mobility of the semiconductor channel, the gate capacitance per unit area, the threshold voltage, and the dimensions of the transistor, i.e., the W/L ratio, that govern the current. Of these, the mobility of the channel is the material property that is determined at the time of material selection. The capacitance of the gate depends on the κ-value and thickness of the dielectric layer, which is determined from the process flow. The threshold voltage is determined by the material and doping because of the field required for inversion. Finally, the W/L ratio is determined during processing based on the lithography, or patterning technique used. Thus, material selection and process flow are key to the performance of the transistor. Further, these parameters also determine the speed of the gates formed using the transistor (switching speed), thus determining the speed of the entire processor. The last piece of the processor puzzle is the number of transistors (and hence gates) per unit area of the chip, also known as the device density. This parameter determines the complexity and sophistication of the processor architecture. A higher device density leads to more gates per unit area, allowing for more complex architecture, leading to a processor with more functionality. Thus, semiconductor material and process flow are two independent variables that determine the drain current, switching speed, power dissipation, and device density. These parameters, in turn, determine the performance of the processor. However, it should be noted that these parameters are not independent of each other. For example, changing the dimensions of the transistor to increase device density, also changes its current characteristics and switching speed. Hence, it is important to study the cascading effect of changing one parameter on all the others.

As discussed earlier, the success of silicon CMOS has been because of the dimensional scaling of transistors. Let us consider that we change the linear dimensions of all the components in the chip by a constant scaling factor "s" (>1), as shown in Figure 13.1. This leads to a reduction in the width and length of the transistor, and thickness of the dielectric by the same scaling factor, whereas material properties like mobility are unaffected. The capacitance per unit area increases proportionally because of the decrease in dielectric thickness ($C_{ox} \rightarrow C_{ox}s$), leading to a proportional increase in the drain current, because the (W/L) ratio remains constant. The gate capacitance decreases because the area reduction is more than the gain from the reduction in dielectric thickness ($C_g \rightarrow C_g / s$). This reduction in gate capacitance along with an increase in drain current leads to a much better switching time ($T \rightarrow T / s^2$). These changes bode well for the speed of the processor. However, the problem is with the power dissipation, which increases proportionally to s ($P \rightarrow Ps$). This power is for a single transistor and is now dissipated in a smaller area, leading to a much higher power density ($P_d \rightarrow P_d s^3$). The explosion in power density with scaling would have acted as a major problem in the argument for dimensional scaling. However, the way this situation is mitigated is by scaling the voltage proportionally as well, i.e., $V_{dd} \rightarrow V_{dd} / s$ and $V_t \rightarrow V_t / s$. These changes, applied along with the dimensional scaling, cause the drain current to reduce by a factor of s ($I_d \rightarrow I_d / s$), however, this does not cause a slowing down of the processor because of the reduction in drain voltage. Thus, switching time still reduces by a factor of s ($T \rightarrow T / s$). Further, because of the reduction in voltage and current, the power dissipation reduces by a factor of s^2 ($P \rightarrow P / s^2$), while the power density remains the same. Thus, higher switching speed and better device density are obtained even with the

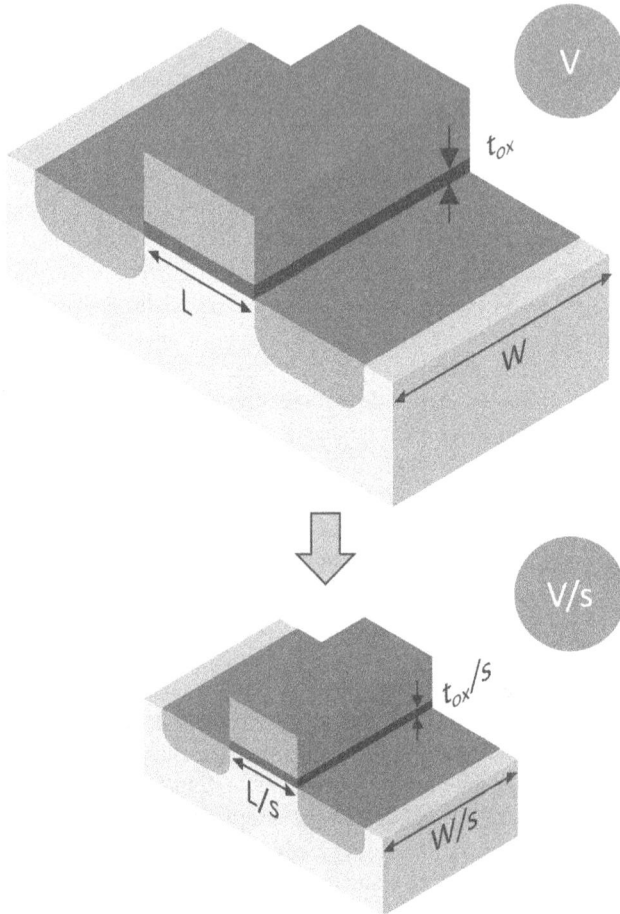

FIGURE 13.1 Dimensional scaling of CMOS transistors reduces all the physical dimensions associated with transistors by a constant factor. The drive and threshold voltages are also reduced by the same amount.

same power dissipation density. This win-win situation is the reason for the success of dimensional scaling of silicon processors. Dimensional scaling of the semiconductor manufacturing process is designated by the size of the minimum feature on the chip. This is known as the technology node or the process node for the particular process. Historically, the minimum feature in a silicon processor was the gate length of the transistor, however, in the past decade, this relationship has not been accurate. Technology nodes are expressed in length terms, for example, 10 μm, 350 nm, 22 nm, and so on. The latest processors are being made using 5 nm node technology.

In the context of flexible electronics, it is important to realize the key enablers for the dimensional and voltage scaling of silicon-based transistors. The first is the use of more advanced lithography processes with every generation of technology. This has led to the continuous improvement in technology nodes from 10 μm in 1970s to 10 nm in 2019. The 1000-fold improvement in size bought improvements in processor speed and the number of transistors in a chip, leading to better functionality. However, to use sophisticated lithography techniques, it is important to have a planar, chemically stable substrate that can withstand several steps of processing. The second enabler has been the use of doping to control the threshold voltage of silicon at a precise point. This generally requires multiple steps of diffusion, ion implantation, and annealing, which is possible with a thermally stable substrate like silicon. It is important to consider these key enablers while designing materials and processes for fabricating complex processors with flexible materials.

13.2.2 CMOS TECHNOLOGY

Transistors are the backbone of all information processing. The state-of-the-art processors are based on CMOS technology, which requires the existence of complementary transistors. In the silicon world, they are referred to as the n-channel and p-channel MOSFETs, depending on the type of carrier conduction taking place in the channel (electrons for n-channel and holes for p-channel). These are complementary to each other in terms of the gate voltage required to switch them on, i.e., p-MOS transistors switch on for a negative gate-substrate voltage (also called gate-source because source and substrate are shorted in most cases) and n-MOS transistors switch on for positive V_{gs}. Thus, if connected together to a common gate, only one of these transistors is on at a given time (Figure 13.2). Thus, if this configuration

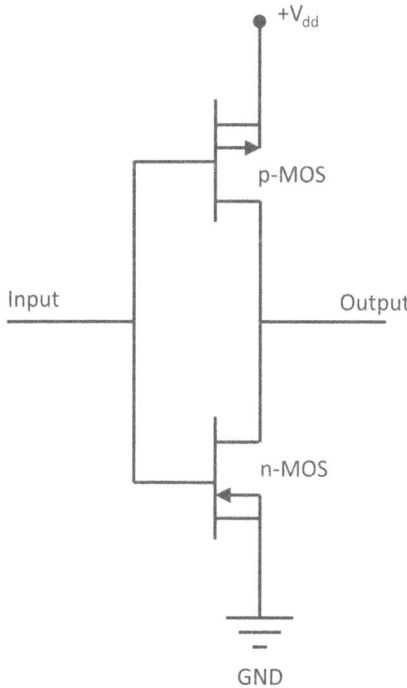

FIGURE 13.2 The CMOS inverter is formed by connected an n-MOS and p-MOS transistor in series with the Vdd and GND. The input is connected to the gate of both transistors, while the drain of both is connected to form the output.

is connected in series between V_{dd} and ground, we can obtain an inverter. The output is connected to ground through the n-MOS when a positive V_{gs} is applied at the gate, whereas it is connected to the V_{dd} through the p-MOS when a negative V_{gs} is applied. Because only one of the transistors is on at a given time, the V_{dd} and GND are never connected to each other. This causes the design to consume very low static power. Power is only needed at the time of switching the output capacitor from low to high and vice versa. Such a configuration can also be formed into digital logic gates such as NAND and NOR, which can then be subsequently interconnected to form large and complex decision-making circuits. Thus, CMOS technology enables the creation of processors, memories, transceivers, and many such circuits using silicon substrates. The key to CMOS technology is the ability of silicon to form stable doped structures that can behave as n-type and p-type semiconductors. The amount of doping can also be well controlled using processes such as dopant diffusion and ion implantation.

13.3   FLEXIBLE TRANSISTORS

From the preceding discussion, it is clear that the key pillars for the success of the processors based on the current electronics manufacturing are dimensional scaling, threshold voltage scaling, and CMOS technology. If a circuit of comparable performance needs to be made using materials other than silicon, say for creating flexible processors, it is important that these three aspects are met. In Chapter 4, we discussed processes to form flexible single-crystal silicon films using techniques such as TPER and spalling, after the wafers have been subjected to standard CMOS manufacturing processes. These flexible silicon pieces obviously support the three aspects mentioned and are comparable in performance with their rigid silicon counterparts. Thus, if we use the same state-of-the-art techniques that are used for rigid processors, followed by a device-first flexible silicon process, a state-of-the-art flexible processor can be obtained. In this chapter, we will focus on non-silicon semiconducting materials, discussed in section II, to access the feasibility of creating high-performance flexible processors.

13.3.1   ORGANIC SEMICONDUCTORS

Popular organic semiconductors rely on the concept of π-conjugation to create electronic molecular orbital states with a certain bandgap. The highest occupied molecular orbital (HOMO) and lowest unoccupied molecular orbital (LUMO) behave in a way similar to the valence and conduction bands in inorganic semiconductors. However, the disordered structure of the organic thin films (both small molecule and polymeric) renders the conduction of charge carriers difficult, leading to hopping as the dominant charge transport mechanism. Further, the energy associated with the LUMO orbitals is high, i.e., the electron affinity of the material is low, leading to free electrons being trapped at reaction sites, and problems with the electron injection barrier being high. Thus, most stable organic semiconductors support p-type conduction at room temperature, however, with advancements in organic electronics, some high mobility n-type organic semiconductors have been discovered to be stable in air at room temperature. Thus, both p-type and n-type conduction are possible in organic semiconductors, which augers well for CMOS integration. For example, polymers based on a (3E,7E)-3,7-bis(2-oxoindolin-3- ylidene)benzo[1,2-b:4,5-b'] difuran-2,6(3H,7H)-dione (IBDF) core have demonstrated very good air stability and n-type mobility of 5-10 cm^2/V-s, which is comparable to p-type organic semi-conductors such as pentacene (Figure 13.3). However, because high electron and hole mobilities are exhibited by different materials, the integration of these thin films onto a single substrate for CMOS device fabrication becomes challenging. This is not the case with silicon, wherein the same material lattice can be selectively doped to form n-type and p-type semiconducting regions.

The non-availability of a single organic thin film for CMOS integration, also makes dimensional scaling challenging because individual organic field-effect tran-sistors (OFETs) need to be fabricated using different thin films. However, because organic polymers and small molecules can be processed into a stable thin film, some lithography processes, such as nanoimprint lithography, can be used to obtain nanometer-scale dimensions. Special care needs to be taken during such processing to make sure that the organic thin film is not sensitive to the chemicals involved. However, the biggest concern regarding the formation of large circuits resolves around the large threshold voltage and turn-on voltage required for OFET transis-tor operation and the dependence of these voltages on the fabrication process. The unreliability in the threshold voltage for an organic thin-film means we cannot scale down the input voltage to the processor reliably, leading to high power density in the circuit. This is particularly problematic for organic transistors because they are tem-perature sensitive, thus, a small rise in temperature because of the power dissipation in the circuit can lead to catastrophic failure. Thus, the high and unreliable threshold voltage, along with the unavailability of a single organic material with tunable con-duction properties are the key challenges in creating large-scale organic processors.

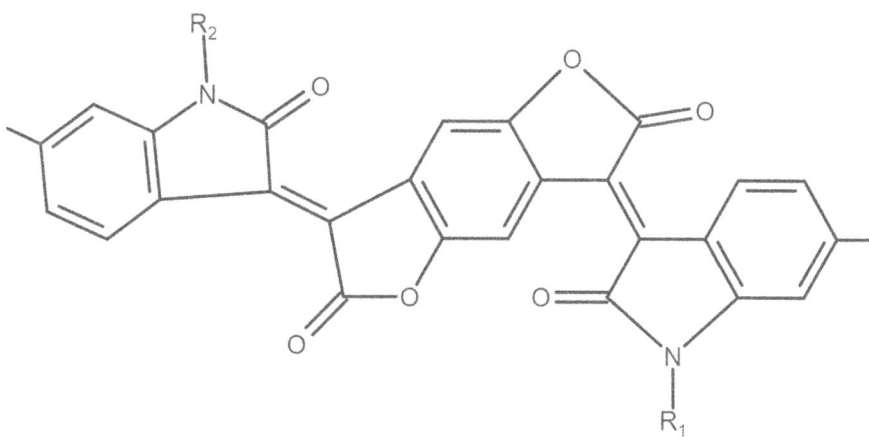

FIGURE 13.3 The (3E,7E)-3,7-bis(2-oxoindolin-3- ylidene)benzo[1,2-b:4,5-b']difuran-2,6(3H,7H)-dione (IBDF) core has been used in several demonstrations of n-type organic semiconductors, with reasonable electronic mobility.

13.3.2 Metal Oxide Semiconductors

Metal oxide semiconductors are promising materials for thin-film transistor fabrication owing to high carrier mobility and better air stability compared to organic semiconductors. The formation of chemically stable thin films using processes such as sol-gel also provides an opportunity to use high-resolution state-of-the-art lithography processes to obtain dimensional scaling. Thus, large-area high-resolution flexible circuits can be obtained using metal oxide thin films as the semiconductors with the processes discussed in previous sections. However, in this case, as well, we struggle to find materials that transport both electrons and holes effectively. Because conduction band minimum in the metal oxide thin film consists mainly of metal s-orbital, the dispersion is better, leading to better electron mobility. Conversely, the valence band maximum consists of the oxide p-orbitals, which are more localized, leading to poor hole mobility. Thus, typical metal oxide thin films such as ZnO, are mostly n-type semiconductors. In recent times, the problem of obtaining high hole mobility in metal oxide semiconductors has been solved by using metals that introduce filled s-orbitals near the valence band maximum energy levels, leading to better dispersion and better hole mobility in the material. Thus, metal oxides such as nickel oxide (NiO_x), copper (I) oxides, and SnO are being studied as promising p-type semiconductors. Metal oxides have low intrinsic carrier concentration compared to crystalline silicon, owing to the fact that the defects in the thin film that serves as donor or acceptor sites have energy levels deep within the bandgap of the semiconductor. The intrinsic carrier concentration in silicon at room temperature, for example, is $10^{10}/cm^3$, whereas that in ZnO thin film is $10^6/cm^3$. This intrinsic carrier concertation can vary depending on the defect density, which depends on the deposition process used for thin-film fabrication. For example, ZnO thin films deposited using atomic layer deposition tend to have lower defects compared to those deposited using sputtering. Further, because these types of energy states (arising out of defects) are embedded deep within the bandgap, they may cause problems of charge trapping during device operation. The inherent randomness of the defects also makes it difficult to obtain the required device performance reliably. Thus, carrier concentration is introduced in metal oxides by doping them with external impurities. This is particularly of interest because extrinsic doping can be used to increase hole carrier concentration leading to p-type conduction in metal oxide thin films. Thus, the same material can be selectively doped to obtain n-type and p-type on the same substrate, opening up the possibility of CMOS integration. Molecular beam epitaxy (MBE) or pulsed laser deposition (PLD) are commonly used to produce doped ZnO thin films in both polarities. While n-type ZnO is easier to produce using access Zn during deposition, or with dopants such as aluminum, gallium, or indium, p-type ZnO has only recently been reported. This is usually attributed to the fact that deep-level donor defect sites compensate for the extrinsic shallow level acceptor site, leading to a charge trap. However, recent studies have shown that group V elements, such as nitrogen, can be used to produce p-type ZnO. However, the mobility of the holes remains low because of the underlying electronic structure of the valence band maximum.

While the dimensional scaling and CMOS integration of metal oxide thin films looks promising, the threshold voltage engineering of these materials has been a key bottleneck in their large-scale use. This is owing to the fact that the bandgap of typical metal oxide thin films is large, thus requiring a higher surface potential at the semiconductor-oxide interface for the onset of drain current. This leads to a higher gate voltage required for switching, thus a higher threshold voltage. With recent advancements, the problem of the high threshold voltage is being solved by engineering the doping levels, deposition processes (defect levels), and film thicknesses. However, a bigger problem with metal oxide threshold voltage is its drift over time. This leads to unreliable circuit operation over the long term, reducing the commercial viability of creating large circuits using metal oxide thin-film semiconductors. The drift in threshold voltage typically comes from charge trapping at the oxide-semiconductor interface. The presence of a large number of defects at the interface leads to more charge traps, leading to the threshold voltage drift. In recent times, the quaternary metal oxide, indium gallium zinc oxide (IGZO), has gained popularity because of high threshold voltage stability, better defect density over a large area, higher electron mobility, and control over composition using the PLD deposition process. However, strategies for obtaining p-type conduction with IGZO channel transistors are still being developed.

13.3.3 Nanostructured Semiconductors

There are several promising semiconducting materials in the nanostructured basket to potentially replace silicon as the preferred semiconductor for processor design. These include silicon nanowires themselves, but also carbon nanotubes, graphene, transition metal dichalcogenides (TMDs), and so on. In the case of nanotubes/nanowire structures, the key challenge is the lack of a strategy for obtaining well-defined predictable deposition on a planar substrate to form the circuit. This can be associated with the problem of dimensional scaling because aggressive scaling is not possible without the semiconductor channel position being predetermined in a predictable way. At an individual strand level, both CNT and SiNW are themselves nanoscale semiconducting channels owing to their structure. However, the way in which CNTs and SiNWs arrange themselves on planar substrates needs to be predictable in order for the transistor structure to be defined using lithography. However, one way of obtaining a reasonable large-scale circuit using CNT-based transistors was showcased in 2013, using aligned CNTs grown on a quartz substrate [7]. The aligned CNTs were transfer printed on a destination substrate and the transistor structure was completed. The metallic CNTs were removed using electrical breakdown. This 1-bit processor consisted of 178 CNT FETs interconnected together. In 2019, a 16-bit RISC-V-based processor with 14,000 CNT transistors was demonstrated using solution-based deposition of CNTs on a pre-patterned substrate [8]. While these processors were not flexible, the processes used to create them can be followed using a flexible destination substrate to yield a flexible CNT-based processor. For CMOS integration, the unique electronic properties of CNTs are leveraged to promote the injection of a specific kind of carrier by changing the source-drain contact metal. Thus, the same nanotube can be used as an n-type or p-type transistor depending on the contact metal and gate stack, providing a facile method for CMOS integration on the same substrate. In the case of silicon nanowires, there is an advantage of creating vertically aligned nanowires that are doped in-situ during the CVD growth process to obtain nanowires ready to be used as transistors. However, reliable transfer printing on flexible substrates along with large-scale circuit fabrication is yet to be demonstrated for silicon nanowires.

Transition metal dichalcogenides are a class of 2D atomic crystal structure materials that hold substantial potential as semiconductors for flexible processor applications. In principle, the atomically thin sheets can be used to create highly scaled circuits without the need for alignment. However, the key issue with TMDs is the availability of a single crystal large area monolayer film for creating such circuits. Graphene, which is also a 2D material, does have a process for the fabrication of single crystal monolayers, however, because of the zero-bandgap, many additional steps such as patterning nanoribbons, etc., need to be completed before it can be used as a semiconductor. On the other hand, the CVD process used for growing TMD materials does not result in single-crystal thin films because of simultaneous nucleations of random orientations on the substrate. This causes grain boundaries to occur when these individual flakes grow and eventually meet. In the absence of a process for creating high-quality TMD monolayers, it is still unclear how large-scale circuits, with thousands of transistors, if not more, can be reliably fabricated. Thus, at present, the work related to TMDs such as MoS_2, WS_2, $MoSe_2$, etc., concerns the fabrication and characterization of single or few transistor circuits to determine important parameters such as mobility, threshold voltage, carrier concentration, and so on (Figure 13.4). These studies, along with advancements in thin-film fabrication processes, can lead to the development of flexible processors based on TMD semiconductors.

13.4 FLEXIBLE CIRCUITS

The final piece of the puzzle for creating a large circuit, such as a processor, is the contact metal used for charge injection into the transistor channel and for interconnecting the transistors to each other. These metal contacts need to form "ohmic contacts" with the transistors so that there is no barrier for the injection of carriers into the transistor. Ohmic contacts are junctions between two disparate conductive mediums such that there can be non-rectifying charge transfer in both directions, i.e., the current-voltage characteristic of the junction is linear. Whether a metal will form such an ohmic contact with a semiconductor depends on the Fermi energy levels of the metal and the semiconductor (their work functions). In the case of n-type semiconductors, the work function of the metal should be less than that of the semiconductor to form Ohmic contacts, whereas, in a p-type semiconductor, the work function of the metal should be more. This may lead us to believe that we will require different metals for contacting the source and drain of n-type and p-type transistors. This apparent dichotomy is solved in the silicon CMOS process by heavily doping the region close to the metal-semiconductor interface to increase carrier concentration and by alloying the silicon at the metal interface with metals such as nickel, tungsten, or titanium. These steps ensure the formation of Ohmic contact with the same contact metal. However, such detailed engineering of contact formation has not been carried out for non-silicon materials that may be used for flexible processor fabrication. This is because the physics of the injection of charges in amorphous

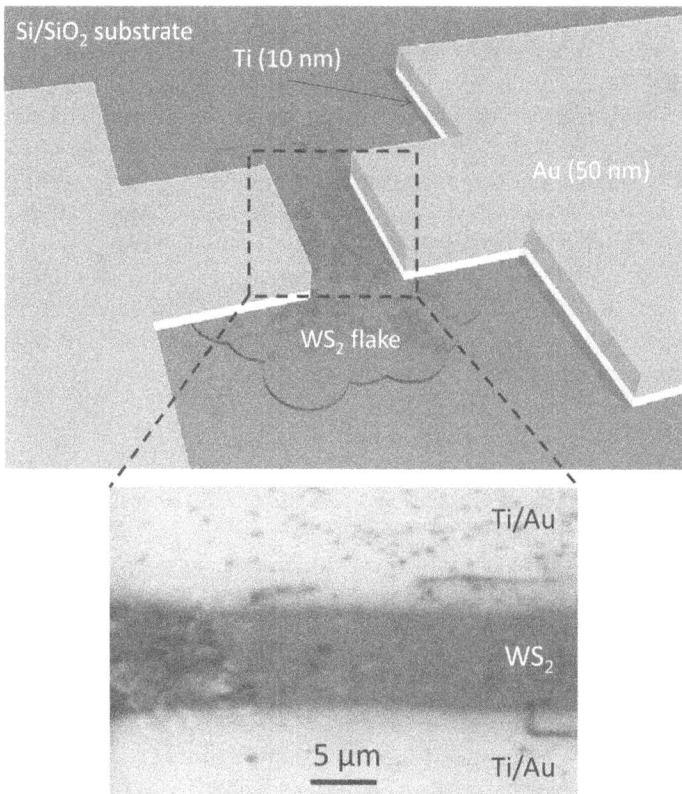

FIGURE 13.4 Illustration and microscope image of TFTs fabricated using WS$_2$ flakes, a member of the transition metal chalcogenide family. The integration of such individual TFTs into a large-area circuit for flexible processor fabrication remains a challenge. (© [2013] IEEE. Reprinted, with permission, from [9]).

semiconductors is not well understood because of the difference in charge transport mechanisms going from crystalline to amorphous semiconductors. This makes it challenging to reliably contact metal oxide or other amorphous semiconductors. The problem persists in the case of organic semiconductors because of yet another charge transport mechanism. Further, the presence of a large number of organic semiconductors with different electronic characteristics leads to the lack of exhaustive experimental data for any metal-semiconductor combination. The problem of charge injection in organic semiconductors is partly solved by involving multiple layers of different organic materials between the metal and the semiconductor to facilitate charge injection. This strategy is particularly popular in the fabrication of organic light-emitting diodes but involves different metals for anode and cathode interconnection. In the case of CNT transistors, the formation of potential barriers at the metal-semiconductor interface is in fact used to create n-type and p-type transistors. This complicates the integration process, but the additional steps required for disparate contact metal deposition are offset by the removal of the doping steps.

We discussed that the success of large-scale flexible circuits made using non-silicon semiconductors depends on the ability of these materials to form dimensionally scaled CMOS devices. Presently, many promising candidates are being investigated as potential replacements for silicon – metal oxides, organics, TMDs, and so on. However, there remain key challenges in the integration of these materials into large-scale CMOS circuits that could challenge the supremacy of silicon. At present, it seems device-first flexible silicon processes are the most promising for obtaining state-of-the-art processor capability in a flexible form. However, non-silicon semiconductors have already found mainstream commercial success in other areas of electronic system design such as memory, displays, photovoltaics, energy storage, and so on. These use cases do not require large-scale CMOS integration, dimensional scaling, or precise threshold voltage control.

EXERCISES

13.1. Explain the effect of dimensional scaling of CMOS transistors, and the voltage, on the drive current, speed, and power dissipation.

13.2. Assume we fabricated a transistor with a semiconductor with mobility of 1000 cm^2/V-s, the dielectric thickness of 2 nm, κ-value of 10, V_{dd} of 3 V, V_t of 1 V, and $W = L = 100$ nm. What is the saturation drive current expected from the transistor? If this transistor is used to drive the gate of another such transistor, how much time is needed to charge it to V_{dd}? What is the power dissipated during switching?

13.3. If the transistor of question 13.2 is scaled by a factor of 2, what are the new values of drive current, charging time, and power dissipation?

13.4. What are the key properties that a material system needs to possess to obtain large-area CMOS circuits? Compare the following material systems on these properties: silicon, organic semiconductors, metal oxide semiconductors, nanostructured material systems.

13.5. Describe the phenomenon of ohmic contact between a metal and semiconductor for n-type and p-type materials.

14 Flexible Memory

14.1 INTRODUCTION

A typical electronic system mainly consists of a processing unit along with some memory to store information relating to the code to be executed and the information being processed. These storage systems are typically designed separately as volatile memory (temporary storage for the information being processed) and non-volatile memory (permanent storage for the program and other critical information). Volatile memory only retains information so long as the power is supplied, whereas non-volatile memory can retain information in the absence of power. Ideally, we could have designed everything as non-volatile, such that there is retention of information at all times, however, a key reason for the use of volatile memory is that non-volatile memories are slow to respond, creating a potential bottleneck in information processing. Further, non-volatile memories have a limited number of switches, i.e., the data written in non-volatile memory cells can only be erased and rewritten a specific number of times during their lifetime. While this number can be in the thousands, it is much higher for volatile memories. Thus, volatile memories are used as a cache for the information being processed because of their quick access and easy switching. Within volatile memories, there are some memory blocks that are positioned very close to the processor for immediate access and information caching, called level 0 (L0). These are generally formed using arrays of registers. These blocks obtain information from level 1 storage, called L1 cache, which is typically made using SRAM cells and is located on the same chip as the processing units. It has a larger storage space, compared to L0 cache, to provide for more information to be buffered for the processor to fetch from. The main memory, also called primary memory, or RAM of the system is the level 2 storage that typically exists as a separate memory chip. Finally, the disk storage, or the non-volatile memory, is the level 3 storage that provides permanent information such as coded instructions, files, data, and so on. These levels typically form a pyramid, as shown in Figure 14.1, because the storage available increases as one goes from level 0 to level 3. For example, a processor contains a register bank of a few hundred bytes at level 0, and a volatile memory of hundreds of kilobytes to few megabytes at level 1. The main memory (at level 2) these days consists of a few gigabytes, whereas the non-volatile disk storage (at level 3) is typically a few terabytes.

With the advancement of manufacturing technology, the density of both volatile and non-volatile memories has significantly increased. Memory density typically refers to the number of bits stored per unit area of the chip. This is particularly relevant for L0 and L1 volatile memories because they are positioned along with the processor inside the silicon chip, thus accounting for a portion of the silicon floor. This will be an important point to remember while designing the integration strategy for flexible electronic systems, particularly from the memory point of view. The ability

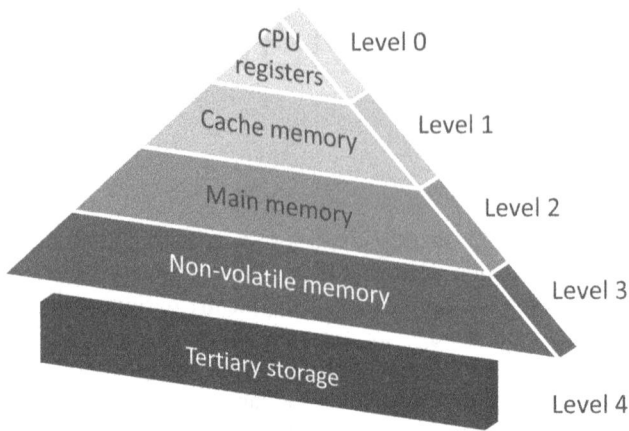

FIGURE 14.1 The levels of memory are typically depicted as a pyramid because the storage size increases going from level 0 to level 3. The memory units are lower levels are smaller, closer to the processor core, and faster.

to integrate L0 and L1 cache depends on the strategy used for creating the processor because the same integration process flow also accounts for the creation of these volatile memory blocks. The integration of these blocks then follows directly from the integration of the processing unit. The material system and processes used to create the processor circuit can be used to create the circuitry required for these levels because they are based on cross-coupled gate structures such as latches and flip-flops for information storage. The integration strategy for level 2 and level 3 memory can be different because they are external to the processor chip. Hence, in this chapter, we will discuss the level 2 and level 3 memory, and the integration strategies for making them flexible.

14.2   VOLATILE MEMORY

14.2.1   Conventional Volatile Memory

The colloquially used term for the volatile "main memory" module is RAM or random-access memory. The name comes from the fact that any word in the memory can be accessed at any time, as opposed to direct access for some of the level 3 memory technologies such as magnetic tapes, hard disks, and CDROMs, where the time of access depends on the physical location of information storage. RAM memory chips are typically based on SRAM (static RAM) or DRAM (dynamic RAM) cells that are used to store a single bit of information and form the basic building blocks of these chips. SRAMs are based on latching circuitry typically consisting of several transistors in a single cell, whereas DRAMs are based on a single capacitor whose charge indicates the state of the information stored and is usually controlled using a transistor. Thus, a DRAM cell is much smaller than an SRAM cell, but needs to be continuously refreshed (hence the name dynamic) because leakage in the capacitor can cause the information to be lost, even when power is available. Because of their smaller size and hence higher memory density, DRAM architectures have become very popular for modern RAM chips. However, the charge leakage and continuous refresh cause DRAM cells to consume more power compared to SRAM, limiting their application in power-constrained systems.

It may seem that memory structures involve similar components (transistors and capacitors) compared to a processor chip, however, the processes involved in fabricating memory chips can be quite different. For example, DRAM cells require deep trench capacitors (DTCs) that are specifically designed to be placed close to the DRAM cell and provide high capacitance per layout area. The capacitance and leakage current of the capacitors are key in determining the refresh rate and dynamic power needed for the DRAM cell, thus, making the integration of an efficient capacitive structure critical. The capacitance of a capacitor can be increased by increasing the area of the plates and decreasing the distance between them, both of which can be achieved efficiently using a DTC. A standard DRAM cell is shown in Figure 14.2. Writing information onto the DRAM capacitor is a relatively simple operation. The word line is activated to connect the capacitor to the bitline, which can charge or discharge the capacitor, which is equivalent to writing a 1 or 0 bit in the cell. To read from the cell, the bitline is precharged to a voltage halfway between the two logic levels (say $V_{dd}/2$). The precharge circuit is disconnected and the wordline is then activated to connect the capacitor to the bitline. If the capacitor had been charged (thus storing the value logic 1), charge flows from the capacitor to the bitline, while if the capacitor had been discharged, charge flows from the bitline to the capacitor. The capacitance of the bitline is typically much larger than that of the cell capacitor, thus, this flow of charge changes the voltage on the bitline by only a few

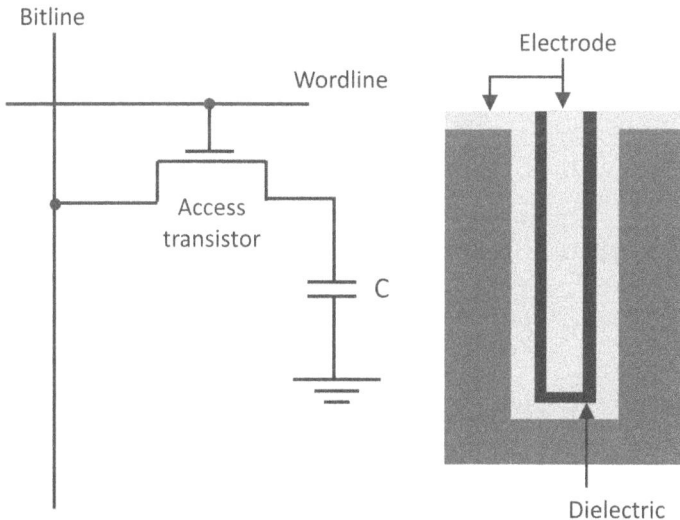

FIGURE 14.2   Structure of a typical DRAM cell consists of a capacitor, which stores the binary information as its charge state and an access transistor. The capacitor is typically fabricated as a deep trench capacitor (DTC) using special integration techniques.

tens of millivolts. This small change in bitline voltage is detected by sense amplifiers at the end of the line, converting this information into logic levels 0 or 1. The read operation discharges the DRAM capacitor; thus, it needs to be rewritten after every read operation. This is in addition to the periodic refresh required to counter the charge leaking from the capacitor. This process makes the DRAM interface circuit much more complex compared to that of SRAM, however, the increased cell density in DRAM makes up for the chip area lost to the interface circuitry.

A typical SRAM cell is shown in Figure 14.3. It consists of six transistors – four transistors are organized as two inverters connected in a positive feedback loop, with the remaining two transistors serving as access transistors for read and write operations. The inverter feedback loop is a bistable configuration that can keep the voltage at the output stable, as long as power is provided to the circuit. The configuration also provides immunity from electrical noise because the voltage levels for change in inverter output are typically higher than the noise voltage. The write operation is performed by activating the wordline and driving the output of the bistable system to the desired value through the bitline. This requires careful design of the bitline driver, the invertors, and the access transistor so that the bistable system can be successfully overturned by the bitline. To read the stored bit, the bitlines are precharged at V_{dd} voltage (logic 1), following which, the precharge circuit is disconnected. The wordline is then activated to connect the bistable system to the charged bitline. Because one of the outputs of the system is zero, the charge on the corresponding bitline decreases, which can be detected using a sense amplifier connected to the bitlines. The wordline is then deactivated and the bistable configuration recovers the charge lost to the bitline. Thus, this configuration does not require refresh circuitry

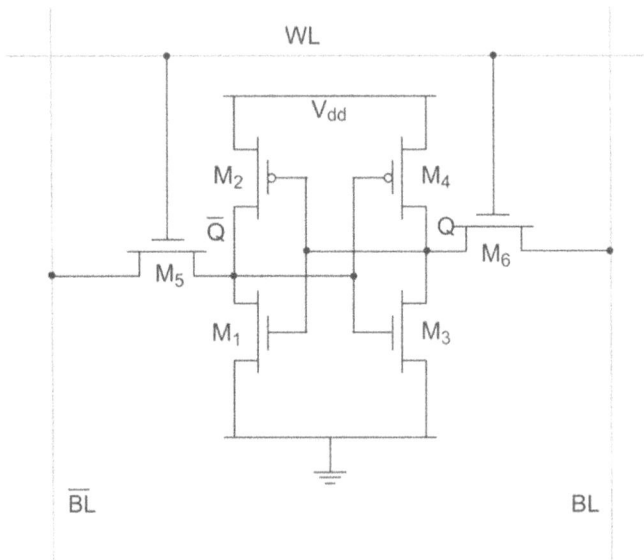

FIGURE 14.3 Circuit diagram of a six transistor (6T) SRAM cell. (Reprinted from Wikimedia Commons under the Creative Commons Attribution-Share Alike 3.0 Unported.)

or rewriting of information after the read cycle, and consumes less power compared to DRAM, however, the number of transistors per cell, and hence the area of the cell is much higher.

14.2.2 FLEXIBLE VOLATILE MEMORY

While the individual cell is a relatively simple construction in both SRAM and DRAM, the complexity of the read/write circuit creates a problem while creating flexible versions of volatile memory. In particular, the read operation in both SRAM and DRAM cells requires the use of high precision sense amplifiers, which can be difficult to integrate on a flexible platform. Hence, reports in the literature pertaining to demonstrations of random-access memories only focus on the implementation of the cell, particularly as a technology demonstration for a novel thin film or integration process, without implementing the interface circuits. For example, in 2011, Fukuda et al. fabricated an SRAM cell using Dinaphthothienothiophene (DNTT, $C_{22}S_2H_{12}$) as the semiconductor [10]. This and other such demonstrations have been important first steps, following which larger arrays of SRAMs were fabricated on flexible substrates. In the few cases when large arrays or interface circuits have been demonstrated along with the memory cell array, the integration has been performed using thin-film transistors (TFTs) based on large-area uniform thin films such as metal oxides, polymers, and polysilicon. In these integrations, the thin film is typically deposited on a flexible substrate using a low-temperature process such as sputtering. The memory cell and the interface circuitry are then fabricated using standard lithography processes. For example, Ebihara et al. demonstrated a flexible 16 kb memory using TFTs based on low-temperature polysilicon as the semiconductor [11]. The devices were fabricated on a glass substrate and transferred onto a 0.2-mm thick flexible plastic substrate. This system, consisting of 110,000 transistors, included the interface circuit on the flexible platform. Another such example is the demonstration of a 128-bit SRAM array (16x8), along with the interface circuit, fabricated using amorphous indium gallium zinc oxide (a-IGZO) as the semiconductor on a flexible polyimide substrate [12]. Flexible volatile memory demonstrations typically feature SRAM cells because of the simplicity of the interface circuit compared to DRAMs, and the absence of a refresh circuit.

14.3 NON-VOLATILE MEMORY

14.3.1 CONVENTIONAL NON-VOLATILE MEMORY

Non-volatile memory (NVM) refers to the level 3 of memory hierarchy that deals with permanent storage, i.e., the stored information is retained even with a loss of power. This generally involves causing some permanent "damage" to an aspect of the memory. For example, pits dug inside a plastic substrate (CD ROMs) or magnetization of a thin film (hard disk drives), can serve as permanent storage of information. These are generally decoded using techniques specially designed for the particular medium of storage. For example, optical drives are used to read from CD ROMs and an actuator arm with a magnetic head is used to read from HDDs. These methods have given way to modern methods that do not require moving parts for read/write operation and are entirely contained inside the chip architecture. These solid-state drives (SSDs) are gaining popularity because of their memory density and speed, along with better reliability (no moving parts). These memory architectures rely on the same electronic components as volatile memory, such as transistors, capacitors, resistors, along with a special material whose properties can be changed to store information permanently. This change to the thin film should follow some ground rules: it should be brought about using standard voltage and current levels; it should be reversible to obtain an erasable and rewritable memory; it should be permanent, i.e., once inflicted, it should not reverse on its own; and we should be able to read the damage done as information using standard electronic components. Some of the technologies satisfying these rules include resistive RAM (ReRAM), ferroelectric RAM (FeRAM), phase change RAM (PCRAM), and flash memory. It should be noted that "RAM" here refers to the fact that information can be retrieved non-sequentially, unlike in the case of CDROM or HDD, however, these are not to be confused with volatile memory systems that are also colloquially referred to as RAM. These technologies provide the benefits of fast access, permanent information storage, and low power consumption, along with the possibility of integration with state-of-the-art silicon CMOS processors. Hence, we may see penetration of SDD NVM technologies from level 3 to level 2 of the memory hierarchy in the future.

All the NVM technologies work on the principle of causing a permanent change in a particular material thin film for information storage. In resistive RAMs, or memristors, the resistance of an oxide thin film is changed by causing a soft breakdown by applying a set voltage. The application of reset voltage reverses the condition of the thin film. A high resistance oxide is read as logic 0, while a low resistance is read as logic 1. A similar approach is followed with FeRAMs, the key difference being the switching of the magnetic polarization state of a thin film to store information. This can also be achieved by applying appropriate voltages across the thin film. In the case of phase change RAM, the phase of a thin film is changed from amorphous to crystalline and back by applying pulses of current, or through optical excitation. Flash memory has the same structure as that of a MOSFET, with the difference that there is an additional isolated thin film in the gate stack acting as the floating gate (FG). This conductive thin film is isolated from the channel using a tunneling oxide layer and from the gate metal using a blocking oxide layer. The threshold voltage of the transistor depends on the number of charges trapped inside the FG layer. Information stored in the transistor is changed by causing charges to tunnel into the conductive layer through the semiconductor channel and change the threshold voltage of the underlying transistor. Flash memory is the most commonly used NVM technology for manufacturing solid-state drives, memory cards, USB drives, and so on. These technologies are schematically illustrated in Figure 14.4.

Several different architectures can be employed for creating a non-volatile memory from any of the aforementioned technologies. These most often include a transistor as the access transistor controlling the access to the storage cell, and the NVM technology element in the form of a separate resistor/capacitor or acting as the gate of the transistor. Further, even with the same memory cell structure, there

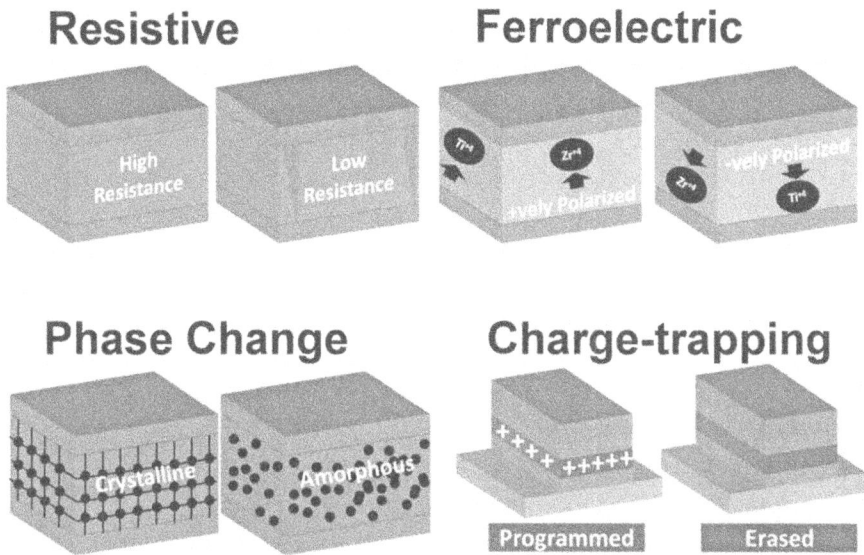

FIGURE 14.4 Various technologies used for creating non-volatile random-access memories. (Adapted under the Creative Commons CC BY License from [13].)

can be differences in the way they are connected together to form the memory array. For example, FG flash memory cells can be connected in series between the bitline and ground, or each cell can be individually connected to the bitline and ground terminals. In the former case, the bitline is only pulled low if all the wordlines are high, resembling NAND logic operation, hence it is called NAND flash architecture. In the latter case, when one of the wordlines is brought high, the entire bitline is pulled low, resembling NOR logic operation, hence it is called NOR flash architecture. Finally, there is the crossbar architecture, where the bitlines and wordlines form a crossbar array with the NVM technology element, such as the resistive oxide, at the cross-over points. These architectures are schematically illustrated in Figure 14.5.

Given the diversity in NVM technologies and their architectures, it is important to understand the figures of merit used to compare them. One of the most important parameters is the memory density, i.e., the number of bits that are stored per unit area of memory. For a successful NVM technology, the ability to write, erase and rewrite information reliably is an important attribute. This is captured through the endurance and retention of the technology. Endurance is the number of write/erase cycles that the memory can be subjected to before significant damage sets in, whereas retention is the amount of time for which the information, once written, is stored in memory. The reliability of the read/write operation is also related to the memory window, which measures the differentiation between the two logic levels stored in memory. Finally, it is important to have fast memory operation, which is represented using the read, write, and switching speed of the memory technology.

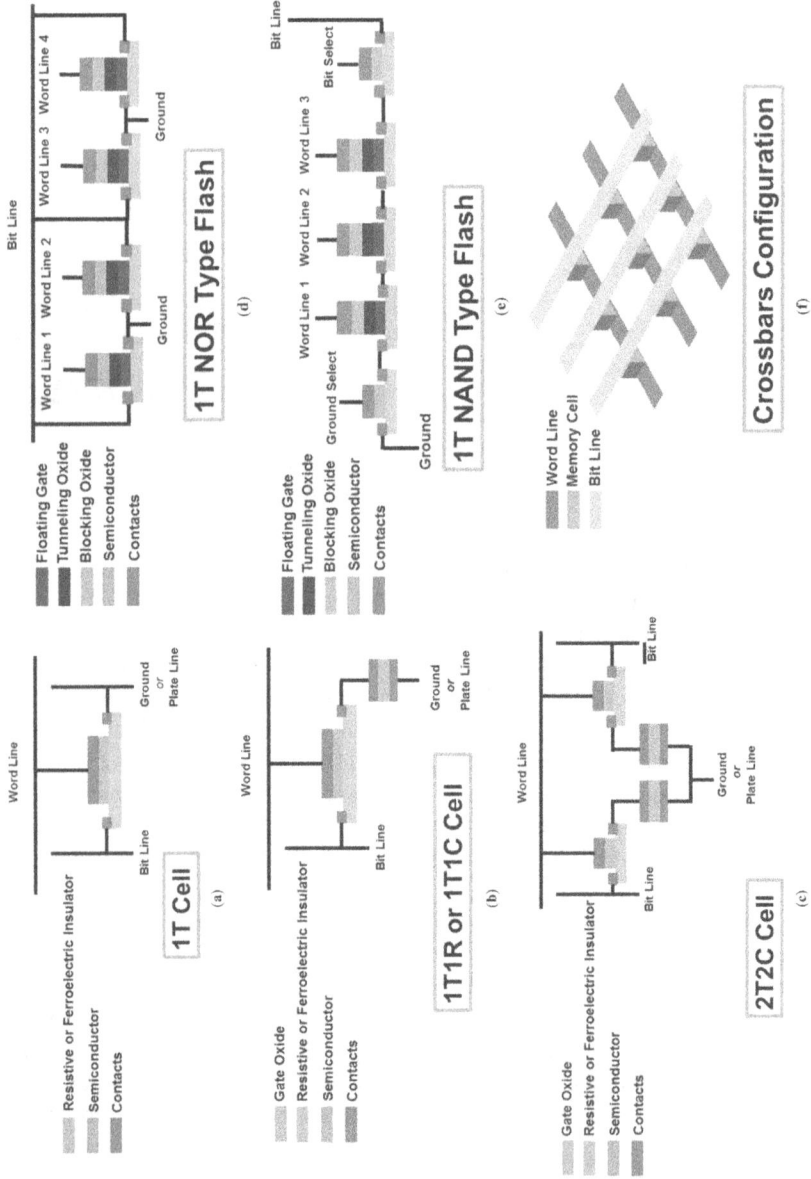

FIGURE 14.5 Commonly used non-volatile memory architectures. (Reprinted under the Creative Commons CC BY License from [13].)

14.3.2 FLEXIBLE NON-VOLATILE MEMORY

14.3.2.1 Flexible ReRAM

Resistive RAMs or memristors are non-volatile memory elements that store the logic levels using two resistance states – high and low. These states are set or switched during the write cycle and sensed during the read cycle. The memristor array can be implemented in a crossbar architecture, providing very high data density. The simplicity of this design has caused memristors to attract a lot of attention from the research community in the past decade. Further, the variance in resistance of the memristor element is similar to the behavior of brain synapses. It is well known that highly used pathways in the brain offer lower resistance to signal transfer, providing scientific credence to the phrase, "practice makes perfect". The potential use of memristor elements as brain synapses can lead to the design of a structure similar to that of the brain cortex. These potential applications have further sparked interest in creating flexible memristive structures.

The thin film being used as the memristor element in ReRAMs is the most essential component of the device, defining most of its performance parameters. A crossbar architecture consisting of orthogonally arranged bitline and wordline conductors is used to read/write onto the memory. The memristive element is sandwiched between the bitline and wordline conductors. Such an architecture is called the 1R configuration because each memory cell contains a single resistive element. To change the resistance of the element, a large potential is applied across the bitline and wordline conductors. The applied electric field causes structural changes in the thin film, causing its resistance to change permanently. The resistance is sensed by applying a voltage across the bitline and wordline conductors, however, the read voltage is not sufficient to cause changes in the film structure. To switch the written bit, an inverse electric field is applying across the element. An example of a crossbar architecture-based flexible ReRAM was demonstrated by Ji *et al.* in 2010 [14]. In this work, an 8×8 memristor array was fabricated using a composite of polyimide (PI) and 6-phenyl-C61 butyric acid methyl ester (PCBM) as the active element, with aluminum conductors on a PET substrate. The voltage for setting the bit, or switching the resistance of the thin film to a low resistance state, was reported to be 3.5 V. At this voltage, we have transport states causing the injection of charges from the metal electrodes and their transportation through the film. The voltage for resetting the bit, or switching to a high resistance state, was reported to be 5 V. At this voltage, the deep level charge traps are filled, leading to the buildup of space charge that opposes charge injection. This charge buildup is stable and remains even after the electric field is removed, leading to the non-volatile behavior of the thin film. A low voltage level of 0.3 V was used to read the state of the memory element without causing changes to its state. This general mechanism pertaining to space charge formation and charge injection is also valid for other organic and inorganic memristive thin films.

The simple crossbar architecture does have a problem, called the sneak path current problem, that refers to the current that flows through three adjacent cells when a particular wordline/bitline combination is activated (Figure 14.6). Because the direction of current in at least one of the sneak path cells needs to be reversed

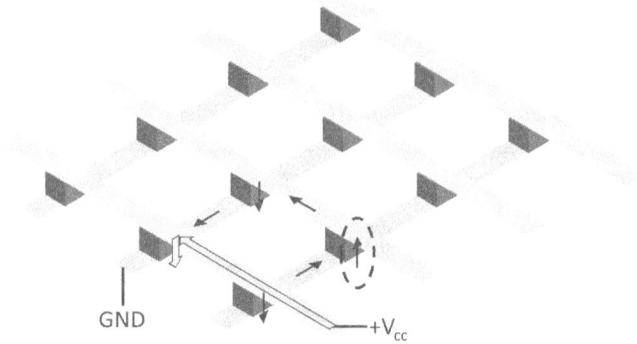

FIGURE 14.6 In a crossbar architecture, apart from the primary current going from the Vcc to ground through the resistive element being probed, there can be a "sneak path" current that goes through three (or more) different resistive elements. The current through one of the elements needs to be in a direction opposite to the expected one, for the sneak path current to form.

compared to that in the desired cell, a simple solution to the sneak path problem is to have a diode in series with every cell element, limiting the current in the reverse direction. This architecture is known as the 1D1R configuration. An example of a flexible ReRAM based on the 1D1R configuration was presented by Yoo *et al.* in 2014 [15]. In this work, copper oxide, deposited using ICP-RIE system at room temperature, was used as the memristive element with aluminum and copper contacts. The diode was fabricated using single crystal silicon transferred from an SOI substrate onto the Kapton substrate. The silicon was doped to form a p-n junction diode to resist sneak path current. Silicon doping was completed before transfer to the plastic substrate because it is a high-temperature process. The copper oxide layer undergoes formation and rupture of conductive filaments, upon application of an electric field, giving the low resistance and high resistance states of the resistive RAM. The set and reset voltages were reported to be 2.5 V and 1.8 V, respectively, with a reading voltage of 1 V.

14.3.2.2 Flexible FeRAM

FeRAMs operate by changing the magnetic polarization of the NVM element using applied electric fields. The key advantage of FeRAMs over memristors is the endurance and retention of the stored data. This pertains to the physics these devices use for data storage. In the case of memristors, the resistive state is dependent on the space charge in the deep level charge traps or the oxide vacancies created due to the applied potential. These states are supposed to be permanent, however, charges can migrate over time to flip the state, creating retention issues. Further, the switch in the resistive state requires a soft "breakdown" of the thin film or its reversal, reducing the endurance of the memristor device. In the case of FeRAMs, the applied electric potential realigns the magnetic regions of the ferroelectric thin film, which provides better retention and endurance. Hence, they can be integrated into a simple conductor/ferroelectric/conductor structure to obtain a bistable change in magnetic polarization in the ferroelectric material upon application of electric field through the conductors. These ferroelectric capacitors can form the crossbar architecture discussed previously, usually with an access device such as a transistor to avoid the sneak path current problem. An example of a flexible ferroelectric RAM integrated into a ferroelectric capacitance architecture (1C configuration) was demonstrated by Ghoneim *et al.* in 2015 [16]. In this work, the ferroelectric thin film used was lead zirconium titanate ($Pb_{1.1}Zr_{0.48}Ti_{0.52}O_3$, PZT) deposited on silicon substrate using the sol-gel process, while platinum thin films, deposited using sputtering, were used as the conductors. After device fabrication, the silicon substrate was finally flexed using the trench-etch-protect-release (TPER) process. The fabricated devices showed endurance for 10^9 read/write cycles, compared with typical values of 10^6 cycles in memristive devices.

Ferroelectric thin films with readily reversible polarities are difficult to obtain, thus, most of the FeRAMs demonstrated in the literature rely on the same thin films. Apart from PZT, an organic material known as PVDF-TrFE is commonly used as a ferroelectric thin film for FeRAM applications. PVDF-TrFE, or poly(vinylidene fluoride-trifluoroethylene), is a copolymer consisting of vinylidene fluoride (VDF) and trifluoroethylene (TrFE) monomers. The copolymer contains randomly oriented dipoles because of the presence of a heavily electronegative fluorine atom. These dipoles can be aligned with the application of an external electric field, resulting in a remanent polarization. An example of a flexible FeRAM based on PVDF-TrFE as the ferroelectric thin film was reported by Yoon *et al.* in 2011 [17]. In this work, the structure was based on a transistor configuration (1T), wherein the ferroelectric material thin film is used as an addition to the gate stack. The transistor was fabricated on a flexible PEN substrate using ALD-deposited ZnO thin film as the semiconductor and Ti/Au/Ti as the source/drain contact metal. In such a structure, information is stored in the form of polarization of the ferroelectric gate insulator layer, which changes the turn-on voltage of the underlying transistor. This change in V_{on} can be used to sense the stored information. The wordline is typically connected to the gate, whereas the source and drain are connected to bitline and ground. When wordline is activated, the current in the biltine depends on the polarization of the ferroelectric layer, which is sensed to determine its state. For the write operation, the polarization is set by applying a specific electric field across the gate and the channel.

14.3.2.3   Flexible PCRAM

Phase change memories rely on the change of crystalline state of a thin film from amorphous to crystalline, or vice-versa, to store information. These materials exhibit high resistance in the amorphous state and low resistance in the crystalline state, which is used to represent the two digital states of binary information. The architecture used can be the same as memristors, because the primary property being changed is the resistance of the thin film. The transitions from one state to another are achieved by applying a temperature pulse to the thin film. The highly localized nature of the phase change leads to very high integration densities. An example of a flexible phase change memory was demonstrated by Yoon *et al.* in 2012 [18]. In this work, a glass-fabric reinforced composite substrate was used with the phase change thin film $In_3Sb_1Te_2$ (IST) deposited using sputtering. The lack of options for high-quality phase change thin films and the complexity associated with their integration into flexible substrates results in fewer reports in the literature for flexible PCRAMs. However, the potential for high integration density is the main driver for research in PCRAMs.

14.3.2.4   Flexible Flash Memory

Flash memory technology is the most commonly used NVM technology at present. The basic structure is that of a transistor that includes an additional layer in the gate stack that traps charges and changes the threshold voltage of the underlying transistor. The layer can be either conductive or insulating depending on the integration strategy. The injection of charges into and out of the charge-trapping layer occurs from the transistor channel through the thin dielectric layer called tunneling oxide. Although rigid flash technology is very well established, the progress towards flexible flash memories has been restricted by the requirement of a transistor architecture and strict constraints on the properties and size of the tunneling oxide. For example, using the usual CMOS process-based crystalline silicon/silicon oxide/polysilicon stack can provide a defect-free interface and thin oxide for charge injection into the polysilicon layer. Such a combination of properties is not present with materials used in flexible electronics fabrication. Further, creating a memory array based on flash memory technology requires the interconnection of several such structures, as opposed to the simple crossbar architecture that can be used in other NVM technologies. Nevertheless, an example of a flexible flash memory was reported by Jeon *et al.* in 2012 [19]. In this work, a silicon nanowire (SiNW) based transistor was fabricated with a gate stack consisting of Al_2O_3 as the tunneling oxide and Pt-nanocrystal (Pt-NC) thin film as the FG. The silicon nanowires were fabricated using crystallographic wet etching of doped silicon wafers, followed by transfer to a polyethersulfone (PES) substrate. The tunneling oxide layers were deposited using ALD to obtain precise control over thickness, while the Pt-NC layer was deposited using sputtering.

There has been significant progress in the field of flexible memories, both volatile and non-volatile, over the past decade. The integration strategy used for flexible volatile memories, owing to their proximity to the processor, is dependent on the integration strategy used for fabricating the processor. On the other hand, the increased use of NVM SSD technologies in electronic devices has increased emphasis on research in rigid NVM technologies, leading to progress in their flexible forms as well. The progress is particularly clear in the various rigid flash memory architectures, however, progress in flexible flash memory has been slow. The crossbar architectures afforded by ReRAM, FeRAM, and PCRAM provide a facile strategy for fabricating flexible versions of their memory cells. However, for the memory module to be complete, the interface circuit should also be flexible, requiring flexible CMOS transistors, gates, amplifiers, and flip-flops to be fabricated. Thus, the progress of flexible memory modules is eventually tied with the overall progress in flexible electronics fabrication.

EXERCISES

14.1. Explain the different levels of memory hierarchy. Why are they typically represented as a pyramid?

14.2. What are the key challenges in creating flexible volatile memory modules? Give examples of a few methods that can be used to create flexible volatile memories.

14.3. Describe, in brief, the working principle of the following non-volatile memories:
   • Resistive RAM
   • Phase change RAM
   • Floating gate flash memory

14.4. Describe how an organic polymer material, PVDF TrFE, behaves as a ferroelectric thin film. How can this property be used for fabricating a non-volatile memory? What are the challenges associated with it?

14.5. Why haven't several flexible versions of flash memory been reported in the literature, even though the rigid versions are the most mature NVM technology?

15 Flexible Displays

15.1 INTRODUCTION

Flexible displays are, at present, the most promising flexible electronics technology. They have already been commercialized in many applications and several consumer products such as the foldable phone from Samsung, rely on the flexible display technology. Displays have evolved greatly in the twentieth century, going from vacuum tube-based greyscale bulky and power-consuming devices to slim and flexible full-color OLED panels. After the turn of the century, display technology has primarily focused on increasing the pixel density to accommodate high-resolution screens in a small area, particularly with the rise of the smartphone industry. The ability to make thinner displays has led to the push for and the development of flexible displays. This has been augmented by the explosion of the internet and social media around the world. The growing need for information access at all times translates into the demand for digital screens everywhere – in our palms, on walls, on hoardings, on street signs, and even in our eyeglasses. This demand needs to be partly fulfilled using flexible displays because many of the surfaces requiring screens are not flat and rigid. Further, such a potentially high demand also entails the development of low-cost solutions to make sure the digital divide is minimized. The displays of the future, if they are going to be ubiquitous, also need to be environmentally friendly and easy to discard. Hence, it is important to study the integration strategies available for creating conventional displays and their flexible counterparts.

One of the key reasons flexible displays are ahead of the curve from a technology readiness level (TRL) point of view, is that displays can be a stand-alone product. In some applications, a display signboard only displaying a few directions can be installed at public events. Such displays do not require information processing, network connectivity, or large memory capacity, greatly simplifying the system design and integration. Even in the case of connected displays, the processing and networking can be taken care of separately using rigid modules, keeping the display flexible. Further, in most devices, the display takes up the largest surface area. For example, a 65-inch TV will always have a 65-inch screen, even if the support electronics can now be accommodated in a one-square-inch area, owing to the dimensional scaling of electronics. Thus, in the display industry, larger is better, whereas, for all other electronic components, smaller is better! This unique aspect of the display industry entails fundamentally different integration strategies compared to silicon electronics. In this chapter, we will take a look at various integration strategies that have evolved for the fabrication of flexible displays. However, before we can discuss flexible integration strategies, we need to understand the fundamentals of conventional (rigid) displays.

DOI: 10.1201/9781003010715-19

15.2   CONVENTIONAL DISPLAYS

Displays, like all things electronic, have undergone a sea change in the twentieth century. The ancestors to current displays worked on the principle of electron emission in a vacuum tube. The electron beam in the cathode ray tube (CRT) was accelerated toward a fluorescent screen to create temporary luminance upon impingement. These devices were bulky, to provide space for electron beam generation, focusing and guiding modules, and power-hungry, to account for the creation and acceleration of the electron beams. Further, they were susceptible to image persistence, meaning an image displayed for a long time would "burn in" (leading to the creation of screen savers!). The dominance of CRT displays ended with the evolution of the liquid crystal display (LCD) technology, which uses the light polarizing property of liquid crystals. LCD pixels act as a gateway for light, inhibiting or allowing light to pass through depending on the applied electric field. It is important to note that LCD pixels cannot produce light but can only modulate the passage of light. Thus, LCD screens depend on another light emission technology as the source of light, generally acting as the backlight panel. An LCD panel is composed of three basic layers: two glass sheets with polarization orthogonal to each other, sandwiching a layer of liquid crystals, as shown in Figure 15.1. The glass sheets also need to have transparent conductors for the passage of electric current through to the liquid crystal pixels. Further, the glass sheets are coated with a polymer with grooves aligned to the direction of polarization. Thus, in steady state, the liquid crystal molecules gradually twist from one orientation to another. The operation of LCD pixels depends on the fact that liquid crystal molecules can polarize light passing through in a particular direction. The light from the back panel enters the first glass sheet and gets polarized in a particular direction. The direction of polarization of light is then gradually changed as light passes through the liquid crystal. When light exits the liquid crystal film, its polarization is aligned with that of the second glass sheet, allowing it to pass through. However, if an electric field is applied, the orientation of the molecules inside the liquid crystal gets disturbed. This causes the light exiting the liquid crystal to be of a different orientation than that of the second glass sheet, thus, no light passes through the pixel. This control of whether light is allowed to pass, at an individual pixel level, allows for the creation of LCD displays.

The source of light can be a CFL (compact fluorescent lamp) or LED panel. Color displays based on LCD technology operate by using a high-energy, blue/white light backlight panel and color filters in each pixel to create red, green, and blue pixels. The filters are usually fabricated as layers of material on top of the backlight panel, with each pixel having a specific color filter deposition. These pixels are arranged in a staggered configuration to create an illusion of a color display. However, the inefficiency in filtering can allow for some light to pass through even when the pixel should be "off" leading to black levels that are not perfectly black. Recently, semiconductor quantum dots (QDs) (zero-dimensional nanostructured materials) have been introduced as the functional constituents of the filter layer. We know that as the dimensions of a semiconductor particle approach nanometer scale (<100 nm), its bandgap can be tuned by changing its dimensions. This property arises from the confinement of charge carriers leading to a higher dispersion in the available energy

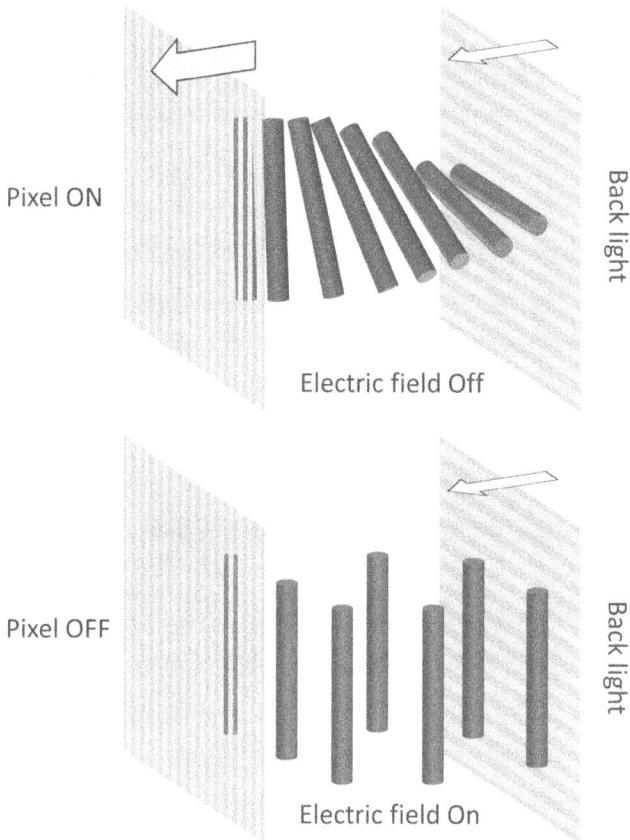

FIGURE 15.1 The liquid crystal molecules change the polarization of the incoming light such that it is able to exit from the front panel. The application of an electric field aligns the molecules such that light polarization does not match the front panel, leading to an off state of the pixel.

states. Thus, the emission wavelength of a QD can be finely tuned merely by changing its dimensions. The QDs, used for creating the color filtering thin films in LCD displays, are sized such that they produce almost purely monochromatic red, green, and blue colors of a specific wavelength. These particles absorb the incoming blue light from the back panel and emit the specific wavelength they are tuned for. Thus, the LCD displays based on QD thin-film color filters provide better color gamut coverage and perfect black levels. They have been commercialized as QLED screens.

The development and proliferation of the LCD technology were closely followed by LED-based displays. Both emerged in the 1960s and got popularized by simple applications such as calculators and public signboards. Pixels based on the LED technology have the capability of producing light on demand, whereas LCD pixels only obscure light from the back panel when switched off. This key difference provides a much better contrast ratio (the difference between on and off intensity) in LED displays compared to LCDs. Early LED displays were based on the use of III-V semiconductors for fabricating LED pixels, whereas nowadays, it is common to find organic semiconductor-based LED (OLED) displays in commercial smartphones. The working principle of an LED display is the same as that of a single LED, however, the strategy to array the pixels together and control each one separately is the key challenge, particularly for very high pixel densities. With the development and commercialization of QD thin films for color filtering, there have been efforts to create emissive displays using semiconductor QD thin films, i.e., instead of being excited by light from the back panel (photo-emissive), the thin film can be excited by the application of electric current (electro-emissive), as is the case for LED displays. However, the technology is only at a nascent stage.

For both LCD and LED displays, the two main strategies used for pixel control are the passive matrix and the active matrix strategies. In the passive matrix strategy, the pixels are addressed directly using control lines corresponding to the particular row and column, without the use of a switch (Figure 15.2). The electronic structure is similar to the crossbar architecture discussed for memory applications. When an electric potential is applied between a particular row and column line, the pixel in the corresponding row and column is activated. This strategy provides for simple fabrication because of the absence of an electronic select switch. However, the lack of the switch also creates problems like crosstalk and sneak path current issues. The problem is typically solved using the active matrix strategy involving an electronic select switch (typically a thin-film transistor) for controlling each pixel. The presence of the switch resolves the crosstalk problem resulting in better optical contrast;

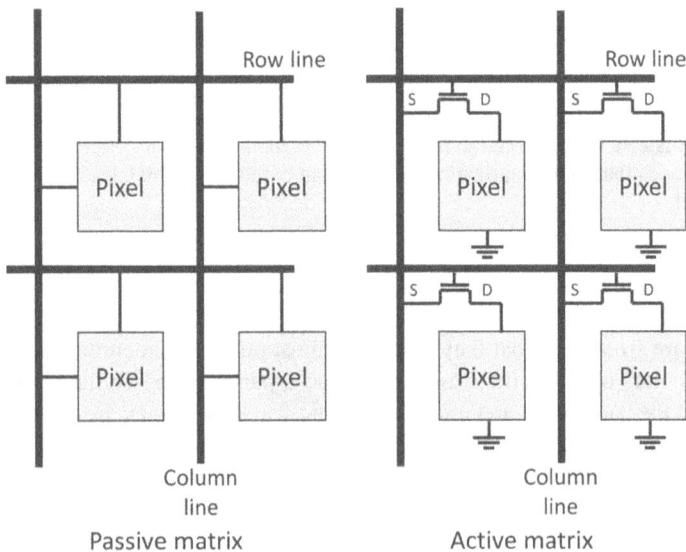

FIGURE 15.2 The passive matrix connection strategy does not involve an active switching components. The pixels are directly connected to the row and column lines. In active matrix configuration, a switching component is used to control the state of the pixel, reducing sneak path current and crosstalk problems.

however, it requires the integration of an additional component in the pixel, leading to a more complex and expensive fabrication process. The advantages of the active matrix strategy outweigh these problems, leading to their widespread adoption in the commercial display industry, including television screens, computers, laptops, and mobile phones.

Another display technology gaining traction in the commercial space is the electrophoretic display (EPD) technology. As the name suggests, the technology is based on the principle of electrophoresis, which is the motion of dispersed charged particles in a fluid medium upon application of an electric field. For creating a display, a liquid film consisting of charged color pigments is encapsulated between two electrodes. When a particular electric field is applied, the color pigment particles with opposite charges are attracted toward the front of the display, leading to the appearance of that color. In case the opposite field is applied, the other color pigment particles move to the front (Figure 15.3). There are many variations of this process, however, all EPD technologies involve the use of charged pigments dispersed in a liquid medium, being manipulated by electric fields. In most cases, the color pigments used are black and white to create the most contrast between the two states of the display. A large high-resolution display can be created by operating each pixel in the same way and controlling the state of the pixels using active or passive matrix strategies. Colored displays can also be created using color filters for each pixel. The key difference between EPD and other display technologies is the use of reflected ambient light for display, which can reduce the power requirement significantly. Such technology is also less stressful for the eyes because humans have naturally evolved to observe through reflection (very few things in nature are emissive). The most popular commercial example of the use of EPD technology is the display for the Amazon Kindle series.

Other display technologies being used for niche applications include electrowetting displays, thermochromic displays, and electrochromic displays (ECDs). Electrowetting displays change the intensity of a pixel (using backlight) by changing the wettability of the surface, which changes the size of an opaque bead of liquid on

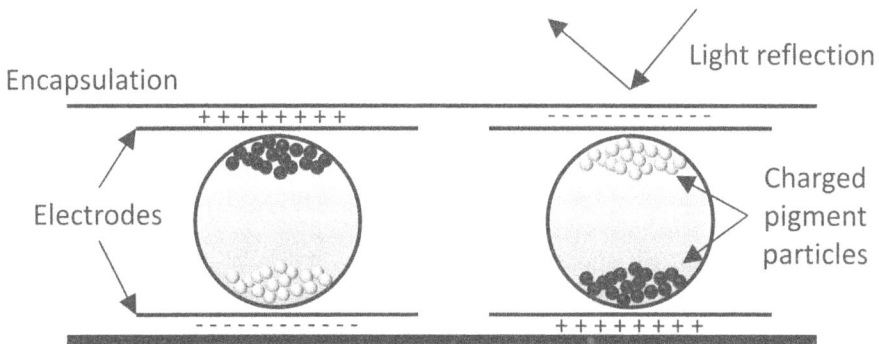

FIGURE 15.3 Electrophoretic displays consist of encapsulated liquid with charged pigment particles in each pixel. When an electric field is applied, the particles with opposite charge get attracted to the top electrode creating a reflective display of that color.

the surface. The change in surface wettability can be obtained reversibly by applying an electric field. Thermochromic displays use the phenomenon of change in the absorption spectrum of materials because of a change in temperature. The change can occur due to phase change, crystal structure reorientation, or dissociation/formation of chemical bonds upon application of thermal energy. Whereas ECDs utilize an electric field to create the change in optical properties of a thin film using ion-exchange reactions driven by the applied field.

15.3 FLEXIBLE DISPLAYS

As discussed in the previous section, the key commercial display technologies depend on regulating either emission (LED), transmission (LCD), or reflection (EPD) of light using electric fields. Of these technologies, LED technology-based displays have shown the most promise for flexible display fabrication because of the reduced number of layers and fabrication complexity. Both LCD and EPD technologies rely on multi-layered structures with encapsulated "moving" components for operation. While challenging, there has been some progress toward the development of flexible displays based on LCD and EPD technologies. These are discussed at the end of this section.

15.3.1 ORGANIC LEDs

Conventionally, LEDs are fabricated using high-quality III-V semiconductors. The fabrication process involves the epitaxial growth of multiple layers of highly crystalline thin films on III-V substrate. The most commonly used III-V substrate is gallium arsenide, GaAs. The composition of the epitaxial layers grown for light emission is balanced such that the lattice parameter closely matches that of the underlying thin film to reduce stress; and that the required wavelength of light is obtained. The thickness of the film stack can also be used to fine-tune the bandgap of the semiconductor because the electrons are confined in a specific region of the device, creating a dispersed energy band structure. The operation of LEDs based on III-V materials is dependent on high-quality crystalline lattice structures because defects in the crystal structure create deep-level charge traps that can be used by excited charge carriers to recombine non-radiatively. Thus, amorphous or polycrystalline thin films of III-V materials deposited using low-temperature processes such as sputtering, or e-beam evaporation cannot be used for the fabrication of LEDs because of the defects at the grain boundaries. This leads to the development of methods such as the TPER process for creating flexible thin films usable for light-emitting applications (details in Chapter 7).

Because of the challenges associated with creating highly crystalline flexible thin films using III-V materials, we turn our attention to the organic light-emitting diodes (OLEDs). Organic materials have been known to have electroluminescence for decades, however, practical organic LEDs have only been made in the 1990s. The basic working principle of OLEDs is the same as that of LEDs based on III-V materials, i.e., a semiconductor with a certain bandgap is injected with excess charge carriers, which recombine radiatively to create photons. The intensity of the light emitted is proportional to the current (amount of charge carriers injected) and the wavelength is dependent on the energy bandgap of the semiconductor. In the case of organic semiconductors, the energy bandgap is the difference in energy of the highest occupied and lowest unoccupied molecular orbitals (HOMO and LUMO). Charge injection into the semiconductor is done through metal contacts that form Ohmic contacts with the semiconductor, which requires their Fermi levels to be close. However, because the Fermi energy of organic semiconductors cannot be tuned as easily as III-V semiconductors (say through doping), the injection of charges into light-emitting organic semiconductors can be challenging. The way this charge injection is achieved is by creating a multilayer structure such that the HOMO and LUMO energy levels of each layer are precisely engineered to provide a small barrier to charge injection into the next layer. These layers are called hole injection layer (HIL), hole transport layer (HTL), emitting layer (EML), electron transport layer (ETL), electron injection layer (EIL), and so on (Figure 15.4). In many cases, a single material is used to function as one or more of these layers.

The goal of the structure is to provide an efficient pathway for charge carriers to enter the EML where they recombine to form photons. Because the wavelength of the light emitted is dependent on the energy gap between HOMO and LUMO in the

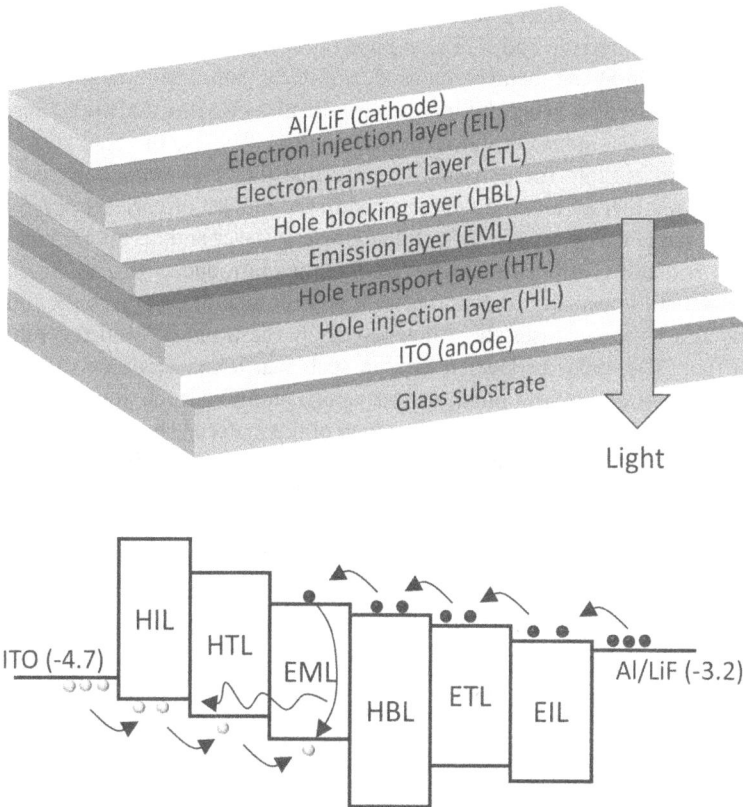

FIGURE 15.4 The structure of an organic LED consists of many layers with specifically tailored LUMO and HOMO energy levels to obtain Ohmic contact with the metal electrodes and for providing a low energy barrier for charge carriers.

EML, it is important to tune its electronic properties. This is particularly important for display applications where a highly monochromatic light of the correct wavelength is desirable for producing true color screens. The structure can be formed into a pixel and addressed using passive or active matrix strategies for the creation of an OLED display screen. The electronic band structure of the layers used for creating the OLED pixel is finely tuned to obtain the correct values of ionization potential (energy needed to take an electron from HOMO to vacuum level) and electron affinity (energy needed to take an electron from LUMO to vacuum level). The metals at the cathode and anode are required to have the correct work function such that the injection of electrons and holes respectively, can take place smoothly. For example, poly(p-phenylenevinylene) can be used as an emissive layer with ionization potential and electron affinity of 5.1 and 2.7 eV respectively, providing photon energy of 2.4 eV, which results in a bright green color emission. If aluminum, with a work function of 4.3 eV, is used as the cathode, the energy difference between its work function and electron affinity creates a large barrier for the injection of electrons. Thus, if calcium, with a work function of 2.9 eV, is used instead, the electron-injection barrier is only 0.2 eV, allowing for a more efficient structure. However, calcium is unstable in air and cannot be used as a stand-alone thin film reliably, hence, different layers, such as electron injection and ETLs, are used to bring the electron affinity close to the work function of the metal to be used. Similarly, hole injection and hole transport layers are used to bring the ionization potential close to the work function of the metal in smaller steps (as shown in Figure 15.4). The use of these layers brings the charge carriers into the emission layer so that they can radiatively recombine. However, given that an electric field is causing the motion of carriers, it is possible that some charge carriers pass through the emission layer and do not recombine. To limit this possibility, a blocking layer is sometimes introduced next to the emissive layer such that the carriers transported to it are not able to escape to the following layer. Such a layer increases the efficiency of charge transport and recombination even further. The emissive layer is chosen such that it can facilitate the transport of both electrons and holes (ambipolar transport) along with the ability to facilitate radiative recombination. Further, the bandgap of the material should be such that the wavelength of light emitted should be in the visible region for display applications (1.75–3.1 eV). Finally, the emissive layer material should be able to form a uniform and stable thin film. In the case of materials that satisfy the first two properties and lack good film-forming properties, they can be used as emissive dopants dispersed inside conductive organic thin films. Examples of such materials include Nile red (red), rubrene (yellow), N,N'-Dimethylquinacridone (DMQA, green), and perylene (blue). Further, metal complexes, such as osmium complexes, platinum complexes, and iridium complexes have been reported as phosphorescent emissive dopants for OLEDs. They are dispersed in a host material with good film-forming ability.

There are several advantages of forming light-emitting devices using organic semiconductors. Because light originates from the pixel itself, OLED displays do not require a back panel, allowing for a very thin and potentially flexible integration. The thin-film stack deposited as cathode, anode, emissive layer, and other charge transport layers can all be accommodated in a few hundred nanometers thickness. Hence, the total thickness of the structure, including that of the substrate can be as low as sub-one-micron, leading to truly flexible OLEDs. Further, in the case of color screens, the red, green, and blue light can be obtained from three different finely tuned emission layers, without the need for an additional layer of color filtering. This leads to sharper image quality and true black levels. Finally, the use of all organic layers for creating the OLED stack leads to the opportunity to use room temperature fabrication processes such as sputtering or solution processing, thus providing an opportunity for integration on any plastic substrate. This is not possible with inorganic semiconductors (III-V semiconductors) that require high-temperature deposition processes such as epitaxy. The OLED device stack can also be integrated using roll-to-roll production processes leading to high throughput and low manufacturing costs.

Indeed, there are some challenges associated with creating flexible large-area displays using organic semiconductor-based light-emitting devices. The structure of the OLED is such that the metal contact layers form the top and bottom layers of the structure. Obviously, if we require light to leave the device from at least one side, one of these metal layers needs to be transparent for light in the visible region. In the rigid OLED displays, indium tin oxide (ITO) is used as the transparent metal electrode. It is typically deposited prior to the deposition of the organic layers and is annealed at a temperature above 200°C for better electrical conductivity. However, ITO deposited on flexible plastic substrates has been reported to develop fractures leading to conductivity failure upon application of strain. Hence, several transparent conductive thin films have been reported as a replacement for ITO, such as carbon nanotubes/graphene, metal nanoparticles/nanowires/nanotubes, and ultrathin metal films. In particular, ultrathin metal films have shown promise because of their high conductivity and transparency, along with mechanical stability during strain application. Another challenge with the fabrication of large-area OLED displays is the encapsulation of the organic layers. This is a critical feature of all OLED displays because the organic materials used for their fabrication deteriorate in the presence of oxygen or moisture. Thus, transparent encapsulation layers which act as barriers against oxygen or moisture ingress are typically added to the fabrication process to protect the underlying organic layers (refer to Chapter 10 for details). Exposure of the OLED stack to the atmosphere can cause dark spots of non-emission to appear on the surface of the display, which grow, eventually leading to the disintegration of the complete display. To avoid this catastrophic failure, the acceptable water vapor transmission rate of the encapsulation is less than 10^{-6} g/m^2/day. The encapsulation layer also needs to be thin, flexible, and transparent in order to maintain the flexibility and light emission properties of the device. This problem is typically solved using a multibarrier encapsulation layer consisting of successive layers of inorganic and organic thin films. With the problems associated with OLEDs solved and given the advantages of OLEDs for flexible integration, there has been a boom in OLED-based large-area flexible displays.

15.3.2   FLEXIBLE AMOLED DISPLAYS

The success of flexible displays based on organic semiconductors has led the platform from research laboratories to commercially available products. These products are the first flagship commercially available flexible electronic products. Examples of these include the foldable phone from Samsung and the rollable TV from LG. The race for creating better flexible OLED displays has evolved from proof-of-concept demonstrations to increasing pixel density, stability, and reliability of the flexible display. The active matrix (AM) strategy for controlling displays based on OLED technology has been in use for the past two decades for conventional (rigid) laptops, smartphones, and television displays. The so-called AMOLED technology has been responsible for the high density achieved by state-of-the-art smartphone displays. The key in AMOLED integration is the use of an active electronic switching element to control the state of an individual pixel. This switching element is typically realized using thin-film transistors (TFTs). The inclusion of the TFT adds a few more layers in the OLED display stack, namely the gate metal, dielectric, and semiconductor used for creating the transistor structure. In order to achieve flexibility with AMOLED integration, it is important to make sure the layers comprising the TFT structure are also flexible.

The conventional rigid AMOLEDs are manufactured using glass as a substrate and LTPS (low-temperature polysilicon) as the TFT semiconductor material. LTPS has high carrier mobility and can be created using low-temperature CVD deposition of amorphous silicon followed by laser annealing for film recrystallization. The OLED films are then deposited to make the pixels followed by a glass superstrate for encapsulation. However, in the case of flexible polymer substrates, high-temperature annealing or film densification for the LTPS-TFT fabrication is challenging. This is why one of the first flexible AMOLED displays was demonstrated using a stainless steel substrate [20]. Subsequently, the polyimide substrate, due to its temperature stability, became a popular substrate choice. The process used for the recrystallization of amorphous silicon into polysilicon is the excimer laser annealing process. This process provides a controlled amount of thermal excitation to the top surface of silicon for recrystallization, with minimal heat transfer to the substrate. In some cases, barrier films of SiO_x and SiN_x are deposited between PI substrate and Si to provide water vapor ingress protection and to reduce heat transfer to PI substrate. There are still concerns of overheating, particularly outgassing, of polyimide substrate during the process, requiring special design of the process equipment to detect outgassing. The process is generally carried out on a glass support substrate by first depositing the PI layer followed by the barrier, TFT and OLED layers. This is followed by the deposition of the superstrate for encapsulation. The complete structure along with the flexible substrate and device layers is then removed from the rigid glass support and (optionally) transferred to a flexible substrate for better encapsulation and barrier properties. Hence, although the fabrication process for flexible AMOLED displays has undergone great progress, there are still many challenges preventing mass adoption of large-area flexible AMOLED displays.

15.3.3 Other Flexible Display Technologies

There are several challenges in creating flexible versions of LCD displays. Because the display is based on a backlight obstruction mechanism, it can be challenging to create LCD pixel with a small bending radius, because the light entering from the back panel may get misaligned with the front end polarized layer, leading to dead pixels. Further, the liquid crystal layer depends on the movement of molecules to create the change in polarization required for the pixel to work. Flexing such a structure can cause issues with the reliability of the movement and light polarization. Nevertheless, flexible LCD displays have been reported in the literature using organic TFTs, instead of amorphous silicon TFTs that are used in the commercially available rigid LCDs. These organic TFT-based LCDs (called OLCDs) can be fabricated using low-temperature processes, opening up the possibility for integration on flexible plastic substrates. Meanwhile, black and white EPDs have been integrated into flexible versions on plastic substrates. Their color counterparts have only recently been announced as proof-of-concept demonstrations. However, because of their low power and reflective nature, flexible EPD technology seems poised to take over the wearables industry. The flexible displays in both LCD and EPD technologies are also yet to be demonstrated for large-area applications such as computer and television screens.

The mass manufacturing of flexible AMOLEDs on PI is still in its infancy, while great progress has been made in conventional AMOLED display manufacturing. However, the success of rigid AMOLED displays bodes well for their flexible versions, because lessons from the manufacture and use of the rigid structure can be used for creating better flexible displays. Further, with the rise of the internet of things and wearable technologies, the demand for flexible displays is on the rise, because displays, in general, consume the largest surface area for any electronic device.

EXERCISES

15.1. Describe the working principle of the liquid crystal display (LCD). What are the challenges in fabricating a flexible LCD?

15.2. How is the emergence of quantum dot (QD) technology helping create better displays?

15.3. Why are organic semiconductors better suited for fabricating flexible displays compared to inorganic ones?

15.4. What is the key difference between passive matrix and active matrix configurations? Why are active matrix displays more popular than passive matrix?

15.5. We need to design a flexible OLED stack for the visible region. We have the following materials: (a) ITO [WF: 4.7 eV]; (b) Ca:Al [WF: 2.9 eV]; (c) Tris(4-carbazoyl-9-ylphenyl)amine (TCTA) [IP: 5.7 eV, EA: 2.4 eV]; (d) 2,9-dimethyl-4,7-diphenyl-1,10-phenanthroline (BCP) [IP: 7 eV, EA: 3.5 eV]; (e) CuI [IP: 5.3 eV, EA: 2.1 eV]; (f) 4,4'-(9H,9'H-[3,3'-bicarbazole]-9,9'-diyl)bis(3-(trifluoromethyl) benzonitrile) (pCNBCzoCF3) [IP: 5.8 eV, EA: 2.9 eV]. Which of these should be used as the anode, cathode, hole injection, hole transport, emission, hole blocking, and electron transmission layers. Explain your choices. What will be the colour of the LED?

16 Flexible Energy Generation and Storage Devices

16.1   INTRODUCTION

The Sun is the source of all energy on our planet. Be it from the conversion of sunlight into chemical energy by plants (photosynthesis), or the wind energy that arises because of the pressure changes due to heating of the atmosphere, or the hydroelectricity generated because of the evaporation of water to higher ground, or from the direct conversion of sunlight into electrical energy (photovoltaics). Even fossil fuels such as natural gas, coal, and crude oil contain chemical energy from plants and animals, concentrated over millions of years. A possible exception to the rule is nuclear energy that originates from the suboptimal arrangement of elementary particles in the nucleus of large atoms. Indeed, the Sun itself gets its energy from atomic nuclei – from the fusion of atomic nuclei of small atoms. This burst of energy manifests in terms of photons of a large range of frequencies released outward from the Sun. We "see" these photons as the visible light and the associated warmth received from the Sun. In modern industrial society, electrical energy is becoming increasingly important. With computers, laptops, and personal electronic devices becoming ubiquitous, the demand for electrical energy has been growing relentlessly. Now, with the advent of electric vehicles (EVs), the demand for electrical energy is expected to grow even faster. We have employed many methods for generating electrical energy from other energy sources, including but not limited to gravitational energy (hydroelectricity), wind energy (wind turbines), chemical energy (gas/coal and other fossil fuels), nuclear energy (nuclear fission reactors), and so on. In most of these cases, the source of energy is first used to create kinetic energy in the form of a rotating turbine, which is then used to generate electricity. The carbon emissions from the burning of fossil fuels, the problem of disposal of nuclear waste from fission reactors, and the ecological damage caused by constructing large dams for hydroelectric power generation have led to global efforts in developing environmentally sustainable technologies for the generation of electrical energy. Hence, since the end of the twentieth century, there has been great interest in using sunlight to directly create electrical energy (photovoltaics), because the Sun is the most abundant energy source and electrical energy is the most useful form of energy for us. This has led to increased research interest in photovoltaic technology, resulting in better efficiency at the cell and module levels. Further, the rise in demand for eco-friendly sources of electrical energy has increased the volumes of solar cell production, driving down costs per energy output. The global adoption of solar technology has been such that it accounted for 3% of the total electrical energy generated in the World in 2019 (720 TWh).

DOI: 10.1201/9781003010715-20   **233**

The relevance of photovoltaic technology for flexible electronics applications is clear from the fact that it is one of the few electricity generation technologies that does not involve moving parts and can be fabricated using thin-film technology. Other electricity generation technologies such as hydroelectric, nuclear, wind power, gas turbine, and so on, utilize the source energy to create mechanical motion, which is converted into electricity using electromechanical drives. Further, photovoltaic devices can be scaled based on the power output required, simply by changing the area of exposure to sunlight. Thus, PV devices are available for electricity generation in all sizes, ranging from tiny solar cells with an output of a few milliwatts, to giant solar farms with an output of gigawatts. This makes the use of flexible PV devices for low-power wearable devices feasible (it is hard to imagine a tiny turbine powering a smartwatch!). Furthermore, solar cells produce DC output that can be directly used for powering modern electronic circuits or for energy storage. Other electricity generation technologies based on thin films include triboelectric, thermoelectric, and piezoelectric generators. The generation of static charge because of relative motion between two surfaces is known as the triboelectric effect. Triboelectricity is the buildup of static electric charge encountered in everyday experiences such as brushing a plastic comb through the hair or rubbing glass with silk and so on. In fact, early experiments with electricity mostly involved triboelectricity, and the word "electricity" comes from the Greek word *elektron* for amber, which when rubbed against wool, acquires static electric charge. The phenomenon of triboelectricity can be used to generate electrical power from any kind of mechanical motion or vibrations. Piezoelectricity is also used to generate small amounts of electrical power from stray mechanical vibrations or everyday movement such as walking, etc.; however, the physics of conversion of motion into electronic energy is completely different from triboelectricity. Piezoelectricity is the property of certain materials to produce electric charges due to the application of mechanical strain which causes a change in electric polarization of the material mainly due to the reorientation of internal molecular dipoles. The change in polarization results in a buildup of surface charge to balance the resulting electric field. These charges have an electromotive force associated with them and can be used to operate external electrical loads. Thermoelectricity is the generation of electromotive force because of the difference in temperature across a junction. The junction, called a thermocouple, is typically made using metals that have different temperature dependence of work functions. When two metals are contacted, electrons from one metal move into another creating a change in the vacuum electrostatic potential, equating the Fermi levels of both metals and restricting the flow of current. However, if the temperature of one junction is changed with respect to another, a potential difference is developed in the circuit because there is a change in the relative Fermi level, causing an electric current to flow as long as the temperature difference is maintained. This electrical current can be used to perform work.

The thin-film technologies discussed so far for electricity generation have one common factor – all of them are intermittent sources of power, i.e., power flows for a certain time depending on the external conditions, for example, if there is sunlight or mechanical motion, etc., but does not remain continuously available throughout the operation of the generator. This is not the case for conventional electromechanical generators, because the fuel source used for obtaining the mechanical motion can be stored beforehand and consumed at a required pace. Such storage is only possible in photovoltaic and other thin-film electricity generators *after* the electrical power has been generated. For example, one cannot store sunlight or vibrations to be used later, as one can store coal or gasoline. The storage of electricity after its creation is generally done using electrical batteries that typically convert the applied electrical energy into chemical energy for long-term storage. The battery then acts as a source of energy and can be summoned to supply electrical power, when the primary source of electricity is not in operation. The simplest example of this method is the use of batteries to store extra energy from photovoltaic cells during the day so that it can be used during the night when there is no sunlight for PV generation. Further, energy storage solutions also play a key role in boosting the power output of a system that has low-power output over a large period of time. For example, a triboelectric energy harvester can be used to trickle power into a battery overnight so that the available energy can be used over the course of a few hours in the morning.

The most commonly used process for creating energy storage solutions involves the use of electrical energy to create a chemical reaction, such as redox or ion exchange. This reversible reaction is not spontaneous in a particular direction and requires electrical work to be done in order to proceed in that direction. This stores energy in the electrochemical cell in the form of chemical energy, which can be released spontaneously when a transfer of electrons is facilitated from one end of the cell to another (typically through an electrical load). A simpler way to store electrical energy is to store the charge in a capacitive structure. This does not require conversion from electrical to chemical energy, thus saving on the energy losses due to conversion inefficiencies. However, capacitors typically leak charges through the dielectric causing the charge to neutralize over time and the energy to dissipate in the form of heat. This is a major challenge in the use of capacitors for long-term, large-scale energy storage. Nevertheless, because capacitive structures are very simple to fabricate using thin-film technologies, there is an interest in the use of specific types of capacitive structures, so-called supercapacitors, for energy storage applications. Other methods for storing electrical energy for later use involve mechanical or gravitational structures. For example, a heavy flywheel can be set into motion using the excess energy produced from photovoltaic cells during the day, and the stored rotational energy can be converted into electricity during the night. Another example is the use of hydroelectric structures to store energy – water is pumped upstream using excess energy during the day, and the gravitational potential energy in the water can be used to create electricity at night. These are interesting energy storage solutions, however, because they involve large mechanical parts, they are typically ignored from a flexible electronics point of view.

16.2   ENERGY GENERATION

16.2.1   Photovoltaic Devices

The generation of energy using photovoltaic devices depends on the use of semi-conductors to capture the photon and excite an electron to a higher energy state, i.e., create an exciton (an electron-hole pair). Although the creation of electron-hole pairs upon absorption of photons happens in almost all materials, in most materials this pair is allowed to recombine either radiatively (creating light) or non-radiatively (creating heat). In photovoltaic materials, a semiconductor junction is used to separate the electron and hole from each other and are then forced to pass through an external circuit before being allowed to recombine. This passage of charges results in electrical work being done. The use of impinging photons for the excitation of electrons and subsequent use of the electric potential energy stored in them is not a new concept. In fact, plants have been using the same concept for photosynthesis for literally billions of years. The only difference is that the ions formed in the chlorophyll pigment are then used to carry out a chemical reaction to store energy in the form of chemical bonds.

In its simplest form, a photovoltaic device consists of a p-n junction formed using a semiconducting substrate, as shown in Figure 16.1. Because silicon is the most used and widely studied semiconductor material, it is only logical to use silicon as the semiconductor for the fabrication of photovoltaic devices. When a p-n junction is created in a semiconductor, the excess electrons from the n-side and excess holes from the p-side migrate across the junction. This diffusion of carriers depletes the region adjacent to the junction of charge carriers. The diffusion continues until an internal electric field is developed opposing the migration. The internal electric field is called the built-in potential. If a photon impinges and excites an electron inside the depletion region, the electron and hole thus created are pulled to the opposite side of the junction because of the built-in potential. They then have to pass through the external circuit to recombine, leading to the generation of photocurrent.

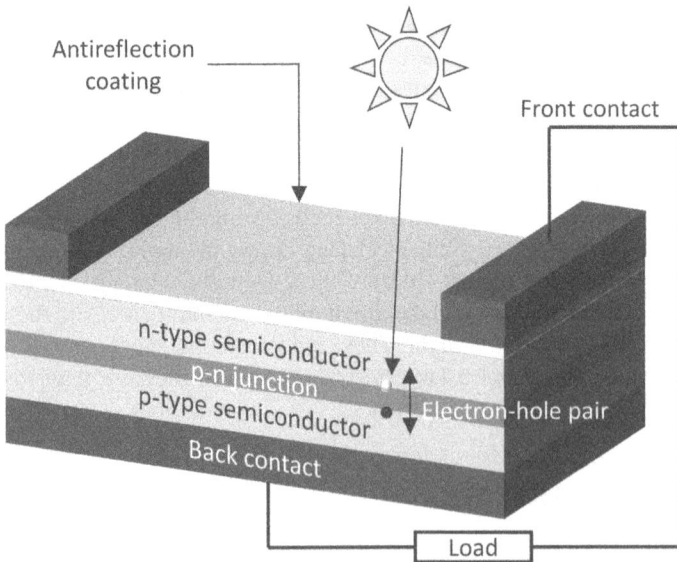

FIGURE 16.1 A solar cell consists of a p-n junction that helps separate the electron-hole pair created because of the impinging photon. The carriers are collected by the metal contacts and passed through the external load to perform electrical work.

The bandgap of the semiconductor used for photovoltaic applications plays a major role in determining the amount of energy captured in the form of usable electrical energy. This bandgap also depends on the distribution of energy in the solar spectrum, i.e., the relative intensity of each wavelength in sunlight referred to as spectral solar irradiance ($W/m^2/nm$). If the bandgap is chosen to be high, only high-energy photons, such as those in the ultraviolet region, will be able to excite electrons in the semiconductor. This will lead to a lesser number of electrons being excited for a given spectral irradiance pattern. On the other hand, if a semiconductor with a low bandgap is used to fabricate a photovoltaic device, it will be able to create a large number of excited electrons for the same spectral irradiance pattern. However, these electrons will lose most of their energy (above the bandgap energy level) in the form of heat, limiting the amount of energy converted into useful electrical energy. This fundamental trade-off in the selection of bandgap, along with other losses associated with electron-hole recombination probabilities, leads to a theoretical limit to the amount of electrical energy that can be produced using the solar spectrum. These theoretical considerations, first calculated by William Shockley and Hans-Joachim Queisser in 1961, show that the efficiency of single-junction solar cells varies with semiconductor bandgap, giving a maximum efficiency of around 30% at 1.1 eV (Figure 16.2). State-of-the-art solar cells based on single-crystal silicon (bandgap of 1.16 eV) have reached close to 25% efficiency. With multiple junctions, the probability of photon capture increases, and the efficiency of the cell can be significantly increased. Solar cell efficiencies of 47% have been reported in the literature for multi-junction solar cells based on III-V semiconductors using light concentrators.

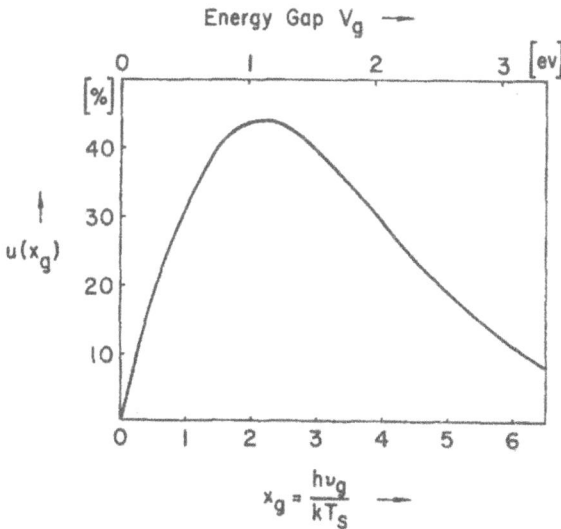

FIGURE 16.2   The Shockley-Queisser limit is the maximum efficiency a single-junction solar cell can attain, based on the bandgap of the semiconductor used. Reprinted from [21], with the permission of AIP Publishing.

Once the photon is captured and charge carriers are separated for conduction, the next consideration in the operation of solar cells is the amount of external load to be connected for maximum power transfer. Individual cells are typically connected together to form a module before being subjected to an external load. Solar cells act as DC cells in the presence of sunlight. When individual cells are connected in series, the output voltage increases, while the current capacity of the system remains the same, whereas cells connected in parallel provide higher current output at the same voltage. In practical solar modules, the connections are a combination of series and parallel connections to boost both the voltage and current at the output. However, the question of applying an appropriate load still holds. If no electrical load is applied, there is a large voltage built up across the module, called the open-circuit voltage (V_{oc}), but no current, leading to zero power delivered to the load. On the other hand, if the module is short-circuited (to mimic infinite load), the voltage across the module drops to zero, with a finite current output, called the short circuit current (I_{sc}). The V_{oc} and I_{sc} represent important parameters associated with the module, however, the power transferred to the load at both these extremes is zero. For the maximum power to be transferred, a specific voltage and current need to be maintained at the output depending on the I-V characteristics of the solar cell (Figure 16.3). The ratio of the maximum power output to the product of V_{oc} and I_{sc} is referred to as the fill factor (FF) of the cell. These parameters, V_{oc}, I_{sc}, FF, and efficiency are frequently used to compare solar cell performance.

16.2.1.1 Flexible Inorganic Photovoltaics

The structure of a solar cell is such that it can be easily manufactured using thin-film fabrication processes. Because the absorption of the photon and creation of the electron-hole pair happens inside the depletion region, which is typically a few microns

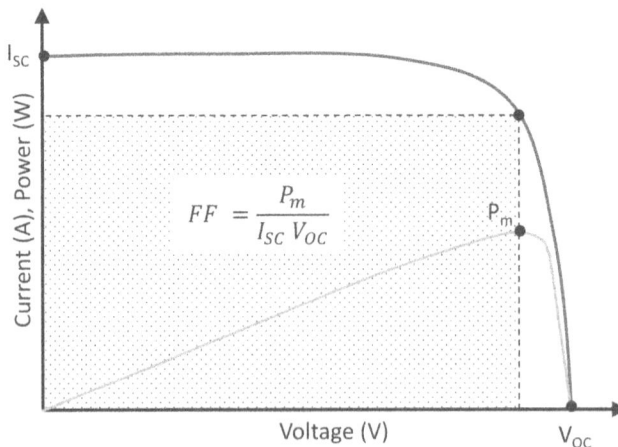

$$FF = \frac{P_m}{I_{SC} V_{OC}}$$

FIGURE 16.3 Typical IV characteristics of a solar cell for a given level of solar irradiance. The power delivered to the load depends on the current drawn, with a maximum occurring at a specific point on the IV curve. This determines the "fill factor" and efficiency of the solar cell.

wide, the thickness of the structure can be significantly reduced without compromising performance. The compatibility with thin film processing allows for solar cells to be created on flexible substrates using room temperature deposition processes such as vacuum deposition, solution processing, transfer printing, and so on. The performance of the resulting solar cell and module then depends on the material chosen as well as the deposition process used. In this section, we discuss some of the promising inorganic thin-film semiconductors, and their deposition processes, for the fabrication of flexible solar cells.

The success of crystalline silicon for large-area rigid solar cells has led to the development of amorphous silicon (a-Si) or low-temperature polysilicon (LTPS) for the fabrication of thin-film flexible solar cells. In particular, a-Si can be deposited at room temperature using physical vapor deposition techniques like sputtering on a flexible plastic substrate. The use of a-Si instead of crystalline silicon also reduces the cost of production per unit area of solar cells. The deposition process and substrate material can be tuned to create large-area solar cells using roll-to-roll production process, giving high throughput, and reducing the cost further. However, the key disadvantage of the use of a-Si for solar cell fabrication is the defect density in the thin film. These defects arise from the "dangling bonds" associated with unsatisfied silicon atoms in the thin film. The defects create localized electronic states inside the bandgap of the semiconductor, which act as charge traps and recombination centers. The existence of these states increases the probability of recombination of charges and reduces the diffusion length of free carriers. Thus, the number of electrons available for conduction through the external circuit reduces for a given number of photons absorbed, causing the efficiency of the solar cell to drop. Thus, even though a-Si solar cells can be manufactured at a lower cost compared to crystalline silicon, the requirement of more solar modules and larger areal coverage for the same power output, offsets the cost advantages of using a-Si. The problem of defects in a-Si thin films can be solved to some extent by annealing the thin film in the presence of hydrogen. The resulting thin film, called hydrogenated amorphous silicon (a-Si:H), has significantly less defect density compared to a-Si. Thus, the efficiency of solar cells based on a-Si:H is higher compared to a-Si solar cells. However, solar cells based on a-Si:H suffer from a drop in their efficiency over the first few months of operation. The exact cause for this, known as the Staebler–Wronski effect, is still debated, however, it is ascribed to the degradation of bonds in the Si-H matrix because of exposure to light, leading to the creation of defect sites. The structure of the a-Si solar cell is typically of the form p-i-n, with a layer of intrinsic silicon sandwiched between p-type and n-type layers. This is done to increase the area where photon capture will lead to the dissociation of excitons.

The defect density in a-Si:H thin films, and the resulting low cell efficiency, have led to the development of several compound inorganic thin-film semiconductors for flexible solar cell fabrication. These materials include cadmium telluride (CdTe), copper indium gallium sulfide (CIGS), copper zinc tin sulfide (CZTS), and metal oxide nanoparticle thin films. Of these, CdTe is most widely available in commercial thin-film solar cells, particularly for semitransparent PV windowpanes. CdTe is a semiconducting material with a direct bandgap of 1.5 eV. The typical structure of the junction includes a thin layer of cadmium sulfide (CdS) as the n-type semiconductor, while a relatively thick layer of p-type CdTe is used for photon capture. The solar cell is constructed in an inverted manner, i.e., a transparent substrate is used to deposit the bottom conductor, the n-type CdS layer, the p-type CdTe layer followed by the top conductor. The solar irradiation enters the device through the transparent substrate. CdTe thin-film solar cells have found commercial success because of their relatively high efficiency of ~20% compared to a-Si:H solar cells. Further, because of the high photon absorption of CdTe, a thinner film of the material is required to obtain reasonable efficiency leading to higher flexibility. CdTe thin films can be deposited using several methods such as sputtering, evaporation, electrodeposition, laser ablation, and so on. However, in most cases, a thermal treatment is required to recrystallize the thin film and improve film morphology. Such a thermal treatment is not possible with plastic substrates. The use of flexible metal substrates results in the device fabrication steps being reversed because light cannot enter from the substrate side. This has led to the use of polyimide (Kapton) as a thermally stable substrate for the fabrication of high-quality flexible CdTe thin-film solar cells using the usual through-substrate structure. Another popular material used for solar cell fabrication is copper indium gallium selenide (CIGS). It is a compound semiconductor with a bandgap ranging from 1 eV to 1.7 eV depending on the relative proportion of the constituents. The absorption coefficient of the material is high leading to the requirement of a very thin film for solar light absorption. Further, the typical structure of the solar cell involves light entering the absorber layer from the front side, alleviating the constraint on the substrate and back contact metal to be transparent. This provides the opportunity to create highly flexible solar cells using any flexible substrate. Several processes can be used for depositing CIGS layers on flexible substrates. For example, all four elements in the material can be deposited together using co-evaporation. This allows for precise control over the thin-film stoichiometry and thickness, and is a process that can be done at low substrate temperatures. CIGS can also be deposited using reactive sputtering of copper, indium, and gallium in a selenium environment, or by first depositing the three metals followed by a "selenizing" step which involves annealing in a selenium environment. The latter process is expensive (creating a selenium vapor environment in the entire chamber) as well as involves high substrate temperature. High photon absorption and tunable bandgap enable very high-efficiency CIGS thin-film solar cells and the versatile deposition processes allow for integration on flexible substrates. However, the key disadvantage of CIGS (and that of CdTe) is the material cost involved, particularly because of the use of rare and expensive materials such as indium, selenium, and tellurium, even if the thin film deposited is only a few microns in thickness. Hence, other compound semiconductors such as copper zinc tin sulfide (CZTS) have gained popularity. CZTS is composed of earth-abundant and non-toxic elements and offers similar material properties such as high photon absorption and tunable bandgap, as compared to CIGS. Further, the deposition strategies used for CIGS can also be applied for obtaining CZTS thin films. However, solar cell efficiencies using CZTS have not reached CIGS levels because higher defect density in CZTS reduces the carrier lifetime.

The development of solution-based processing techniques, such as sol-gel, spin-coating, spraying, inkjet printing, etc., has led to the development of solar cells fabricated using solution-processed thin films. These deposition techniques offer flexible solar cell thin films, that can be deposited on large area substrates at room temperature. Thus, the manufacturing scalability of these processes, compared to vacuum deposition, can lead to very low-cost flexible solar cells. For example, particular "ink" chemistries have been developed to obtain thin films of well-known solar cell materials such as CIGS and CZTS. Typically, these inks are prepared using nanoparticles of the semiconductor that are separately fabricated, purified, and dispersed in a suitable solvent. This can be extended to create solutions of nanocrystals of group IV, III-V, or II-VI semiconductors (quantum dots). The solution can then be deposited on a substrate using any of the solution processing techniques, followed by a low temperature, post-deposition anneal for solvent evaporation, and film densification. Such a process is scalable, leading to low costs, however, the thin films obtained using this process have higher defect density and well-defined grain boundaries (at the edges of the nanoparticles) leading to lower conversion efficiencies compared to vacuum deposited thin films.

Finally, single crystal silicon, which is the most widely used material for solar cell fabrication, can also be used to fabricated flexible solar cells. The processes discussed in Chapter 4 can be used to create flexible pieces of monocrystalline silicon, which can then be used to create high-efficiency flexible solar cells. Apart from these processes, a unique process has been introduced by the Australian National University to produce flexible monocrystalline silicon solar cells, called Sliver® cells. In this process, a thick silicon wafer, around 1-2 mm thickness, is used. Rectangular sections are etched all the way through the silicon using laser scribing, dicing saw, or deep reactive ion etching. These sections are patterned such that thin "slivers" of silicon remain connected to the rest of the wafer after the etching process (Figure 16.4). The slivers are then further processed to create solar cells, detached

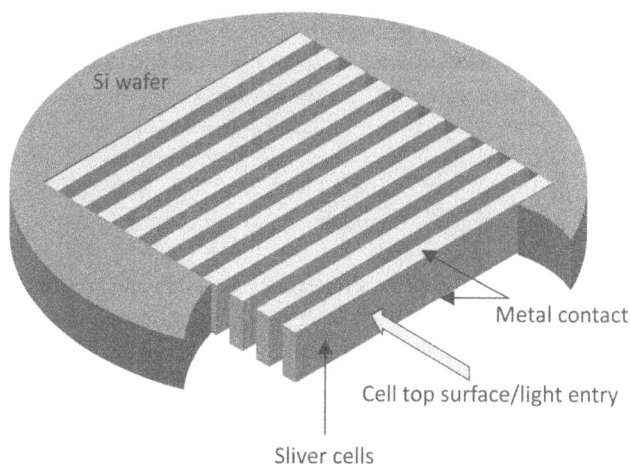

FIGURE 16.4   Sliver cells are created by pattering silicon wafers into thin slivers such that individual cell thickness is small enough for flexibility. The doping and contact formation are completed before creating the slivers.

from the original wafer, and integrated into modules. The thickness of these slivers, which determines their flexibility, can be controlled using the patterning process. This process helps reduce the amount of single-crystal silicon used for a solar cell because most of the silicon substrate used in rigid solar cell production is only there for handling purposes. The efficiency of Sliver® cells is reported to be comparable to that of rigid monocrystalline silicon solar cells.

16.2.1.2 Flexible Organic Photovoltaics

Organic photovoltaic solar cells (OPVs) are a class of solar cells that utilize organic thin films as the photon absorption layer. Like the inorganic thin-film solar cells discussed in the previous section, OPVs provide a way to obtain flexible solar cells through the use of low-temperature thin-film deposition processes that give an opportunity to use flexible plastic substrates. Another advantage of the use of organic semiconductors for flexible solar cell fabrication is the inherent flexibility provided by the use of polymer thin films. This provides for reliable solar cell operation under strain and after strain cycles. The principle of operation of OPVs is similar to that of inorganic solar cells, i.e., a photon is absorbed by the organic semiconductor and an electron-hole pair (exciton) is generated. As discussed in Chapter 5, an organic semiconductor consists of molecules with conjugated π-bonds, leading to molecular orbitals. The highest occupied molecular orbital (HOMO) and the lowest unoccupied molecular orbital (LUMO) behave like the valence band and conduction band in an inorganic semiconductor, respectively. The impinging photon excites an electron from HOMO to LUMO across the energy bandgap, from a bonding (π) to an anti-bonding (π^*) orbital, creating an exciton.

The key difference in organic semiconductors is that the binding energy associated with these excitons is much more than that in inorganic semiconductors. This is because inorganic lattice semiconductors typically have a higher dielectric constant compared to organic semiconductors, which leads to a reduction in the Coulomb attraction between excited electron-hole pairs due to dielectric screening. The binding energy of the exciton is a few tens of meV in inorganic semiconductors, whereas it is of the order of hundreds of meV in organic semiconductors. Thus, the excitons created in organic semiconductors are harder to dissociate into individual free charge carriers. If they do not dissociate in time, excitons typically recombine and give out the stored energy in the form of radiation or heat. This can be great for fabricating LEDs, because there is a greater probability for electron-hole recombination and photon emission, however, for photovoltaics, this is a cause of concern, because the conversion of a photon into electric current requires dissociation of the electron-hole pair. This problem is solved in OPVs using the layered structure of organic thin-film semiconductors. The absorption layer is divided into electron donor (ED) and electron acceptor (EA) layers such that the difference between their HUMO and LOMO energy levels is more than the binding energy of the exciton. This promotes spontaneous dissociation of the exciton into electron and hole – the electron is captured by the EA layer while the hole is captured by the ED layer. These charge carriers are then collected by the metal contacts for conduction in the external circuit. This structure is similar to the p-n junction in a silicon solar cell, however, there is no built-in potential or depletion at the OPV layer junction, which implies that there is no force making the carriers drift toward the junction. The excitons created in the OPV cell can only diffuse randomly toward the junction and get separated into individual charge carriers. The diffusion length of excitons inside organic semiconductors is very short, thus, the thickness of the EA and ED layers cannot be large. Otherwise, the excitons formed inside these layers will recombine before finding their way to the junction. However, the reduced thickness of the semiconductor reduces the absorption of a photon inside the semiconductor layer, creating a trade-off that needs to be carefully balanced by the choice of organic semiconductors and their thicknesses. This issue can be overcome by blending the EA and ED materials into a single layer, such that the junction between the two materials is spread throughout the layer. In such a design, called the bulk heterojunction (BHJ) structure, the exciton diffusion distance is reduced without reducing the thickness of the absorption layer (Figure 16.5). However, practical difficulties with contacting the individual layer separately for free carrier capture, make manufacturing challenging and reduce the overall efficiency. Commonly used EA materials include fullerenes and their derivatives because of their high electron affinity, for example, phenyl-C61-butyric acid methyl ester (PCBM). Electron donor (ED) materials have a wider variety, but typically include polymeric derivatives of polythiophene, polyfluorene, and PPV.

The choice of electrodes for OPV is also an important design problem because there are not many opportunities for controlling the Fermi level of organic semiconductors through doping. Thus, contact metals are chosen such that the work functions provide the electromotive force required for the separation and conduction of respective carriers at anode and cathode. Further, low carrier diffusion lifetime and the thin nature

Bilayer organic PV   **Bulk heterojunction organic PV**

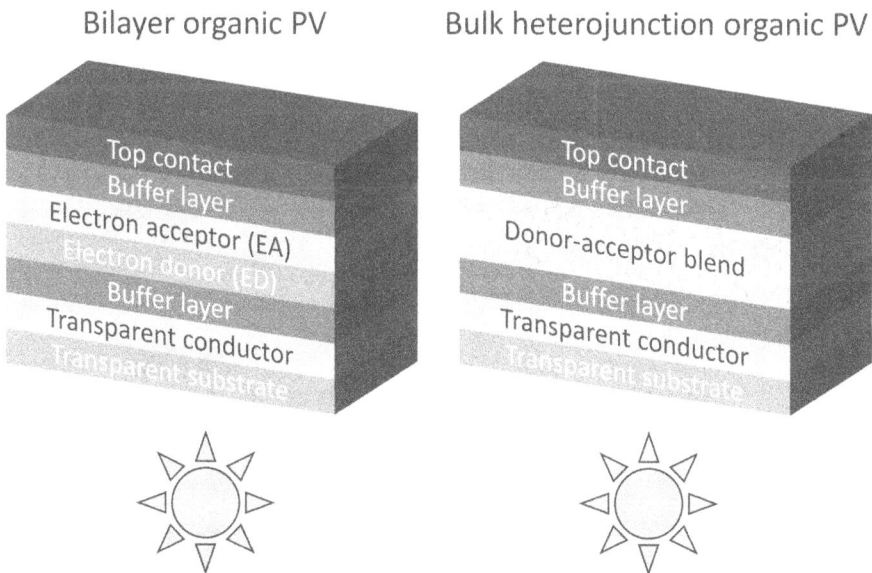

FIGURE 16.5 Organic solar cells are made using layers of thin films such as electron donor (ED), electron accepter (EA), buffer layers, and metal contacts. In some configurations, conversion efficiency can be improved by blending the donor and acceptor layers into one.

of the organic semiconductors necessitate the complete coverage of the semiconductor with metal electrodes. Thus, apart from the constraint of work functions, one of the electrodes needs to be transparent to allow for sunlight to enter the device. These constraints reduce the number of electrode materials available. A popularly used transparent metal electrode is indium tin oxide (ITO), because it has high transparency and conductivity, along with a relatively high work function of 4.7 eV. This makes ITO suitable for anode contact because the high work function leads to the Fermi level of the metal close to the HOMO level of the organic semiconductor in contact, which helps in the capture and transport of holes, out to the external circuit. For the construction of cathode, a metal with a low work function is required, so that the Fermi level of the metal is close to the LUMO level of the organic semiconductor. However, most metals with low work functions such as lithium (2.9 eV), potassium (2.3 eV), sodium (2.4 eV), cesium (1.9 eV) are highly reactive and unstable in the atmosphere. In most cases, a bilayer of aluminum (4.2 eV) and calcium (2.9 eV) is used, such that calcium is in contact with the organic semiconductor, while the aluminum layer protects the calcium layer from atmospheric exposure. The metal electrode layers are typically deposited using physical vapor deposition processes such as sputtering or evaporation.

The structure of the OPV depends on the strategy used for light penetration. If solar light is to enter the organic semiconductor through the substrate, the transparent conductor (ITO) is deposited on the substrate first as the anode, followed by the ED layer, EA layer, followed by the cathode. This structure is known as the "conventional" structure. As the name suggests, this has been the conventional structure used for OPV manufacture. However, the "inverted" structure can also be fabricated with the light entering the OPV from the top electrode. In this case, the cathode is deposited on the substrate, followed by the EA layer, ED layer, and the transparent anode. This structure is gaining traction because it allows for the reactive cathode layer to be hidden between multiple layers of materials on one side and a relatively thick substrate on the other side.

The structure of OPVs also includes additional buffer layers for various purposes. These layers are typically between the metal electrode layers and the photoactive layers. These can be electron or hole block layers that allow the passage of the correct charge carrier to the electrode while impeding the other type. This is typically done by selecting materials with the right HOMO and LUMO levels. Buffer layers can also be used to act as a barrier for contact metal diffusion into the photoactive layer or to protect the photoactive layer from damage during processing. The layered structure of the OPV, thus, resembles that of an OLED, with the key difference being the material selection is such that exciton dissociation is encouraged in OPVs, whereas exciton recombination is encouraged in OLEDs.

16.2.2 ENERGY HARVESTING DEVICES

The rise of wearable and IoT devices has created a use-case for technologies harvesting usable electrical energy from no particular source, i.e., from ambient energy. These include the use of piezoelectric generators to capture energy from movement, triboelectric generators to capture energy from stray charges, and thermoelectric generators to capture energy from temperature differences. These energy sources

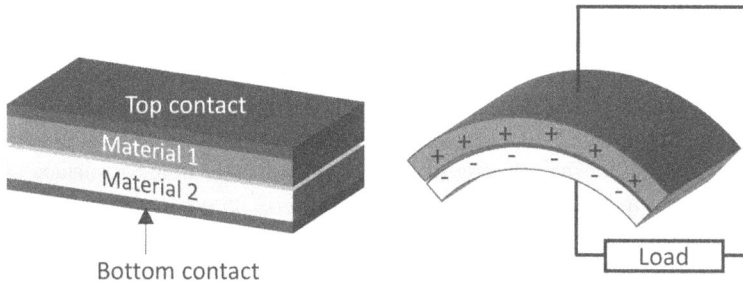

FIGURE 16.6   A triboelectric generator is based on the chargers developed because of friction between two dissimilar materials. The current in the external circuit is caused because of the induction of chargers on the metal electrodes.

are already present in the ambient environment of the user. Because the use-case for these energy harvesters typically exists in the wearable, IoT, and biomedical fields, there has been significant effort in making flexible versions of these devices.

The structure of triboelectric generators involves the use of two thin films touching each other (Figure 16.6). These films can be either inorganic or organic polymer materials and are typically held together only at the edges, without the use of an interlayer binding material. Metal electrode thin films are deposited on either side of the bilayer to connect the device to the external circuit. When the two active layers touch or rub against each other, opposite charges accumulate on their surface. This leads to an electric field being generated that induces opposite charges on the metal electrodes. The change in the capacitance of the system, because of bending, causes a current to flow in the external circuit, performing electrical work. When the system is released, a reverse current is induced in the external circuit. Thus, the electrical current generated by the system resembles an alternating current generation system for periodic bending cycles. The key advantage of the use of triboelectric generators is the simple design of the system. Commonly available low-cost organic polymer materials such as polyimide (Kapton), polytetrafluoroethylene (PTFE), PDMS, PMMA, PET, etc., have been shown to be promising active material layers for the fabrication of triboelectric generators. Triboelectricity also has the potential to produce very high voltages, of the order of hundreds of volts open-circuit voltage (V_{oc}). Further, because the system is based on thin films attached together, it can be made to be completely flexible, providing an opportunity for attaching to any surface for harvesting energy. Examples include the use of everyday mechanical motions such as walking, driving, cycling, rotation of tires, and the motion of doors and windows panes. The major disadvantage is that the amount of power generated per cycle of motion is extremely small, leading to the need for extensive structure optimization and parallelization, and for the use of energy storage solutions. However, while the power generated may not be sufficient to power other systems, a practical use-case for triboelectric systems is to act as self-powered sensor systems for detecting motion and touch. Further, because triboelectric systems are based on mechanical touch and friction to create surface charges, these systems are susceptible to mechanical wear, leading to a reduction of the output over a period of time.

Piezoelectric energy generators make use of certain material systems that generate surface charges when mechanically strained. This is different from the induction of surface charges due to triboelectricity because a single material is involved in charge creation. In a piezoelectric material, the application of mechanical strain changes the alignment of the internal electrical dipoles, leading to a change in polarization. This leads to the creation of an electric field that induces surface charges that balance the induced electric field (Figure 16.7). The physics of piezoelectricity may differ from that of triboelectricity; however, the applications of flexible thin-film energy generators are very similar – the use of mechanical energy from everyday motion to produce electrical energy. However, unlike triboelectricity, there are few thin-film materials that can be used to create a piezoelectric energy harvesting system. Nevertheless, a key advantage of the piezoelectric effect is that it is reversible, i.e., the application of electric field across the piezoelectric film produces a mechanical strain. Thus, piezoelectric systems can be used to produce energy generators, sensors as well as mechanical actuators. In practice, piezoelectric energy harvesters are created using a fixed-free cantilever structure made using piezoelectric thin films, because such a structure can produce large mechanical strains during vibration. A mass is attached at the tip of the cantilever to bring the resonant frequency of the cantilever close to the expected actuation frequency. A vibration

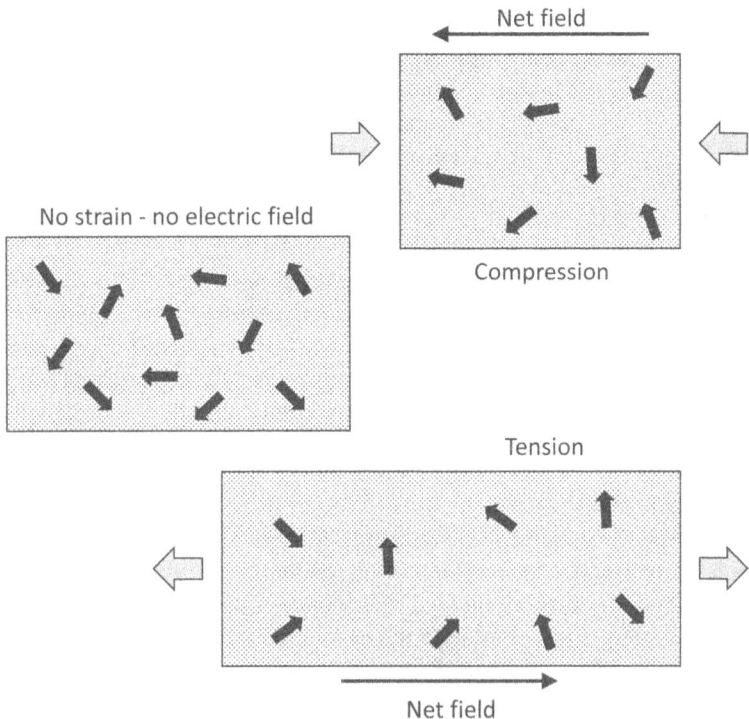

FIGURE 16.7 The piezoelectric generator produces a net electric field in the direction of strain because of the realignment of internal electrical dipoles. The strength and direction of the field depend on the strength and direction of strain.

induced in the cantilever produces an alternating electrical current in the attached electrodes. There is a large variety of materials available for use as piezoelectric materials, however, many of them, such as piezoceramics are not easily obtained as thin films. This is mainly because piezoelectricity is highly dependent on the crystal structure and defects can cause significant degradation in the performance of such materials. Nevertheless, lead zirconium titanate (PZT) is a ceramic piezoelectric material widely reported in flexible electronics literature, that can be deposited using low-temperature techniques such as sol-gel and sputtering. Apart from PZT, barium titanate ($BaTiO_3$), a perovskite material, has received great interest from the scientific community because of its excellent piezoelectric coefficient (electric flux density per unit applied stress). Another commonly reported ceramic thin film for piezoelectric applications is aluminum nitride (AlN). In terms of organic materials, a well-studied piezopolymer is polyvinylidene fluoride (PVDF) and its copolymer with trifluoroethylene (TrFE), PVDF-TrFE. Although organic materials have lower piezoelectric coefficients compared to their inorganic counterparts, these materials provide a much larger range of strain. Further, organic materials are easier to form into thin films, are lighter, flexible, and require less processing. In both organic and inorganic thin films, it is important to note that piezoelectric properties are highly dependent on the lattice structure or the orientation of the material domains in the thin film. Hence, the deposition process, post-processing, and optimization play a critical role in determining the functionality of the system.

The thermoelectric effect, discovered by Thomas Johann Seebeck in 1821, can also be used to obtain usable electrical energy from waste heat energy. The basic principle behind the "Seebeck effect" is the diffusion of carriers because of the applied temperature gradient. This develops an electrostatic potential that depends on the materials involved and the magnitude of the temperature difference. Typically, the voltage produced by thermoelectric generators is low, of the order of a few mVs at best, hence they are stacked together or interconnected in order to obtain a high output. The applications for flexible thermoelectric generators (TEGs) are not as vast as those for triboelectric or piezoelectric energy harvesters because there are not many sources of constant or predictable temperature gradients in everyday life. The materials of choice for TEG applications have been tellurides of bismuth (Bi_2Te_3), antimony (Sb_2Te_3), and lead (PbTe), because of their high Seebeck coefficients, which is defined as the voltage produced per unit temperature difference. However, these materials have a relatively high thermal conductivity that causes heat to flow through the TEG material, reducing the temperature gradient. Thus, the thermoelectric figure of merit, which indicates the performance for a given material as a TEG, has been defined as, zT:

$$zT = \frac{\sigma S^2 T}{k}$$

where σ, S, and k are the electrical conductivity, Seebeck coefficient, and thermal conductivity respectively, and T is the temperature. Clearly, reducing the thermal conductivity of a material can increase the power output, however, this is not very straightforward because the electrical conductivity of the material should be high at the same time. Generally, the materials that have high electrical conductivity, like metals, also have a high thermal conductivity because free charge carriers can diffuse and carry heat across the material quickly. Hence, the challenge in TEG design is to obtain a material system that will have high electrical conductivity and low thermal conductivity simultaneously. Further, because of the rarity of tellurium, the cost of traditional TEG materials is too high for any large-scale commercial application. Hence, efforts are being made to create alternative material systems to the traditional inorganic TEG materials based on metal tellurides. Interestingly, organic conductive polymers such as polyacetylene, polyaniline, polypyrrole, PEDOT, etc., have low thermal conductivity because of the lower free charge carrier concentration compared to metals. Thus, they become candidate materials for high zT thin films. However, for obtaining a thermoelectric generator, both p-type and n-type conduction is required. Whereas p-type conductive polymers, such as PEDOT:PSS, are common, it has been difficult to obtain n-type conductive polymers that are stable in air. This has led to the use of nanocomposites of inorganic materials with organic polymers. In this case, conductive polymers are used as a matrix to incorporate nanostructures such as nanoparticles, nanowires, and nanorods of inorganic thermoelectric material, such as tellurium, bismuth telluride, and antimony telluride. In such cases, the inorganic materials provide a high Seebeck coefficient, while the polymer matrix reduces the thermal conductivity of the thin film. Further, all carbon nanocomposites such as CNT/PEDOT have been reported as effective thermoelectric materials.

The application of energy harvesters in wearables and implantable electronics promises to usher in a new era of biomedical devices, which will not require a battery change or recharging. However, it is important to understand that energy harvesters are not creating energy out of nothing – that is impossible because of the Law of Conservation of Energy. Energy harvesters simply utilize the energy that would have gone to waste to produce minute amounts of usable energy. While this is exciting, one should remember to check whether the source of energy for an energy harvester is *actually* wasted energy. For example, once fossil fuels are burned inside an internal combustion engine to produce motion, the heat produced is completely incidental, and can be used to produce electricity using thermoelectric generators. However, using TEGs to produce electricity using the difference between the temperature of a human body and ambient can be problematic, because the TEG essentially "loads" the human body, causing it to do more work to maintain the said temperature difference. Similarly, stray mechanical vibrations from industrial equipment may be used for electricity generation using piezoelectric generators, however, placing such a device on the heart can effectively load the free motion of the heart, ever-so-slightly, causing the most important muscle in the human body to do more work.

16.3   ENERGY STORAGE

The biggest concern with the thin film source of energy such as photovoltaic, triboelectric, piezoelectric, and thermoelectric is that they are intermittent sources of power. Thus, the power being generated through these methods may not be available at the time when the electrical work needs to be performed. Further, the quality and quantity of power produced using these methods may not match the requirements of the load. This problem can be solved using energy storage solutions. Indeed, even at the grid level, the push for renewable energy sources requires the development of substantial energy storage capacity to compensate for the intermittent nature of solar and wind power. Typically, the electrical energy is converted into another form of energy for storage and then converted back to electrical energy when needed. A simple example is the use of electrical batteries (electrochemical storage) for storing power generated using solar cells during the day, so as to be used during the night. An energy storage device, such as a battery, ideally provides constant or on-demand power, while being charged using an intermittent and as-available power supply. Because of the recent push for renewable energy sources, there has been a great interest in developing large-scale energy storage solutions. These include diverse avenues for storing the energy, such as mechanical energy (flywheel technology), gravitational energy (water pumping, stacking of concrete blocks), chemical energy (batteries), electrical energy (capacitors, supercapacitors), thermal energy (chilled water), and so on. There are advantages and disadvantages associated with all of these technologies. From a flexible electronics point of view, we are interested in technologies that can be implemented in a thin-film form. These include the electrical cell (or battery), where energy is stored as electrochemical potential; a capacitor, where energy is stored in the form of surface charge density; and a supercapacitor, where both electrochemical and surface charge storage is utilized. The various technologies used for energy storage are quantitatively evaluated using parameters such as total energy capacity, energy density (energy capacity per unit volume or energy per unit mass), power density (power output per unit volume or energy per unit mass), and energy leakage over time. Apart from these, cost and scalability are always important considerations for the commercialization of any technology.

Off the thin-film technologies used for flexible energy storage, capacitors are the simplest to fabricate. They can be produced as two films of metal separated by a dielectric and can be integrated with any fabrication process flow. The MIMcap (metal-insulator-metal capacitor) structure can be fabricated using a variety of materials for both metal and dielectric. The limited constraints on fabrication and integration make capacitors ideal for storing electrical energy. Further, capacitors can be charged at a much faster rate compared to other battery technologies. This has paved the way for their use in applications requiring instant bursts of energy, such as camera flashes. Because capacitors store energy directly in the form of electrical energy, there are no losses due to the transduction of electrical energy into another form for storage and back into electrical energy for usage. However, there are some key issues limiting the usability of thin-film capacitors for energy storage. First, the energy stored in a capacitor can quickly dissipate because of the current leakage through the dielectric. This is particularly the case when large area thin-film dielectrics are used because the short distance between the electrodes reduces the leakage resistance. Second, because the energy stored in the capacitor is directly proportional to the capacitance, a very large capacitor is required to store meaningful amounts of energy. Large area MIMcaps are challenging to fabricate because a defect anywhere in the dielectric can cause the entire capacitor to short. This is typically mitigated by fabricating separate capacitors and connecting them into a capacitor bank, however, this strategy increases the integration complexity. Finally, the dielectric layer used for capacitance fabrication can break down under continuous electrical stress because of energy storage. Thus, the reliability of the dielectric layer, particularly under mechanical bending, needs to be studied for long-term energy storage using capacitors.

16.3.1 FLEXIBLE BATTERIES

Electrical batteries can produce an electric current using chemical reactions and vice versa. The fundamental principle of operation in most rechargeable batteries is a redox reaction occurring because of the application of electric current (electrolytic cell) and the same redox reaction leading to the generation of electric current (Galvanic cell). The basic components of any battery are the anode and cathode electrodes and an electrolyte that conducts ions from anode to cathode and back. In practical implementations, a separator film is inserted between the cathode and anode. It is a permeable membrane used to prevent an electrical short between the cathode and anode, while allowing ion exchange through the electrolyte. Current collector electrodes are also connected to the cathode and anode to provide a way to connect the battery to the external circuit. The choice of cathode, anode, electrolyte, and separator material depends on the electrochemistry being used for cell operation. In the case of flexible batteries, all these basic components of the battery need to be flexible for the complete battery to be flexible.

The most commonly used rechargeable battery technology in modern electronics is the lithium-ion technology (Li-ion). The battery operates through the exchange of lithium ions from the anode to the cathode and back through the electrolyte. Both anode and cathode are made using materials that can reversibly incorporate lithium atoms in their lattice structure. The process of incorporation of lithium is called intercalation and the process of removal of lithium is called deintercalation. A typical implementation of Li-ion cell consists of lithium cobalt oxide ($LiCoO_2$) as the cathode and graphite (C_6) as the anode. During charging, the positive lithium ions from $LiCoO_2$ travel through the dielectric, combine with the supplied electrons and intercalate into the graphite electrode as lithium atoms. During discharging, the lithium atoms at the anode dissociate into positive lithium ions and electrons. The electrons are collected by the current collector electrodes, travel across the external circuit, and combine with the positive cobalt ions at the cathode, while the positive lithium ions travel through the electrolyte and intercalate in the cobalt oxide lattice. Electrical work is done by the electrons during this process. The structure is schematically illustrated in Figure 16.8. The Li-ion batteries provide high energy and power density, low energy leakage, and scalability (Li-ion batteries with capacities ranging from 1 mAh to 150 MWh have been produced), making them the darling of the consumer electronics industry. Thus, Li-ion technology finds applications from toy cars to real cars. A flexible version of this chemistry is possible if all the components are integrated into a thin-film form on a flexible substrate.

Electrolytes are one of the key components of any battery technology. They provide a medium for ion exchange between cathode and anode while discouraging electron exchange. However, they are also the hardest to obtain in a solid-state,

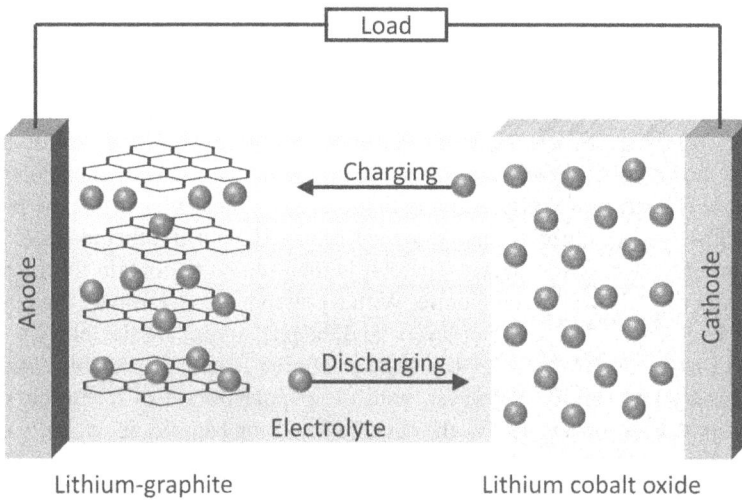

FIGURE 16.8 Lithium ion cells operate through the exchange of lithium ions between anode and cathode. During charging, lithium ions move from cathode to anode and collect the supplied electron, while during discharging, lithium ions lose an electron and move from anode to cathode.

thin-film form, marking a key challenge in flexible battery integration. Traditionally, liquid electrolytes are used to provide high ion mobility and good contact with electrode material (wetting). However, the use of liquid electrolytes makes the integration of batteries challenging because of electrolyte encapsulation and leakage issues, particularly upon bending. Hence, solid electrolytes have received increased attention in the past decade. These electrolytes include solid-state inorganic thin films, polymer gels, and composite materials. For example, lithium phosphorous oxynitride (LiPON) is an amorphous solid that has been used as a solid electrolyte in Li-ion integration. With the use of LiPON, the complete Li-ion battery can be fabricated using depositions of the current collectors, anode, cathode, and electrolyte thin films. If this deposition is done on a flexible substrate with low-temperature processes, a flexible Li-ion battery can be realized. The use of LiPON to create a completely solid-state battery has been demonstrated in many reports in the literature. Other solid-state electrolytes include phosphates and silicates of lithium. A key disadvantage of the use of solid-state electrolytes is that they are inherently brittle materials, requiring very low thicknesses to achieve the desired flexibility, which can lead to a higher rate of Li-ion leakage. Hence, a popular alternative to these electrolytes is the use of polymer composites. Lithium polymer (LiPo) batteries are based on the Li-ion chemistry, using a polymer electrolyte, instead of liquid or solid-state electrolytes. The polymer electrolytes consist of lithium salts incorporated into a polymer matrix. The lithium salts provide Li-ion mobility while the polymer matrix provides flexibility and wettability to the electrolyte. The batteries based on polymer electrolytes are typically lighter than traditional Li-ion batteries, thus providing better energy and power density. The polymer matrix can be based on polyethylene oxide (PEO), or polyvinylidene fluoride (PVDF), while lithium salts with relatively low dissociation energies are used to facilitate ion transport. Such thin films provide much higher inherent flexibility compared to inorganic thin-film electrolytes. Further, they can be deposited using low temperature, solution processing techniques, reducing the cost of integration. Some implementations of the Li-ion chemistry make use of polymer gels as electrolytes. These electrolytes are not technically "solid" but are based on the polymer/lithium salt composite structure that solid polymer electrolytes use. These gels are created by trapping liquid organic molecules inside the polymer matrix. They are typically pumped into the Li-ion cell, with all other material layers, including encapsulation, in place. The cell is then sealed to contain the gel inside, completing the LiPo cell manufacture. With a polymer gel electrolyte, the complete cell can be flexible if all the other layers (and the packaging) are flexible.

The Li-ion battery technology has been a defining feature of portable electronics in the twenty-first century. However, with the popularization of renewable energy technologies and electric mobility, the demand for Li-ion batteries is set to sky-rocket. This has some energy experts concerned because lithium is a relatively rare element on earth. In the absence of robust recycling infrastructure, the supply-demand imbalance for lithium metal may make Li-ion technology commercially unviable. Hence, there has been a push to find alternatives to the Li-ion chemistry for rechargeable batteries. One of these promising technologies is the sodium ion battery (Na-ion or NIB) technology. These batteries make use of Na+ ions instead of Li+ ions for energy storage, otherwise, the construction and working principle are similar to the Li-ion batteries. However, graphite cannot be used as an anode because the larger sodium ion does not intercalate into graphite, thus, a non-graphite disordered carbonaceous anode is used. Na-ion batteries can solve the biggest challenge facing Li-ion industry today – the availability of lithium – because sodium is the sixth most abundant element on earth, around $1000\times$ more by weight than lithium. There has been some progress in recent years on electrode and electrolyte construction for achieving flexible Na-ion batteries. Other promising technologies include aluminum ion, magnesium ion, zinc ion, and so on. All of these chemistries rely on a concept similar to that of Li-ion and can provide some advantages, for example, the use of Al^{3+} ions allows for the release of three electrons for every ion of aluminum, increasing energy storage capacity. However, the development of anode, cathode, and electrolyte materials for creating stable and flexible versions of these technologies is still being investigated.

16.3.2 FLEXIBLE SUPERCAPACITORS

Supercapacitors are capacitive structures that use an electrolyte to create much larger electric fields compared to conventional capacitors. The structure consists of the two electrodes in contact with an electrolyte and uses the buildup of ions in the electrolyte upon application of an external electric potential. The buildup of charge on the metal electrode attracts the ions of opposite polarity from the electrolyte, leading to a buildup of the oppositely charged ions at the metal surface. This double-layer capacitance is used to store energy. The chemistry of the metal electrode and the electrolyte is chosen such that no redox reaction takes place at the metal/electrolyte interface. The buildup of charged ions at the metal electrode surface effectively leads to a very thin (one atomic layer) dielectric between the two charge layers, causing the capacitance of the structure to be extremely high (hence the term "super"capacitor). The structure is schematically illustrated in Figure 16.9. The high capacitance exhibited by the structure can be used to store much larger amounts of energy compared to regular capacitors. Thus, supercapacitors bridge the gap between MIMcaps and metal-ion batteries, in that they use electrostatic energy storage, but involve ion transporting electrolytes. The key advantage of supercapacitors is that their charge/discharge rate is much higher than Li-ion batteries. Further, the absence of chemical reactions in their operation leads to much better lifetimes compared to Li-ion batteries. However, a major drawback in the use of supercapacitors for energy storage is their energy density compared to Li-ion batteries. This is mainly because energy is stored only on the surface of the electrode, leaving the bulk volume of the electrode wasted. This is one of the reasons why the creation of thin film-based flexible supercapacitors can be significant. In the electric double-layer supercapacitors, the key factor determining energy storage capacity is the area of the electrodes. Hence, many state-of-the-art implementations of supercapacitors utilize graphene, CNT, and other conductive carbonaceous material electrodes to enhance the electrode surface area (high surface-to-volume ratio). These electrodes can be fabricated in thin-film form using several low-temperature deposition processes on a variety of flexible substrates. However, as in the case of batteries, the key problem for the fabrication of flexible supercapacitors is the integration of an electrolyte. Approaches similar to those used in Li-ion batteries have been demonstrated, for example, solid-state (inorganic and organic), liquid (aqueous acidic or alkaline solutions), and polymer gel electrolytes have been used to fabricate flexible supercapacitors.

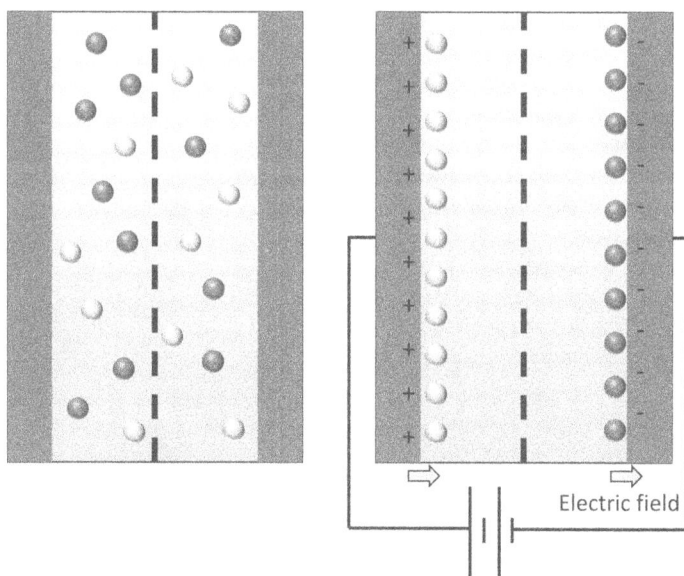

FIGURE 16.9 A supercapacitor consists of an ionic electrolyte that produces a layer of charged ions when the capacitor is charged. This structure stores energy in the form of the electric field produced.

The significance of energy generation and storage in the realization of truly free-form electronics has paved the way for innovations across designs, processes, architectures, technologies, and materials. Wireless communication makes electronics free of data links, however, electronics can only be truly portable if they are also free of the requirement of frequent charging. This is why it is important to develop new and efficient methods for energy generation and storage. There have been significant advances in the past few decades, for example, the invention and proliferation of Li-ion batteries have made many of our current devices possible. However, the battery capacity, weight, and volume continue to be the main bottleneck in the creation of wearable devices. With all the other modules of an electronic system such as processor, memory, display, transceiver, and antenna available in thin-film form, it is important to make progress toward a highly flexible battery to achieve the next generation of wearable devices.

EXERCISES

16.1. We plan to deploy a solar cell to power a system that requires a constant power of 2 W. The average daily solar irradiation is 1000 Wh/m^2. What should be the area of the solar cell if the module efficiency is 20%? What should be the capacity of the battery connected to this solar cell so that constant circuit operation is ensured? What will be the area and battery capacity if the solar cell efficiency is 10%?

16.2. Why is the bonding energy of an exciton higher in organic materials compared to inorganic? What effect does this have on organic photovoltaic cell design?

16.3. What are the major energy harvesting technologies involving thin-film fabrication technology? What are their relative advantages/disadvantages?

16.4. How does the lithium-ion cell work? What is the main challenge in the development of flexible Li-ion batteries?

16.5. What is the difference between capacitive energy storage and supercapacitors? Provide the pros/cons for each technology, from a flexible electronics point of view.

17 Flexible Sensors and Actuators

17.1 INTRODUCTION

For a majority of IoT applications, sensing and actuation are the primary purposes of the electronic system. We have discussed the materials and integration strategies that are involved in creating flexible processors, memory, display, and power supply modules. Now, we are at the "End Game" of the fabrication of a flexible electronic system. There is a digital world inside the memories and processors of our computers, laptops, smartphones, and other electronic gadgets that is dominated by digital signals – 0 and 1. Everything, including books, paintings, sounds, and images, is stored, communicated, and processed in the digital world using binary information. However, the physical world we live in is not based on digital and discrete states. Time and space seem to be continuous for the most part (unless we zoom into the quantum world), and different types of information like sights, sounds, and smells seem to be encoded differently. Sensors and actuators help bridge this gap between the digital signals of the electronic circuit and the physical signals of the real world (Figure 17.1). Both are essentially transducers that convert one form of energy into another. Sensors convert physical signals into electrical responses, while actuators convert electrical signals into physical responses. Even in the natural world, we find the transduction from the physical world into electrical signals for processing and storage, when we examine the way humans and animals interact with their environment. Eyes, ears, skin, tongue, and nose continuously monitor and sense information about the environment and transduce them into electrical signals for the brain to process. Similarly, our muscles actuate based on the electrical information received from the brain. The architecture of modern electronic systems takes a similar form, with sensors and actuators being the means to interact with the environment. The physical principles and materials used for sensing and transducing each type of physical signal are different. This also implies a change in the integration strategy, both in its rigid and flexible form. Hence, it is important to understand the underlying physical principles behind the transduction to be able to design integration strategies for flexible sensors and actuators.

The electrical signals produced by sensors after transductions are typically very small. This is because the signals are produced by small changes in sensor properties because of a variation in the physical parameter being sensed. In fact, in some cases, the electrical signal being produced by a sensor may not be voltage but a change in another electrical property such as resistance or capacitance. In these cases, the change is first converted into a proportional electric voltage or current signal. These small electrical signals are then amplified and filtered to obtain a usable electrical output. This output is finally converted into digital form using analog-to-digital

DOI: 10.1201/9781003010715-21

255

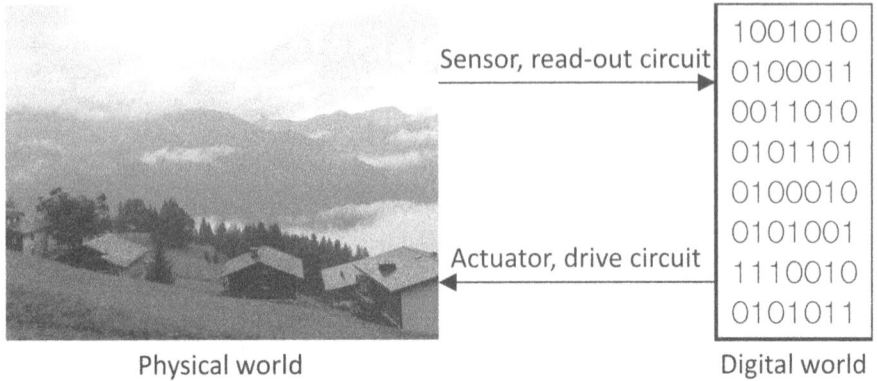

FIGURE 17.1 Sensors and actuators are transducers that convert physical signals into electrical signals and vice versa. They are typically accompanied by read-out and drive circuits to perform amplification, filtering, digitization, and so on.

circuits. The complete circuitry used to convert, amplify, filter, and digitize the transduced signal are called sensor readout circuits. Similarly, while going from the digital world to physical actuation, it may be important to amplify the power of the electrical signal and convert it into an analog form. The circuits tasked with this work are called actuator drive circuits. Because most of these circuits are based on transistors (MOSFETs or TFTs) that have already been covered in this book, we will focus on the transduction part in this chapter.

17.2 FLEXIBLE SENSORS

17.2.1 TEMPERATURE SENSORS

A temperature sensor produces an electrical output corresponding to the temperature of a particular surface. It is one of the most important sensors in wearable devices for biomedical applications. Temperature sensors also play a key role in controlling many industrial and manufacturing processes. Some of these applications, such as sensing the temperature of the human skin, require conformality with the sensed surface. This has led to the development of thin-film temperature sensors on flexible substrates. Change in temperature affects various properties of materials, such as physical dimensions, resistance, dielectric constant, elasticity, and so on. A temperature sensor can be created by monitoring any of these properties to obtain an output readable by an electrical circuit.

A commonly used technique to sense temperature is to monitor the electrical resistance of a material. These sensors, known as resistive temperature detectors (RTD), typically use metals such as platinum (Pt), nickel (Ni), or metal alloys such as nichrome (Ni-Cr), as the sensing material. These metals typically experience an increase in resistivity on account of an increase in temperature. The physical principle at play, in this case, is the increase in the scattering of electrons by the lattice atoms that vibrate more intensely because of an increase in temperature. This scattering decreases the number of charge carriers crossing over a given cross-section of the metal (current density), for a given electric field, reducing the conductivity of the metal (and increasing resistivity). The response of resistance to temperature is generally given by the temperature coefficient of resistance (TCR) as:

$$\frac{dR}{R} = \alpha \, dT$$

where α is the TCR (measured in K^{-1}), R and T are the resistance and temperature, while dR is the change in resistance because of dT. An increase in resistance because of an increase in temperature, as is the case in metals, leads to a positive TCR. The solution to this expression is an exponential rise in resistance with temperature, however, for a small range, the temperature response is typically linearized as:

$$R = R_0 \left(1 + \alpha \left(T - T_0\right)\right)$$

where R and R_0 are resistances at temperatures T and T_0. Metals such as platinum, nickel, and nichrome are commonly used because they have a very stable and linear change in resistance with temperature, particularly in the temperature range close to room temperature. Further, these metals do not corrode or rust, thus remaining functional for a long time. While these metals can be expensive, because the dimensions of the sensors can be very small, they do not add significantly to the device cost.

Thermistors are also a class of thin-film materials that are widely used for temperature sensing. These materials, typically semiconductors or metal oxides, have a much stronger resistivity change in response to change in temperature (sensitivity). In semiconductors, an increase in temperature reduces the resistance offered by the material. This effect, opposite to that in metals, occurs because an increase in temperature increases the number of charge carriers available for conduction through thermal excitation. The scattering due to increased lattice vibrations, as in metals, also occurs, however, its effect is normally overpowered by the increase in charge carriers. These competing phenomena cause the linearity of the temperature response of thermistors to be limited; however, over a specific limited range, thermistors are able to provide better sensitivity, accuracy, and precision compared to RTDs.

17.2.2 STRAIN SENSORS

A strain sensor measures the mechanical strain in a system. This is a fundamental sensor because it can be modified to measure many other mechanical parameters such as motion, stress, pressure, force, acceleration, and so on. In the flexible form, strain sensors can be attached to the human body to monitor muscle and joint movement for various biomedical and physiotherapy applications. Strain sensors are primarily based on the change of an electrical characteristic of material upon application of strain. Resistive strain sensors can simply be a metal thin film that brings about a change in resistance upon application of strain because of the change in geometric dimensions. Further, the applied strain changes the relative position of atoms in a metal lattice (ever-so-slightly) causing a slight change in the material resistivity, however, the change in geometric dimensions is the dominant phenomenon, leading to a change in resistance according to the equation, $R = \rho l\,/\,A$. The relative change in resistance of a strain sensor, per unit applied strain, is called the gauge factor:

$$GF = \frac{\left(\dfrac{dR}{R_0}\right)}{\varepsilon}$$

where R_0 is the resistance at zero strain and dR is the change in resistance upon application of strain ε. A higher GF value denotes better sensitivity to strain leading to more accurate measurements. The value of GF depends on the material of choice, the deposition process, electrodes, plain of strain application, and so on. Resistive strain gauges are widely used in mechanical engineering applications in the form of load cells, wherein strain gauges are attached to a metal structure with known mechanical properties. The applied force causes a strain in the metal structure that is detected using a strain gauge. Thus, strain gauges are used to measure a large array of physical quantities such as strain, stress, force, pressure, and displacement. Strain gauges can be constructed using metal or other conductive thin films and are shaped in the form of a coil to increase the contribution of the direction of application strain in the overall resistance of the device (Figure 17.2). The simplicity of strain gauges, and the facile nature of fabrication on any substrate, makes them ideal for flexible electronic applications. However, the resistive strain gauge is sensitive to changes in temperature, which can also change its resistance. To mitigate this problem, the change in resistance of the sensor is typically read out using a Wheatstone Bridge

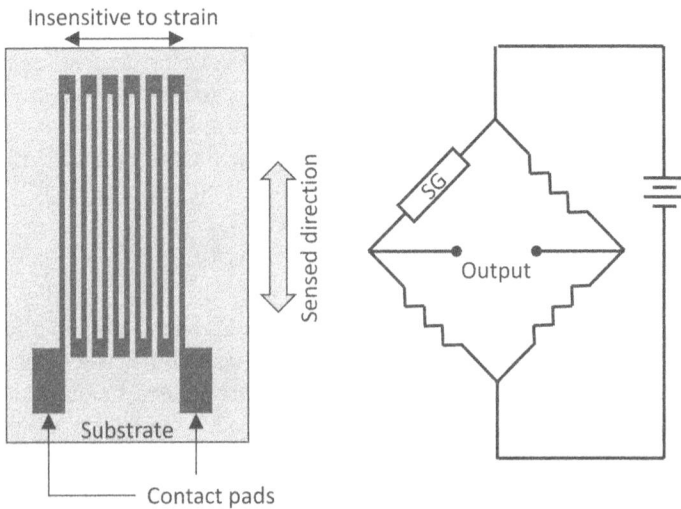

FIGURE 17.2   A resistive strain gauge consists of a simple metal foil structure on a compliant substrate. The structure is typically deployed using a Wheatstone bridge configuration to obtain an output voltage proportional to the strain applied.

circuit configuration, with one arm as the sensor in strain, and another arm as an identical relaxed sensor. Thus, a change in the resistance of the sensors due to temperature affects both the arms equally, canceling out its effect.

Piezoresistive materials are those that exhibit a large change in resistivity upon application of strain. These are normally semiconductors, wherein the electronic band structure is highly sensitive to the interatomic distances. The application of strain changes the interatomic distances, leading to a change in electron transport properties of the semiconductor lattice. This leads to a change in resistivity of the material. The change in geometric dimensions also contributes to the overall change in resistance, however, in piezoelectric materials, the change in resistivity dominates. The relative change in resistivity per unit applied stain is called the piezoresistive coefficient:

$$\rho_\sigma = \frac{\left(\dfrac{d\rho}{\rho_0}\right)}{\varepsilon}$$

where ρ_0 is the resistivity for zero strain, and $d\rho$ is the change in resistivity for applied strain ε. The change in resistivity in semiconductors is much higher than the change in resistance in metals because of variation in geometric dimensions. Thus, strain gauges based on the piezoresistive effect find applications in devices with high sensitivity over a short-range. Further, semiconductor resistivity is also more sensitive to temperature compared to metals, leading to the need for more elaborate temperature compensation.

Despite the difference in the underlying physics and material, both metal-based and piezoresistive strain gauges can be fabricated using thin films on stretchable substrates. The electrodes are connected to the ends of the strain gauge which are typically not designed to undergo strain. A stretchable substrate is required because the entire assembly needs to undergo strain in order for the strain gauge to register a response. Thus, polydimethyl siloxane (PDMS) is a popular substrate for strain gauge fabrication and encapsulation. Further, it is important that the thin film maintains conductivity during and after application of strain, hence, resistive strain gauges based on metal or semiconductor thin films are not used for large strain detection. In such applications, percolating thin films of conductive metal nanowires (Ag, Au nanowires) or carbon nanotubes (CNTs) have been used.

17.2.3 PRESSURE SENSORS

Pressure is defined as the normal force exerted per unit area against a surface. The measurement of applied pressure also involves the measurement of basic electrical parameters such as resistance and capacitance. Resistive pressure sensors are made from materials that undergo a change in resistance due to the application of pressure. These are similar to the metal and semiconducting strain gauges, wherein a change in resistance occurs because of the change in geometric dimensions of the device or change in resistivity, or both. However, whereas strain gauges are based on thin-film techniques, resistive pressure sensors utilize the resistivity of the bulk material to measure pressure more effectively. For example, a conductive foam can be used to form a flexible pressure sensor array simply by sandwiching the material in a metal crossbar structure. The application of pressure decreases the distance between the electrodes causing the resistance between the electrodes to decrease. This change in resistance can be used to estimate the applied pressure. Similarly, the bulk resistivity of semiconductor materials changes due to applied pressure (piezoresistive effect), which can be used to fabricate a piezoresistive pressure sensor.

Capacitive sensing is another method widely reported for measuring pressure. In this method, a capacitive structure is fabricated such that one of the metal plates is free to deflect under applied pressure. The deflection in the plate reduces the distance between the two metal plates, causing the capacitance of the structure to increase. The change in capacitance depends on the deflection, which depends on the applied pressure. However, the relationship between the applied pressure and the change in capacitance is not straightforward and depends on the shape of the capacitive structure, the material, and the integration process. The deflection of a clamped circular/elliptical diaphragm because of the applied pressure is given by:

$$\left[\frac{\partial^4}{\partial x^4} + 2\frac{\partial^4}{\partial x^2\,\partial y^2} + \frac{\partial^4}{\partial y^4} \right] w(x,\, y) = \frac{P}{D}$$

where w is the deflection in the z-direction at point (x,y), P is the applied pressure, and D is the flexural rigidity of the diaphragm. The capacitance of a device with this diaphragm as one of the plates is given by:

$$C_w = \iint_S \frac{\varepsilon\,dx\,dy}{d - w(x,\, y)}$$

where ε is the permittivity of the dielectric between the two plates (typically air gap), d is the original distance between the plates. The integration is over the surface area (S) of the capacitor. For small deflections, the change in capacitance can be approximated as linear. In the case of large pressures, the response is non-linear, but predictable, thus pressure values can be obtained using a device model. For very large pressures, the device architecture is modified so that the deflecting diaphragm is allowed to touch the stationary plate. This can be done without shorting the two electrodes by using a metal/insulator bilayer for fabricating one or both of the plates. The change in capacitance per unit pressure after the diaphragm has touched is much lower, leading to lower sensitivity, however, the range of measurement increases substantially. This structure is called the "touch mode" capacitive pressure sensor (Figure 17.3).

17.2.4 OTHER SENSORS

The measurement of physical quantities such as temperature, strain, and pressure can provide information about many parameters such as force, deformation, heat, phase change, displacement, and so on. However, there are still many physical parameters

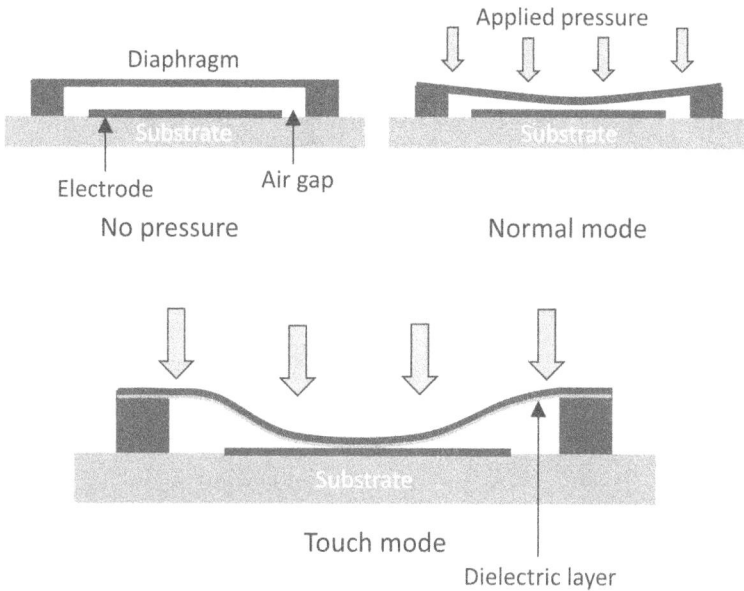

FIGURE 17.3 Capacitive pressure sensors typically feature a movable diaphragm that causes a change in capacitance of the sensor upon application of pressure. In the touch mode sensor, the diaphragm consists of a metal/insulator bilayer and is allowed to touch the substrate to obtain a higher range of sensed pressure.

whose measurement is critical for the success of a system and can even be a matter of life-or-death. Primary among these are gas sensors that inform a user of the presence of hazardous gases in the ambient. Gas sensors are typically based on a thin film that changes its characteristics upon exposure to a specific gas. The change in characteristics is then detected using a read-out circuit to determine the concentration of gas in the ambient. For example, tin oxide (SnO_x) and zinc oxide (ZnO_x) thin films have been used to create sensors for specific volatile organic gases such as methane. These metal oxide thin films adsorb ambient oxygen at the surface which reduces the number of free charge carriers in the material. In the presence of reducing gases, such as methane, the adsorbed oxygen concentration on the surface reduces, increasing the free charge carriers in the thin film, thus increasing the conductivity. The change in resistance of these thin films, thus, provides an indication of the concentration of the reducing gas. Metal oxide nanoparticles are often used to fabricate these thin films to increase the surface-to-volume ratio of the material. In most implementations, the thin films are kept at a specific temperature using microheaters in order to enhance the adsorption/desorption of gases on the surface and increase sensitivity. These sensors can be fabricated and demonstrated in a flexible form because of the thin-film nature of the sensing material, however, a key constraint in commercial deployment is their encapsulation. Because the sensing thin film needs to be in contact with ambient air, the packaging and encapsulation of gas sensors involves significant integration challenges and requires additional processing to create the window of interaction.

Ultraviolet (UV) light sensors are used to detect and warn users of the presence of harmful levels of UV light in the ambient. The basic mechanism of UV sensors is the same as that of solar cells. They absorb photons in a specific wavelength range (100–400 nm for UV) and produce electron-hole pairs proportional to the intensity of the incident light. The wavelength range of light absorption can be limited using the bandgap of the material or a light filter. The free carriers are then collected by the external circuit in the form of an electric current to determine the level of UV irradiation. The materials used for UV sensors based on the photovoltaic effect have a high bandgap (>3 eV) so that they are only responsive to high energy photons in the UV range. These materials include gallium nitride (GaN) and silicon carbide (SiC). Because producing power is not the primary function of a UV sensor, as opposed to solar cells, the integration strategy focuses on maximizing the change in current in response to UV intensity rather than the absolute current output. This parameter is called the responsivity of the sensor:

$$Responsivity = \frac{(I - I_0)}{P_{in}}$$

where I_0 is the output current for zero UV irradiation, and I is the output current for UV irradiation of power P_{in}. It should be noted that for a given UV sensor, responsivity is a function of the wavelength of incident light. Apart from the photovoltaic effect, the performance of UV sensors can also depend upon the surface adsorption of oxygen molecules. In metal oxides such as zinc oxide, the exciton generated because of the photoelectric effect causes the oxygen molecules adsorbed at the surface to release, enhancing the conductivity of the thin film. Thus, UV sensors based on zinc oxide nanoparticle thin films show high responsivity for UV-A irradiation (315–400 nm).

Humidity sensors are gaining popularity owing to the efforts in the regulation of buildings, factory floors, and farm environments at the optimum point. Flexible humidity sensors are commonly deployed for wearable applications to detect the presence or initiation of sweat on the skin surface, to control the ambient environment in the building, or to regulate the humidity of an industrial production line. In this case, as well, the key parameters monitored to determine humidity levels are the capacitance and resistance values of specific materials. In the case of capacitive humidity sensors, a dielectric film is chosen such that it spontaneously absorbs water molecules from the ambient (hygroscopic material). The absorption of water changes the permittivity of the material, which changes the capacitance of the structure. Polyimide (PI) is the most commonly used polymer material for flexible capacitive humidity sensing. The sensor is created using an interdigitated structure of metal lines on the PI substrate. The structure is exposed to the ambient and the capacitance of the structure is monitored to determine the humidity in the ambient. A similar structure can also be created for resistive humidity sensing because the resistivity of some materials changes upon the adsorption/absorption of water molecules. However, capacitive humidity sensors have the advantage of low power consumption, better reversibility, and better temperature stability compared to resistive sensors.

17.3 FLEXIBLE ACTUATORS

Actuators are energy transducers, like sensors, however, they convert electrical input energy into another form of energy. The form being converted to can be light, heat, sound, pressure, force, displacement, and so on. Actuators often help close the "control loop", which includes a sensing element, a processing element, and an actuator to hold the value of a process variable at a target setpoint. The process of actuation involves the use of electrical energy and converting it into another form, however, given that electrical energy is a flow of charges across a potential difference, there seem to be limited ways in which this should be possible. We can convert an electrical current into heat using Joule heating, into light using the recombination of electron-hole pairs, and into a magnetic field using a solenoid. We can convert the potential difference into an electric field, which can then be used to create actuation. However, these seemingly limited options have provided us with a vast array of actuator systems. Further, in some cases, the electrical input power is not sufficient to drive an actuation, in which case, the actuation is conducted indirectly by the applied electrical signal. For example, hydraulic and pneumatic systems are used to create large forces, which can be controlled using applied electrical signals to the valves, which are themselves based on electromagnetic actuation. For flexible electronics applications, we will consider the actuators that can be fabricated in miniature or thin-film form.

17.3.1 MICROELECTROMECHANICAL SYSTEMS (MEMs)

Microelectromechanical systems (MEMS) are a class of systems that are fabricated using micromachining techniques typically used in the CMOS industry. The electro-mechanical actuators based on the MEMS platform rely on the ability to "release" from the substate and perform mechanical actuation. The actuations are performed using electrostatics, electromagnetics, or thermal expansion using Joule heating. The structure is fabricated using sequential deposition and patterning of layers, with the final step being the isotropic etch of the "sacrificial" layer. This releases the structure from the substrate allowing it to move upon application of electrical energy. Because they are fabricated on a substrate using micromachining techniques, the size of these actuators is very small (of the order of a few micron square). Thus, the force and displacement created using these actuators is typically in the pN and μm range respectively. If MEMS devices are fabricated on silicon, they can be made flexible using the standard flexible silicon processes we discussed in Chapter 4. However, low temperature micromachining processes can be used to create MEMS devices directly on a flexible platform, such as polyimide. The most commonly fabricated MEMS structure is a fixed-free cantilever switch structure that is attached to the substrate through an anchor at one end, while being free to more at the other (Figure 17.4). The switch is activated using electrostatic attraction between the gate electrode and the cantilever. When the gate voltage is applied, the cantilever is attracted toward the gate, causing it to bend. This connects the free end of the cantilever to the substrate creating a metallic contact between the source and drain. The principle of electrostatic attraction between two oppositely charged plates is also used to create

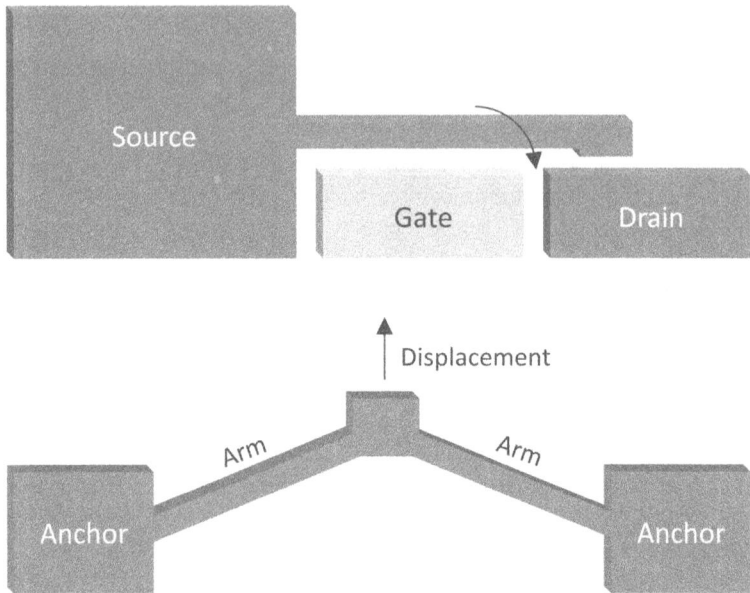

FIGURE 17.4 The design of a simple MEMS cantilever switch. When the gate voltage is applied with respect to source (V_{gs}), the cantilever is attracted to the gate and a connection between the source and drain is established. A chevron thermal actuator's displacement is based on the Joule heating of the individual arms.

a comb drive structure which typically involves the use of interdigitated electrodes to increase the capacitance of the structure. Another commonly used design is based on Joule heating and thermal expansion. In this design, an electric current is passed through a resistive heater to create a local hot spot that causes a rise in temperature of the nearby structures, which leads to thermal expansion. If the structure is free-standing, the thermal expansion can be used to perform mechanical work. Such a structure is normally fabricated as a bimorph design for obtaining bending motion or a chevron design to obtain linear motion. However, the dissipation of heat in the ambient and the non-moving parts of the system can lead to wastage and inefficiency.

17.3.2 ELECTROACTIVE POLYMERS (EAPs)

Electroactive polymers (EAPs) are an important class of actuators particularly for flexible and stretchable applications. These polymers exhibit mechanical actuation (through a change in shape or size) when subjected to an electric field. In some cases, conductive polymers are also included in the EAP category because they respond to an applied electric field with electric current, however, in this section, we will refer to EAPs as those that produce a mechanical actuation upon exposure to electric fields. There are two distinct mechanisms governing mechanical actuation, leading to two different types of EAPs – ionic and dielectric. In ionic EAPs, the displacement is caused by the migration of ions in the polymer thin film. One of the ways of realizing ionic EAPs is using ionic polymer metal composites (IPMCs). These involve the use of ionic polymers or ionomers such as Nafion or Flemion. The structure resembles a capacitor with the ionic polymer sandwiched between two metal electrodes. When an electric field is applied, mobile ions, along with attached water molecules, are attracted toward the oppositely charged electrode. This migration causes one side of the polymer to swell, changing its shape, and causing mechanical actuation. Once the electric field is released, the ions diffuse back into the polymer and the actuation subsides. The fabrication of such an actuator is challenging because the thin films of metal electrodes on either side of the polymer structure need to maintain conductivity with large actuation strains. Further, the process for deposition of the metal electrodes needs to be chosen such that there is good adhesion between the polymer and metal thin film. The actuation of the IPMC structure depends on the surface area of the metal and the concentration of free ions inside the polymer thin film.

17.3.2.1 Dielectric Elastomer Actuators (DEAs)

Dielectric elastomer actuators (DEAs) are a class of EAPs that rely on the electrostatic force between two metal electrodes to produce mechanical actuation. The structure is that of a complaint capacitor with an elastomer acting as the dielectric sandwiched between two metal electrodes. When an electric potential difference is applied to the electrodes, there is an accumulation of opposite charges on the surface, as in the case of an ordinary capacitor. The electrode plates are attracted to each other because of the Coulomb forces between the opposite charges. This force induces a compressive pressure (Maxwell's stress) in the dielectric normal to the electrode surface:

$$\sigma = \varepsilon E^2 = \varepsilon \left(\frac{V}{d} \right)^2$$

where ε is the permittivity of the dielectric elastomer, V is the applied voltage and d is the distance between the two metal electrodes. This pressure causes the dielectric elastomer to compress in the direction normal to the electrodes and expand in the lateral directions (Figure 17.5). This compression and expansion can be utilized to carry out mechanical work. The displacement and the force generated by the actuator are both proportional to the square of the applied electric field, thus, the energy density depends on the fourth power of the applied electric field.

$$U = \frac{\varepsilon^2 E^4}{2Y}$$

where U is the energy density of actuation and Y is the Young's modulus of the elastomer. Thus, it makes sense to increase the electric field to increase actuator performance. However, the physical limit to applying high electric fields to a dielectric is the dielectric breakdown strength of the elastomer, which plays a major role in determining DEA performance. Apart from this, the dielectric constant of the elastomer should be high, while the Young's modulus should be low. The optimization of these parameters is critical for obtaining high actuation energy density in DEAs.

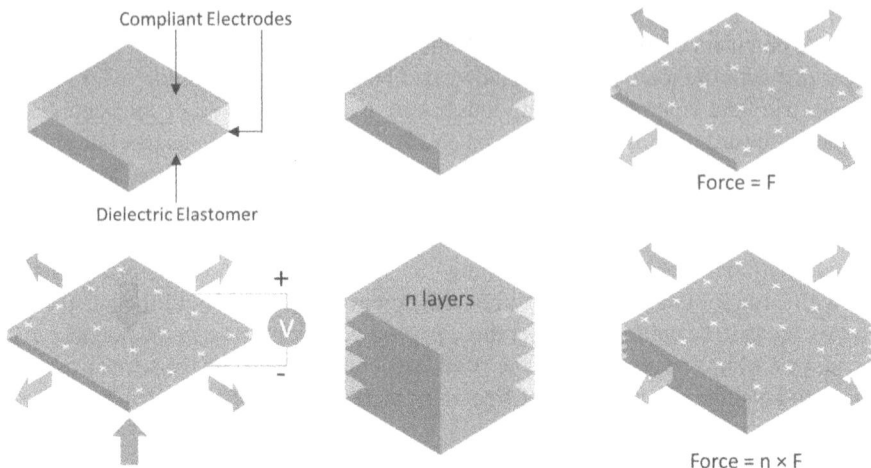

FIGURE 17.5 The dielectric elastomer actuator (DEA) is a complaint capacitor structure. An applied electric field causes displacement in normal as well as lateral directions.
Increasing the number of layers increases the total force output.

Flexible actuators fabricated using DEAs have several advantages. They are
light in weight even for high power output, leading to energy densities comparable
to natural muscles. Owing to the simple underlying physical principle, their fabrication is relatively simple and can be done without using specialized elastomers.
In principle, any elastomer that is non-conducting (dielectric) and complaint can
be used as a DEA. In practice, DEA structures based on many different kinds of
materials, such as silicones, acrylics, urethanes, have been reported. The choice of
an elastomer for DEA applications is based on the consideration that the elastomer
should have low Young's modulus, high permittivity, and high electrical breakdown strength. As for the metal thin films acting as the electrodes, it is important that they maintain conductivity even under large strains. Thus, percolation
thin films of 1D and 2D nanomaterials are often used as electrodes, for example,
thin films of single-walled CNTs. A major drawback of DEAs is the requirement
of high electric fields to create meaningful actuations. The voltages applied can
go up to several kilovolts, even though the power input is still limited because
of the limited current drawn. This limitation can be overcome by creating very
thin elastomer layers, thus leading to high electric fields even for relatively low
voltage. However, this reduces the total output work obtained from the actuator,
which can be resolved by creating multilayer structures. These are created using
alternating polarity electrodes sandwiched between successive dielectric elastomer
layers. The electrodes are connected in an interdigitated manner to the external
power source (Figure 17.5). When a voltage is applied, the potential difference
appears across each dielectric elastomer layer, producing an electric field according to its thickness. The displacement produced by such a multilayer structure
is similar to that produced by a single layer structure, however, the force output

is n-times more, where n is the number of dielectric elastomer layers, leading to higher total output work. Multilayering results in much thinner dielectric elastomer thin films being layered to produce a substantial force output. The reduction in layer thickness helps with achieving the desired electric field with a smaller input voltage. However, a reduction in dielectric thickness also makes them susceptible to electrical breakdown. Further, the effect of nanoscopic defects on dielectric constant and dielectric breakdown strength is enhanced in a thinner dielectric film. Nevertheless, multilayer DEAs have been reported for many soft robotics and artificial muscle applications.

17.3.3 Piezoelectric Actuators

The piezoelectric effect was described when we discussed flexible energy harvesting systems. It is a reversible effect and the inverse effect can be used to create an actuation upon application of an electric field to certain materials. The underlying physics is the realignment of internal electronic dipole regions because of the application of the external field. Examples of inorganic piezoelectric crystals include perovskite materials such as barium titanate ($BaTiO_3$), whereas a well-known organic piezoelectric material is the co-polymer PVDF-TrFE. In a completely relaxed material system (stress is zero), the amount of strain produced is directly proportional to the applied electric field:

$$S = sT + \delta' E$$

where S is the strain developed for an applied electric field E, and δ is the transverse piezoelectric tensor. This mechanical strain can be used for many applications. One of the most common applications of piezoelectric actuators is in speakers to convert the electrical signals into sound signals (in the form of pressure waves in the air). Piezoelectric materials can be easily obtained in thin-film form on flexible polymer substrates using low-temperature deposition processes such as sputtering, sol-gel, spin-coating, and so on.

17.3.4 Shape Memory Alloys

Shape memory alloys (SMAs) are materials that undergo mechanical actuation upon application of thermal energy. This is different from the thermal expansion experienced by all materials when heated. The basic physical principle behind the shape change in SMAs is the rearrangement of atoms in the alloy, at a specific temperature. This rearrangement occurs because the material undergoes a solid-state "phase change". Typically, a phase change is associated with changes from liquid to solid or vapor phases, however, some materials can change from one crystalline phase of the material to another upon application of heat. Such a solid-state phase change can result in physical deformation of the material, even though the material remains solid during the entire process. SMAs are typically metallic alloys of transition metals, most common examples being nickel-titanium (Ni-Ti), copper-zinc-aluminum

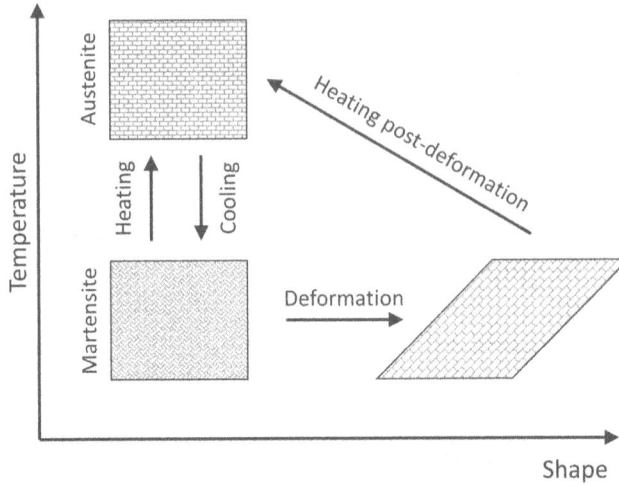

FIGURE 17.6 An example of a shape memory alloy (SMA) is the NiTi that can exist in austenite and martensite states. If deformed in the martensite state, followed by heating, the object recovers its original shape and phase.

(Cu-Zn-Al), iron manganese selenium (Fe-Mn-Se), and so on. In the case of NiTi, the material exists in two different phases – Martensite and Austenite. Martensite occurs at low temperatures, whereas Austenite occurs at high temperatures. When a sample of the material is cooled from the Austenite phase, it changes into the Martensite phase internally without a change in macroscopic physical form. In this phase, the material can be deformed into another shape. If it is heated back to the original temperature, the material then returns to the austenite phase and to the original shape it held before deformation (Figure 17.6). This "memory" of shape can be utilized for one-time deformation and installation in areas such as construction or space applications.

The development of sensors and actuators on the flexible platform has added the final required dimension to the functionality of flexible electronic systems. These help the digital systems interact with the physical world and eventually act as a bridge interconnecting the two starkly different worlds together.

EXERCISES

17.1. The unstrained resistance of a strain gauge is 10 Ω. If it changes to 10.01 Ω on the application of 20 ppm strain, what is the gauge factor? If all the static resistance values on the Wheatstone bridge are equal to 10 Ω, what is the strain for the voltage output of 5 mV for this strain sensor, if the excitation voltage is 5 V.

17.2. The temperature coefficient of resistance of platinum is 0.003927/°C. What is the minimum change in temperature that can be detected using a 100 Ω sensor of this metal, if the resolution of the resistance measurement is 0.1 Ω?

17.3. Describe the structure and working principle of a capacitive pressure sensor. What is the added change in design and advantage of using a touch-mode capacitive pressure sensor?

17.4. A dielectric elastomer has a breakdown field of 10 MV/m, a dielectric constant of 3, and Young's modulus of 10 MPa. We need to create an actuator of area 1 cm^2 that can produce 1 N force using this elastomer. If the thickness of the elastomer is 10 μm, what should be the number of layers in the structure? What is the voltage required for actuation?

17.5. How can we fabricate a flexible gas sensor? Describe the mechanism of sensing volatile organic compounds using this sensor.

Part V

The Road Ahead

18 Stretchable Electronics

18.1 WHY STRETCHABLE?

We have seen many materials, fabrication techniques, and applications for mechanically flexible electronic systems. However, there is now a growing class of electronics that is also gathering momentum – stretchable electronics. The term stretchable electronics is used very broadly to describe any electronic system that can handle some amount of strain without a notable change in its performance. However, it is worth noting that most materials are "stretchable" if you define the term vaguely enough. As we saw in Chapter 2, any material system has a fracture strain. If the system is strained beyond this point, it develops a catastrophic failure and cracks. Thus, all materials can be strained to some degree elastically, i.e., they return to their original shape once the force producing the strain is released. This limit is different for different materials, for example, typical fracture strain values for glass are 1–2%, while those for metals can go up to 50%, and polymers well in excess of 100%. So, should we call all materials stretchable, and by extension all electronics, stretchable electronics? An important point to remember while studying stretchable electronics is that the stretchability needs to be reversible so that we can reliably use the system over many strain cycles. Thus, what we are really looking for is not the fracture strain (the absolute maximum strain a material can withstand before developing fractures/cracks), but the strain at the elastic point (the point which represents the limit of reversible straining). To determine what the reasonable range for elastic point strains should be, we need to consider the reason why stretchability is desirable in electronic systems.

If we take a look around, there is a clear distinction between the intelligent systems made by humans, i.e., the smartphones, computers, server systems, and so on, and the intelligent systems made by nature, i.e., humans themselves. Manmade intelligent systems are rigid, rectilinear, and fixed-form, while animals and humans have a soft, curvilinear, and free-form exterior. This clear distinction makes the integration of these two systems difficult. This is particularly true for wearable electronic systems (on human or animal skin), or implantable inside a living creature. Wearable systems need to have the stretchability of skin so that they do not impair the free movement of the wearer, and implantable systems may need to grow and change shape as the body of the living being changes with time. These unique requirements create a need for electronic systems that are truly free-form – can be flexed, twisted, stretched, rolled, and so on. Thus, to answer the question regarding which systems can be classified as stretchable electronics, it is important to understand the final application for the system. In most cases, if a system can sustain prolonged strains of 20–30% *reversibly*, we can call the system stretchable, mainly because human and animal skin undergoes that amount of strain during normal activities. The term "reversibly" in the previous statement is important because this defines the elastic

DOI: 10.1201/9781003010715-23

point for the material, not the fracture strain. Henceforth, when we refer to a material as stretchable, we will use this range as a reference elastic point. In the case of flexible electronic systems, every component of the system needs to be flexible. Similarly, it is important to have every individual component of a stretchable system reversibly stretchable up to a certain desired strain so that the complete system can strain up to that point. The field of stretchable electronics is only just gaining momentum and many research groups around the world are looking for individual components and material systems that can be reversibly stretchable.

Before we look at some of the means of achieving stretchable electronics, we need to understand the reason why some materials can (and why some cannot) sustain high levels of strains reversibly without any notable change in properties. Also, it is important to understand the difference between flexible and stretchable materials. As discussed in Chapter 2, flexibility of a thin film depends on the properties of the material and the thickness of the film. Given any material, we can keep reducing the thickness of the film to achieve any desired bending radius. This is because, even though the fracture strain of the material remains the same, the actual strain experienced by the film changes with thickness. Thus, the additional parameter of thickness helps achieve flexibility in otherwise brittle materials like silicon (refer to Chapter 4). This is not the case while designing stretchable electronics. In general, it is expected that stretchable electronics will be able to undergo a specific amount of lateral strain reversibly. Hence, we have to fabricate stretchable electronic systems using principles specifically designed for achieving stretchability.

18.2 IMPARTING STRETCHABILITY

To fabricate stretchable electronics, it is intuitive to think of using materials that inherently have large fracture strains. Materials such as natural rubbers, polymers, organic thin films can be used to obtain highly stretchable thin films. However, these materials may not have the desired electronic properties needed to fabricate high-performance electronic systems. While researchers around the world are constantly discovering material systems that have suitable mechanical and electronic properties for the fabrication of stretchable systems, it is also important to consider ways in which stretchable electronic systems can be fabricated with materials that are not inherently tolerant to large lateral strains. Thus, in this section, we will discuss stretchability in two distinct ways – stretchability by material and stretchability by design.

18.2.1 STRETCHABILITY BY MATERIAL

Some material systems are inherently tolerant to large lateral strains. For the purpose of designing stretchable electronics, we generally focus on reversible stretchability, thus, we consider the strain at the elastic point, and not the fracture strain of the material. By that standard, most metals that have a reasonably large fracture strain (around 20–40%) will not qualify as stretchable materials because their elastic point is relatively low (2–5%). In general, many polymeric materials like silicones have the capability to undergo large strains, owing to the molecular structure of the material.

When lateral stress is applied to a crystal lattice, the atomic distance between adjacent atoms increases along the direction of the applied force. As discussed in Chapter 2, the ball-spring model indicates that the interatomic restoration force is proportional to the change in interatomic distance. A steady-state is achieved when the total interatomic distance along a cross-section of the material is equal to the external force (Figure 18.1). Now, this change in the atomic distance also results in potential energy being stored in the material, which, in an ideal case, is equal to the work done by the external force. Fracture takes place when the total potential energy stored in the lattice is more than the energy required to break all the atomic bonds in any cross-section of the material. This is because once fracture takes place, the atoms of the individual pieces are restored to their original interatomic distances, removing the stored potential energy. Thus, at a particular strain, it becomes energetically favorable for the lattice to fracture. This is, of course, in an ideal crystal with perfectly distributed stress along the cross-section of the material and without considering slippage, friction, and thermal losses. In reality, the fracture takes place along the cross-section that has the most defects so that the least potential energy is required for fracture. Further, in many cases, the distribution of stress along the

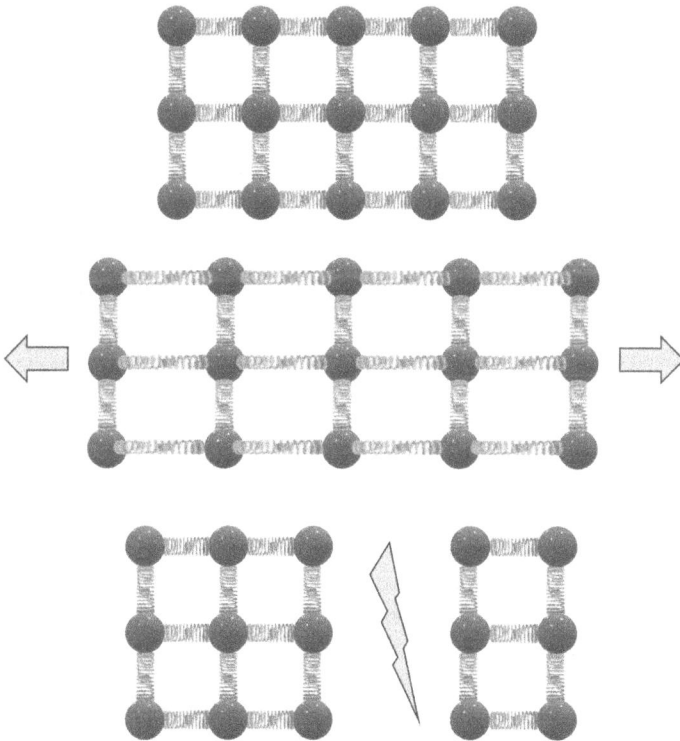

FIGURE 18.1 The fracture of a lattice material can be understood from the ball-spring model. The applied strain causes a potential energy to be stored in the material, if this energy is more than that required for breaking the bonds, it is energetically favorable for the material to fracture.

cross-section of the lattice may not be uniform leading to localized dislocation and fracture propagation. This description holds for materials that form crystal lattices such as silicon, diamond, quartz, and so on. These materials undergo very small strain for large applied forces, leading to very high Young's modulus. Also, the materials fracture when a large enough stress is applied. Thus, the stress-strain curves are straight lines that end abruptly with material fracture. In the case of metals, the stress along the material causes localized slippages to occur resulting in "necking". Thus, most metals undergo a phase of rapid expansion for very small applied stress after the elastic point is achieved. The material finally undergoes catastrophic failure, typically at the place where the necking started.

Polymers, on the other hand, behave very differently when stretched. Polymers are made of large chains of atoms (typically carbon atoms) bonded together into one giant molecule. Each molecule can have thousands of atoms in some cases. To form the bulk material, these molecules are held together because of various forces acting on the chains. One of them is the van der Waal's attraction between the electron clouds and nuclei of atoms in a neighboring polymer chain. This is a very weak attractive force and can be easily overcome, making lateral slippage between polymer chains possible. Another force joining two polymer chains is the hydrogen bond which is the attraction between a lone pair of valence electrons on an electron-rich atom (typically oxygen, nitrogen), and the hydrogen atom in a covalent bond with a highly electronegative atom. This happens because the shared electron cloud in the bond between hydrogen and a highly electronegative atom is held closer to the electronegative atom. This causes the hydrogen nucleus to appear slightly positively charged, which causes it to attract a lone pair of electrons from an atom of another polymer chain. The strength of the hydrogen bond depends on the electronegativity of the atom attached to hydrogen, the availability of lone pair of electrons, and the structure of the polymer chains. Finally, the polymer chains can also be bonded to each other with chemical bonds between different points on the chain. These chemical covalent bonds between different polymer chains are called "crosslinks". These bonds can cause the complete bulk polymer to behave as one gigantic 3D molecule. Depending on the strength and number of one or more of these bonds between polymer chains, the mechanical and chemical properties of polymers are determined. The number of independent parameters affecting properties of polymers, such as the monomer, the number of monomers in a chain on average, the type and number of bonds between chains, leads to an almost uncountable number of stress-strain curves. We can tweak the properties of the polymers by making small variations in the organic ligands within monomers, or in the number of monomers in the chain. This flexibility is important to obtain a material with precise mechanical, chemical, and electronic properties for use in stretchable electronic systems.

Polymer chains are typically twisted and entangled in a random fashion like threads in a stack of yarn (Figure 18.2). These chains may be bonded to one another at some points through hydrogen or chemical bonding. When polymers are stretched along a specific direction, the molecule chains get aligned along that direction. If the polymer chains are not strongly bonded to each other, the polymer can undergo a large deformation for relatively low applied stress (leading to low Young's modulus). However, when released, the stretchable polymer typically returns to its original

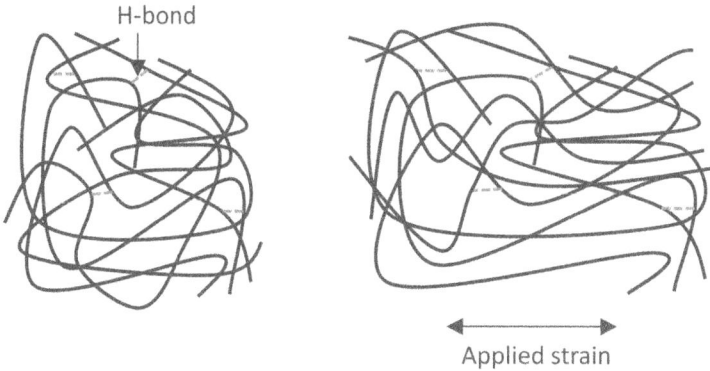

FIGURE 18.2 Polymer materials consist of a network of individual polymer molecules intertwined with each other like strands of noodles in spaghetti. When stretched, some of these strands temporarily align in the direction of applied strain.

state. This elasticity is a result of a fundamental law of Thermodynamics that states that all physical systems tend to increase entropy (the state of disorder). The disorder inherent in the randomly entangled polymer chains is large enough to cause an entropic force that takes the system toward an increase in overall entropy. This entropic force is seen as the restoration force in a stretched polymer material. Because the polymer chains are loosely held together, and because the entropic force responsible for restoration is weak, some polymers (typically rubbers) appear to be very stretchable. Small amounts of lateral stress can create a large deformation in these materials.

If the polymer chains have strong bonding with each other (say because of hydrogen or chemical bonding), a large restoration force is created, because the bonds between polymer chains act according to the ball-spring model and cause it to revert back to the original structure. However, in such polymers, the reversible stretchability is not considerable given that straining beyond a particular point leads to the breakage of the bonds, causing the structure of the material to change irreversibly. Further, the force required to produce a specific strain can be large (high Young's modulus) depending on the number and strength of crosslinks. In many cases, polymers can be made to develop crosslinks to bring about a controlled change in their mechanical properties. This process is called "curing" and is particularly evident in the case of adhesives. In most cases, the original adhesive is a free-flowing liquid because it does not have crosslinks between polymer chains. However, a certain physical process, such as exposure to air, heat, or UV irradiation, can cause the formation of crosslinks in the adhesive, changing it into a strong and hard material after curing. In some cases, exposure to a specific physical process can also lead to the breaking of crosslink bonds, thus altering the properties of the polymer. Thus, the stretchability of a polymer can be defined by choosing the correct polymer chemistry, fabrication process, and deployment conditions. However, even with all these options, there are very few stretchable materials that can be used to replace inorganic semiconductors for processor, memory, or display fabrication. While research is continuously going on to synthesize new stretchable materials with desirable electronic and semiconducting properties, it is important to consider ways in which known inorganic semiconducting materials can be made to stretch reversibly.

18.2.2 STRETCHABILITY BY DESIGN

We can impart stretchability in any material system simply by designing it in a particular shape and form. The stress-strain curve of a system depends on both the materials used and the way they are designed. A classic example of this is the simple steel spring (Figure 18.3). A block of steel of the same dimensions can only stretch reversibly up to 1–2% depending on the type of steel. However, once designed as a spring, the possible reversible lateral strain can be in excess of 100%. This transformation is possible because of the unique structure of the spring – a helical spiral around a linear central axis. The radius of the coil and the pitch play a key role in determining the mechanical properties of the spring. We know that when we apply stress to a piece of metal, say a metal wire, the atoms in the metal are displaced causing the overall wire to strain. This strain is typically small for a large applied force (depending on the Young's modulus of the material). Further, in most cases, the strain is not reversible after a certain point. To overcome this problem, we design the spring in such a way that when stress is applied along its axis, it is uniformly distributed along the helical spiral structure. Naturally, this causes the metal to strain. However, because the length of the coil around a certain distance along the linear axis is much larger (because of the circumference of the radial helix), a small strain in the metal along the helical circumference leads to a large apparent strain along the linear axis. This mechanical amplification of strain in the direction of interest is utilized to obtain large reversible strains using materials that have relatively low elastic strains.

In the case of thin films, the concept of lateral spring structures is used (Figure 18.3). In this case, the thin film is patterned in some form of a coil or meandering structure

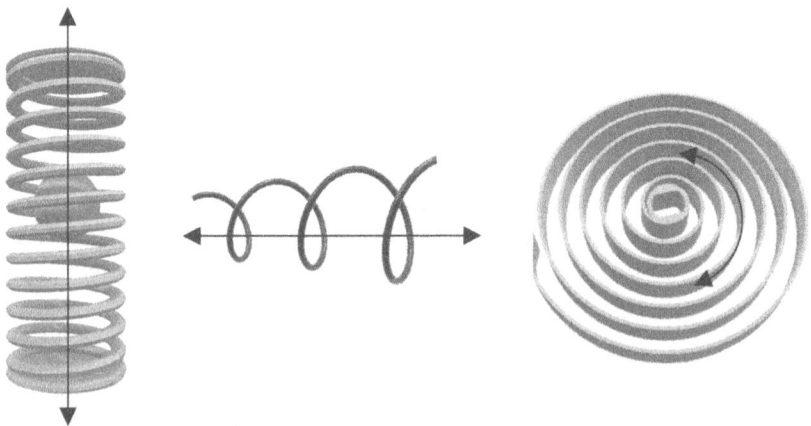

FIGURE 18.3 Various spring designs commonly used in industrial and consumer products take advantage of the distribution of the applied strain along a much larger helical circumference for an apparent strain along the linear axis.

that can "unravel" upon application of strain in a particular direction. Many structures have been reported in the literature for designing lateral spring structures, such as horse-shoe springs, double-spiral coils, self-similar structures, leaf-arm designs, fractal patterns, and so on. A particular structure is generally chosen based on the fabrication technique to be used, the area of the thin film, its thickness, expected levels of strain in the application, and so on. With this approach, even crystalline inorganic semiconductors like silicon have been demonstrated to have reasonable stretchability (up to 450% reversible areal strain).

In cases where patterning the thin film is not an option, stretchability is obtained using a stretchable polymer substrate, while the thin film is deposited on top to obtain the desired functionality. However, if the deposition is done normally, the functional thin film develops cracks when the substrate is strained beyond its yield strain. To avoid this problem, we can use the concept of pre-straining. Pre-straining refers to the concept of applying the desired maximum strain on the stretchable substrate followed by the deposition of the functional thin film. When the substrate is released, the deposited thin film typically forms out-of-plane buckled "wavy" structures to absorb the contraction (Figure 18.4). Now, when the substrate is stretched, these wavy structures act as lateral springs and reduce the strain experienced by any section of the thin film, even for very large substrate strains. The pre-strain in the polymer substrate is obtained using mechanical clamping during deposition or using the thermal expansion during deposition, followed by subsequent cooling. An advantage of this technique is that it can be used with any thin film given a suitable polymer substrate to pre-strain and bond the thin film. However, a major disadvantage of this technique is the compressive strain that the functional thin film undergoes when the prestrained polymer is released. This strain can be particularly problematic in semiconductor thin films like silicon, where the band structure and electron transport properties depend on the strain in the lattice.

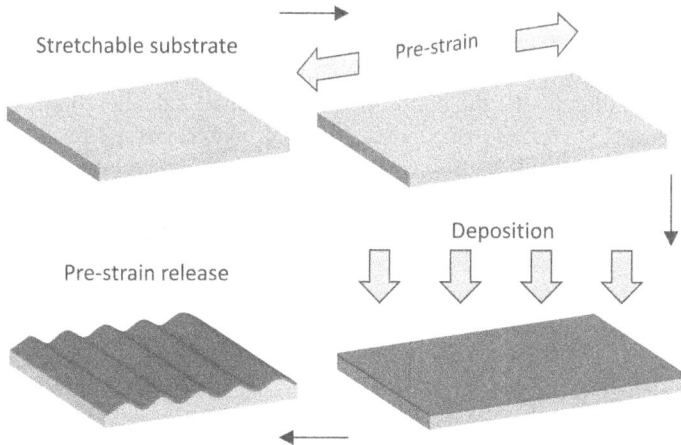

FIGURE 18.4 In the pre-strain technique, a stretchable substrate is pre-strained before deposition of the thin film. The subsequent release creates a structure that can be stretched up to the original pre-strain without loss of conductivity.

18.3 STRETCHABLE CONDUCTORS

One of the most important applications of these concepts of design for stretchability is for making stretchable conductors. In general, stretchable conductors are conductive materials that exhibit reversible stretchability without loss in conduction during stretching. In most cases, however, there is a change in the specific conductivity with strain. These materials are generally used to interconnect islands of functional electronics to create a stretchable system (more on this in Section 18.4), thus it is an extremely important class of materials from a stretchable electronics point of view. There are several ways in which stretchable conductors can be fabricated in practice. The choice of method of fabrication depends on the range of specific conductivity required, cost of fabrication, and other specific requirements of the end application.

18.3.1 Polymer Composites

Composites are material systems that combine two or more different materials into a single system so that certain desirable properties of both materials can be retained. In the case of polymer composites, a polymer material acts as the base or matrix of the composite, while another material is added to provide or enhance a particular property of the material. Polymer composites have been studied for many decades and have been used in applications requiring materials with a high strengthen-to-weight ratio. For such an application, the polymer matrices are generally reinforced with natural or artificial fibers to obtain a lightweight, inexpensive, high-strength, corrosion-resistant material. Polypropylene, epoxies, polyesters, silicones, are some of the commonly used polymers for creating polymer composites. These are reinforced with natural or synthetic fibers such as jute, coconut, glass fiber, carbon fiber, and so on. In particular, carbon fiber composites have found their way in many applications such as aerospace, automobile, sports equipment, etc. where a lightweight and strong material is required. The reinforcement material can be in the form of continuous or discontinuous filaments, unidirectional or random orientation, woven or non-woven. Further, a single polymer composite can have multiple reinforcement materials depending on the end application and the mechanical properties desired.

In the case of stretchable electronics, polymer composites are formed such that they retain the stretchability of the polymer matrix while exhibiting the electronic characteristics of the filler material. The most commonly used filler materials are conductive metal nanoparticles, nanotubes, or nanowires which provide conductivity to the polymer matrix, thus creating a stretchable conductor. In some cases, metal ions are implanted in a polymer matrix after it is cured to obtain conductivity in specific parts of the polymer. The presence of conductive particles/nanowires inside the polymer matrix causes conductive paths to be created through the polymer, thus causing the polymer matrix to exhibit metallic conductivity. This is only possible if sufficient conductive filler is added to the polymer matrix such that some conductive paths are created. However, it is important to understand that the incorporation of metal fillers changes both the electronic and mechanical properties of the polymer matrix. Thus, although incorporating more fillers may increase the probability of finding conductive paths through the material, the polymer matrix may start to stiffen and lose elasticity if too much filler is added. To understand the amount of filler material that should be added to a particular polymer matrix for a given application, it is important to understand the underlying physics of conduction in polymer composites and the theories governing it.

One of the key mathematical concepts used for determining the variation of conduction in a polymer composite with the addition of filler material is the Percolation Theory. The original theory was designed to discern the probability of forming large connected clusters inside hypothetical 3D lattices with a given probability of bonding between adjacent sites. The fundamentals of the theory can be applied to determine if at least one conductive path will be obtained from one end of a polymer composite to another, given the random distribution of a specific concentration of the conductive fillers. This is because conduction in polymer composites takes place primarily because of the network created by junctions between adjacent conductive fillers. These junctions can be direct physical contact between adjacent fillers or a non-contact junction because of the tunneling of current from one filler to another. The latter provides some contact resistance that increases significantly with the distance between the filler particles. Thus, it is clear that an increase in the volume fraction (volume of fillers compared to the total composite volume) of the conductive fillers should increase the conductivity of the composite. However, this relationship between conductivity and conductive filler volume fraction is not a linear function. A typical graph of change in conductivity with volume fraction is shown in Figure 18.5.

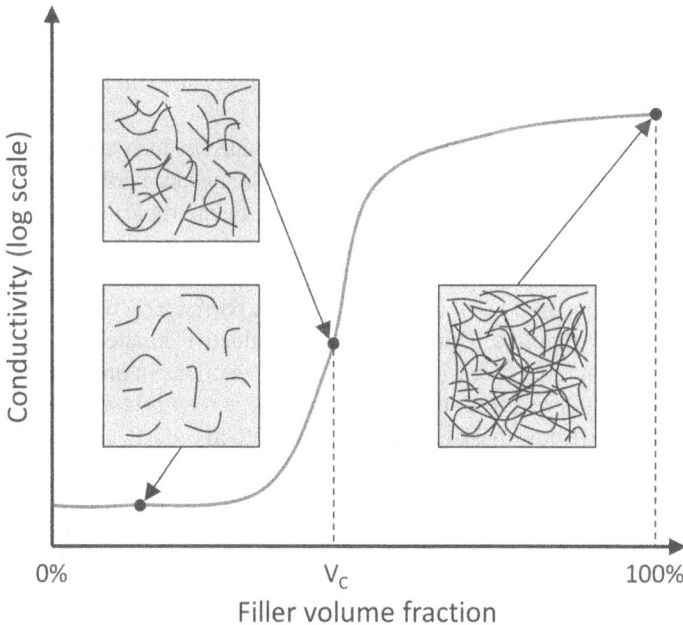

FIGURE 18.5 The conductivity of the polymer composite increases significantly when a specific critical volume fraction is achieved because of the formation of some conductive paths across the polymer.

The conductivity of the composite varies from that of the pure polymer matrix, at zero volume fraction of the filler, to that of the conductive filler material, at 100% volume fraction. It is clear that as the filler volume fraction is increased from zero, there is no significant change in the composite conduction because the conductive filler particles are too far apart to create a conduction path across the polymer. As the filler volume fraction is increased, at a particular volume fraction, there is enough filler material to form direct or non-contact junctions across the polymer and some conductive paths are formed. This volume fraction is known as the critical volume fraction (V_c) and varies with the polymer matrix and conductive filler used. After this volume fraction, there is a sudden increase in conductivity of the polymer composite because some conductive paths have been created across the material, and the conductivity effectively jumps from that of the polymer matrix to that of the conductive filler. A subsequent increase in the filler volume fraction does not cause a discernible rise in conductivity because it only increases pathways for electrons to flow in parallel, not the ease of electron flow.

The slight increase observed is because of an increase in the probability of better direct contact between adjacent filler particles and thus a reduction in junction contact resistance. Thus, for fabricating a polymer composite with a conductive filler, it is important to know the threshold volume fraction for conduction. Further, in the case of stretchable polymer composites, it is important to remember that the percolation threshold changes with strain. As the polymer matrix is stretched, the conductive fillers are separated from each other causing the direct and indirect junctions to break, leading to loss of conduction. In essence, as the volume of the composite increases, the filler volume fraction can fall below the threshold for conduction. Hence, while designing devices with stretchable polymer composites, it is important to take into account the final stretchability required to calculate the amount of filler to be introduced in the matrix.

One of the simplest ways of fabricating stretchable conductive polymer composites is by blending conductive filler material into the uncured polymer solution, followed by curing. Many studies have reported blending microparticles, nanoparticles, 1D structures such as CNT, Ag nanowires, and 2D materials such as graphene sheets, with various polymer matrices to form composite polymers. A key advantage of this method is that, compared to other methods, this is particularly useful for obtaining bulk stretchable conductors. Thus, stretchable conductive components of any shape and size can be obtained by curing these polymer solutions in an appropriate mold. In case a thin film is required, the material can be screen printed and can even be patterned using a shadow mask technique. This method is particularly advantageous with 1D materials because of the low percolation threshold. However, a surfactant is usually required while mixing nanomaterial because of the fast aggregation that takes place in the polymer solution. Surfactants increase the surface interactions between the polymer matrix and the conductive filler to reduce the Thermodynamic favorability of clumping of filler material. This is necessary to obtain a homogenous distribution of the filler inside the composite, leading to conduction at the percolation threshold. However, surfactants on the surface of conductive fillers can increase the junction resistance between filler particles, causing an overall reduction in the conductivity of the composite material.

Conductive polymer composites can also be fabricated by spraying conductive materials onto a stretchable substrate or by infiltrating the polymer into a conductive thin film. Both the techniques result in a thin film of conductive filler material on a stretchable polymer substrate. This is advantageous because of the relatively low amount of filler material used compared to the polymer composite method. Further, because of the low amount of fillers, the physical properties of the substrate such as transparency are better preserved. In the case of spray coating, the conductive filler material is typically dispersed in a volatile organic solvent, often using a surfactant to discourage aggregation. The solution is then sprayed onto a cured polymer substrate or an uncured pre-polymer thin film to make the top surface of the polymer substrate conductive. This technique can be also be used in conjunction with shadow masks to pattern the conductive thin film. In the case of infiltration, the thin film of conductive filler material is first prepared on a solid substrate. The elastomer pre-polymers are then poured onto the thin film and allowed to cure. The resulting polymer thin film, which has a conductive bottom surface, is then peeled off from the solid substrate. An advantage of this approach is that the conductive thin film can be processed on a solid substrate before making a stretchable conductor. For example, annealing a silver nanowire thin film can increase its conductivity because of better direct junctions between adjacent nanowires. Another advantage of this process is that the initial conductive thin film can be patterned into using state-of-the-art lithography techniques because it is on a solid substrate. This can, in theory, lead to micropatterned stretchable thin films. A key disadvantage of both spraying and infiltration is the use of polymer surface for conduction, instead of the bulk as in the case of polymer composites. This makes these thin films susceptible to loss of conduction due to surface damage during straining. Further, while the conductivity for these thin films is higher compared to composites, the current carrying capacity (ampacity) is lower.

Stretchable conductive polymer composites can also be fabricated by implanting a polymer matrix with metal ions. Ion implantation is a well-known process commonly used for doping specific regions of silicon substrates with a specific concentration of dopant ions. The same process can be used to dope cured polymer materials with higher energy metal ions. These ions penetrate the polymer chains and obtain electrons to form metal atoms, which in turn congregate to form metal nanoparticles. If the concentration of these nanoparticles exceeds the percolation limit, the polymer material becomes conductive. The key advantage of this process is that any metal ion can be implanted into any polymer substrate. Further, the depth and concentration of the metal ion in the polymer substrate can be controlled using the acceleration potential and ion current respectively. This provides a great degree of control to create a precisely engineered conductive polymer. The disadvantage of this approach is the formation of metal nanoparticles in-situ inside the polymer matrix. This process is still not well understood, and the shape and size of the resulting metal nanoparticles can have a significant bearing on the conductivity and mechanical properties of the polymer composite.

Finally, it is important to understand that some or all of these processes – mixing, infiltration, spraying, and implantation – can be combined to obtain the desired electrical and mechanical properties in a stretchable conductive polymer composite. For example, metal ion implantation into a polymer composite with mixed conductive fillers may lead to better surface conduction properties and better strain response. Or, a highly conductive silver nanowire thin film infiltrated with a polymer matrix can be made more reliable by spraying the surface with additional Ag nanowires.

18.3.2 Metal Springs

Metallic lateral springs structures are generally used for realizing thin film versions of a metal spring. The structures are generally designed using lithography techniques with patterns depending on the materials involved and the stretchability required. These structures include serpentine shapes, self-similar designs, fractal patterns, and so on. In practice, metallic spring structures are bilayer or multilayer structures with one layer consisting of the metal thin film, while other layers are added for mechanical backing. In particular, a polymeric backing is used to ensure there is enough restoration force when the applied strain is released. This is important because metal thin films tend to behave plastically (such as in the case of aluminum foils) because of the malleability of the underlying metal lattice. The bilayer/multilayer structure is generally patterned throughout, i.e., all the thin films are patterned in the form of the lateral spring structure, to make sure that the applied strain causes "unraveling" of the structure to provide maximum elongation. When used in conjunction with a polymer backing, the metal provides the desired electronic conductivity to the spring structure, while the polymer provides the reversible stretchability. The lateral spring structure reduces the strain experienced by the metal thin film because the strain is distributed along the spring structure uniformly, thus amplifying the apparent strain in a particular direction much beyond the elastic limit of the metal.

The metallic lateral spring structure can be used to connect various components of an electronic system to form a stretchable device. The individual components in this case need not be flexible or stretchable. For example, it is possible to have a decentralized design such that a processor, memory, or display system has disparate components interconnected using stretchable conductors (Figure 18.6). This is similar to the hybrid PCB design approach discussed in Chapter 11. Several examples of stretchable electronic devices have been demonstrated using this technique (Section 18.4). One of the key advantages of the lateral spring structure is that near metallic conductivity is obtained because of the use of the metal thin film, compared to a polymer composite. Also, conductor resistance does not change with strain, in the case of metal springs, compared to polymer composites. This is expected because, in the case of composites, there is a physical separation of filler particles that causes an increase in resistance. However, for metal springs, the requirement to have the spring structure free-standing in order to obtain stretchability limits the integration of these structures into some stretchable designs. In some cases, polymer composites can also be patterned into lateral spring structures for added stretchability (material + design stretchability).

18.3.3 Liquid Metals

The introduction of liquid metal has provided an interesting design tool for creating reconfigurable conductive structures. Mercury has been known as a metal that is liquid at room temperature for centuries. Metals, in general, form strong metallic

Stretchable
interconnects

Unstretchable
islands

FIGURE 18.6 One of the most explored ways of obtaining stretchable circuits is the use of stretchable interconnects to connect disparate unstretchable islands with functional electronic blocks such as battery, processor, memory, and so on.

bonds because of their loosely held electrons, which are readily shared and cause the nuclei to bond together into a solid lattice. However, in mercury, the outermost electrons in the 6s orbital are tightly bonded to the nucleus leading to a lowering of metallic bond strength. This causes mercury to be liquid at room temperature and leads to lower chemical reactivity and low electrical conductivity compared to other metals. However, mercury has not been used in flexible or stretchable electronic devices because of its toxicity. This is particularly relevant for wearable electronics applications. Further, mercury has a very high surface tension that causes it to form spherical droplets to minimize its surface area. Other metals that have melting points below or near room temperature are gallium (29.7°C), francium (27°C), cesium (28.4°C), and rubidium (39.3°C). Of these, francium, cesium, and rubidium are not usable for practical consumer electronic devices because francium is radioactive (with a half-life of only 22 minutes), and all three are alkali metals that react explosively with air. Alkali metal alloy NaK is also liquid at room temperature at the eutectic point (77 wt% K, 23 wt% Na, melting point = −12.6°C), but also reacts violently with air.

This leaves gallium which has a melting point of 29.76°C, so is technically solid at room temperature (25°C), but has very good electrical conductivity. Further, gallium has low toxicity and low reactivity with air, which allow for its use in consumer electronic devices. The melting point of gallium is brought below room temperature by alloying it with other metals such as tin and indium. Two of the most common formulations to this effect are EGaIn (Eutectic gallium indium, 75.5 wt% Ga, 24.5 wt% In, melting point = 15.7°C) and Galinstan (Gallium indium stannum, 68.5 wt% Ga, 21.5 wt% In, 10 wt% Sn, melting point = −19°C). These "liquid metals" have conductivities lower than most conventional metal thin films, however, they are more conductive than polymer composites and are sufficiently conductive for several stretchable electronics applications. Stretchable conductors are generally fabricated by encapsulating these liquid metal droplets inside hollow stretchable elastomer structures. Upon application of strain, these structures expand or change shape and the liquid metal inside them reflows to maintain conductivity. Stretchable conductors made using fiber-enclosed liquid metal show very stable conductivity with strain. However, one key drawback of liquid metals is that they require encapsulation for application in practical electronic devices, which increases integration complexity. Further, they are consumed in higher quantities compared to a metal thin film with a thickness of a few nanometers. This is particularly problematic because of the involvement of indium which is a rare and expensive metal.

18.4   SOME STRETCHABLE ELECTRONICS APPLICATIONS

Some of the most commonly showcased examples of stretchable electronic devices have been fabricated using stretchable metallic interconnects. In this approach, separate, unstretchable islands of electronic circuitry are connected using stretchable metal interconnects. These interconnects can be fabricated using any of the methods discussed in the previous section and can be used to connect many different types of electronic circuits together, such as processing elements, memory elements, energy storage elements, and so on. The interesting thing about this approach is that the interconnected electronic elements need not be stretchable. This lifts a major constraint from the design and integration of the electronic elements, thus allowing for much more optimal design and higher performance. The strategy chosen for the integration of metal interconnects in the design depends on the required stretchability and conductivity and the platform chosen for electronic element integration. Some of the examples are shown in Figure 18.7. For example, a stretchable battery, demonstrated by Xu *et al.* in 2013, was fabricated by interconnecting individual battery elements using metal stretchable interconnects [22]. The metal interconnects were fabricated using patterned copper and aluminum thin films, sandwiched between patterned polyimide layers. The metal interconnects thin films ended with a metal pad that connected to the anode and cathode of every battery element respectively. The complete device was encapsulated using stretchable silicone layers. The stretchable battery was demonstrated to function even at 300% strain. In this design, the complete applied strain was absorbed by the metal interconnects, keeping the individual battery cells unstrained. The silicone encapsulation also made the device usable for wearable electronics applications. Another interesting

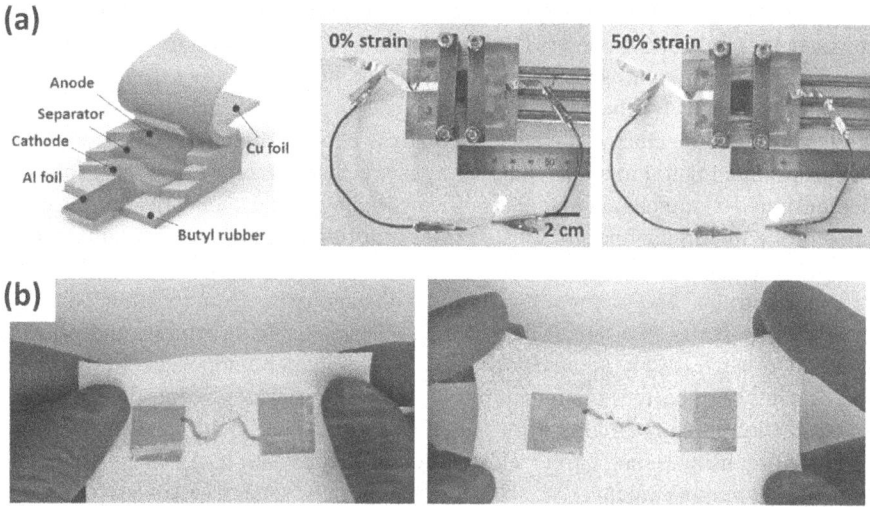

FIGURE 18.7 Examples of stretchable electronic systems: (a) A stretchable Lithium-ion battery based on reentrant micro-honeycomb electrodes and cross-linked gel electrolyte. Adapted with permission from [23]. Copyright 2020 American Chemical Society. (b) A stretchable antenna based on a polyimide/copper bilayer structure, etched using standard lithography techniques.

example of the metal interconnect approach was demonstrated by Song *et al.* in 2013, wherein a stretchable camera was fabricated using individual photodiode elements interconnected using stretchable metal interconnects [24]. In this case, the individual cells consisted of a photodiode and blocking diode circuit fabricated on a silicon-on-insulator (SOI) substrate followed by transfer printing to a destination substrate (refer to Chapter 4). The cells were connected using stretchable metallic interconnects on polyimide backing. Using a similar approach, a stretchable heater was demonstrated in 2015 [25]. In this case, the unit cells interconnected using stretchable metal interconnects were heating elements with a double spiral metal heating coil design. The metal interconnects were fabricated using a copper/polyimide bilayer structure patterned into a lateral spring structure. The structure was fabricated on a silicon substrate and released using amorphous silicon as the sacrificial layer to obtain free-standing horseshoe-shaped metal lateral springs. In this case, as well, the individual heating elements were unstretchable, and the strain during stretching was completely absorbed by the metal interconnects.

Stretchable electronics can also be fabricated using inherently stretchable semiconductors, such as organic semiconductor thin films. In 2011, Yu *et al.* demonstrated a stretchable light-emitting diode (LED) fabricated using inherently stretchable materials [26]. In this work, A luminescent polymer layer consisting of a blue emissive polyfluorene copolymer (PF-B), an ionic conductor (poly(ethylene oxide) dimethacrylate ether (PEO-DMA)), and a salt (lithium trifluoromethane sulfonate, LiTf) was laminated between two polymer composite electrodes. The stretchable composite electrodes were prepared by curing a liquid monomer, tert-butyl acrylate, overcoated on a thin film of single-walled CNT (SWNT). In this structure, a p-i-n junction was formed allowing charge injection from the same electrode (SWNT polymer composite) at both anode and cathode. The stretchable LED was shown to be operational at 45% strain. This and several such demonstrations have shown the potential of organic electronics for the fabrication of stretchable devices because of the inherent stretchability of the materials involved. Finally, we look at an interesting application of encapsulated liquid metal for fabricating a stretchable antenna. In 2009, So *et al.* demonstrated a stretchable antenna fabricated using liquid metal EGaIn encapsulated inside PDMS channels fabricated using soft lithography [27]. The liquid metal antenna was able to maintain electrical continuity and transmission for a wide range of applied strains. The configuration could withstand stress from twisting, bending, or stretching better than metal thin-film-based stretchable structures, and provide much more conductivity compared to stretchable polymer composites. In this case, the device was reported to stretch up to 40% strain, at which point, the elastomer encapsulation tore. Thus, this limit is not inherent to the liquid metal and can be improved by using a more stretchable elastomer encapsulation. An interesting consequence of making antennas stretchable is that because antenna resonant frequency depends on its length, applying strain to a stretchable antenna changes its resonant frequency. Thus, the antenna itself can be used to measure strain or pressure by measuring the change in its resonant frequency.

The field of stretchable electronics has only recently begun to generate interest. With several approaches already developed and demonstrated for fabricating various stretchable electronic devices, the future for large-scale implementation and commercial adaptation looks promising. These demonstrations showcase the fact that stretchable electronic devices are achievable with the choice of the right material system, processes, and integration strategies. It is important to invest in efforts to scale these prototype demonstrations into fully manufacturable technologies. This is particularly important given that stretchable electronic devices can be a bridge between the conventional electronic devices we see around us and the natural animal world, which has many stretchable, flexible, and curvilinear elements. With the vast potential such an integration between the natural and electronic world can provide in terms of applications, it is inevitable that we will see commercial stretchable electronic devices in the future. Further, truly free-form and stretchable systems can give rise to novel devices designs such as a phone that can stretch into a tablet or a television that can be shrunk and rolled up. Indeed, the possibilities are endless.

EXERCISES

18.1. Why is it desirable to have stretchable electronic systems? Describe the various ways in which stretchability can be imparted to electronic circuits.
18.2. How does the conductivity of the composite vary with the volume fraction of a conductive filler in a polymer matrix? What is the physical basis for this?
18.3. Why does a metal spring exhibit reversible stretchability in the linear axis, whereas a metal rod, of the same material, does not?
18.4. Give examples of liquid metals. How are they used in stretchable electronics? Why are mercury and francium not used in stretchable electronic applications?

19 Reliability and Future Outlook

19.1 INTRODUCTION

We have looked at several reasons for the success of silicon-based CMOS electronic systems. The material properties of silicon, the purity of the wafers, the fact that its electronic properties can be precisely controlled using doping, the dimensional scaling of transistors, and so on. However, there is one final reason for the enormous success enjoyed by silicon electronics – reliability. We define reliability as the fact that a specific system or its component will work as per specifications for a specified period of time. The reliability of electronic systems is the reason why we wake up every morning and expect our smartphones to work perfectly. Given that our smartphones have hundreds of components such as microprocessor, memory, transceiver, display, sensors, and in some cases, each component has millions of transistors, it is really a miracle that it keeps working in harmony all the time. The tiny fraction of the time that there is a malfunction, it can be, in most cases, traced back to a human error or mishandling. Further, electronic systems require almost no upkeep or maintenance, such as changing the oil filter in a car engine. The reliability and maintenance-free nature of conventional electronic components and systems partially stem from the fact that there are no moving components, i.e., everything is solid state. Moving components typically experience mechanical fatigue due to the cyclic loading and unloading of stress that can cause the generation and propagation of microscopic cracks through the lattice, eventually leading to catastrophic failure. In the case of solid-state electronic components, there is no mechanical loading, however, these components do undergo electrical loading and unloading, which can lead to electrical fatigue and failure. The points of failure are components that undergo the cycling formation and dissipation of electric fields and electric currents. Electric fields are generated inside capacitive structures such as the gate of a CMOS transistor, while electric currents are generated in resistive structures such as metal interconnect lines. Typical causes of failure can be a sudden surge in electric field/current, damaging a specific section of the component, which can lead to stress localization and eventual catastrophic failure of the entire component. Other reasons include the slow chemical degradation of the material because of interaction with surroundings, for example, formation of native oxide, rust, or leaching into the ambient.

Let us consider the example of a capacitor undergoing electrical load cycles. It consists of two metal plates separated by an insulating dielectric layer. When the capacitor is charged, the dielectric is subjected to an electric field given by $E = V / d$, where V is the applied potential and d is the distance between the two plates. This field causes electrical stress in the dielectric. If it is less than the dielectric breakdown strength, the capacitor should be able to withstand the loading and unloading

DOI: 10.1201/9781003010715-24

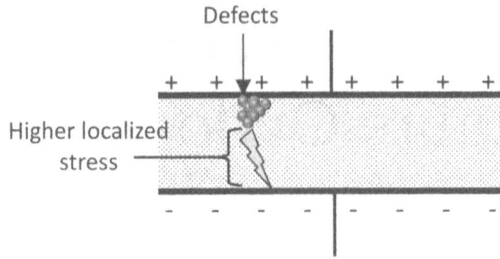

FIGURE 19.1 A microscopic defect in the dielectric of a capacitive structure can cause higher localized electrical stress eventually leading to dielectric breakdown.

of this field infinitely. However, because of the development of microscopic defects and material degradation over time (because of environmental factors), the electrical stress experienced by a localized region in the dielectric can increase, resulting in a localized short. This, in turn, increases the electrical stress of the localized region, resulting in an electrical short that propagates through the dielectric material and eventually causes the capacitor to short (Figure 19.1). This is very similar to the mechanical failure of materials because of loading and unloading cycles. Hence, the electrical breakdown of capacitors is studied using the Weibull statistics, which was originally developed in the early twentieth century to study the breaking of anchor chains in the maritime industry. The core concept is that the breakdown of a larger system (say a capacitor or a chain) depends on the breakdown of the weakest link in the system. In a dielectric, this could be a material defect or a grain boundary, whereas, in a chain, this could be, quite literally, the weakest link in the chain. Because we do not consider the strength or behavior of the statistically average link, we cannot predict the behavior of the system based on averages (Gaussian statistics). In this method, the survival probability of a component is represented by:

$$P = \exp\left[-\left(\frac{E - E_m}{E_0} \right)^m \right]$$

where E is the applied electrical stress, E_m is the stress below which there is zero probability of failure, E_0 is a material parameter, and m is the Weibull modulus. A larger value of m represents lower variability of the strength of the component, which implies a more reliable component. The parameters E_m, E_0, and m can be obtained using experimental data. The failure probability, given by, $F = 1 - P$, is often used in the literature. The Weibull modulus can be obtained by plotting $\ln[\ln\left(\frac{1}{1-F}\right)]$ versus E, resulting in a straight line with the Weibull modulus as the slope.

The static response of a system upon application of a load can be predicted using the three-parameter Weibull distribution, however, the concept of fatigue also involves time as an independent variable. Given enough time or load cycles, components fail even under stress that is much lower than their breakdown strength. In such a case, it is important to predict how long a specific component is expected to survive under given load conditions. This is particularly important in a large system such as the smartphone because the failure of a single transistor (say because of the shorting of gate dielectric) can cause the entire system to stop functioning. The long-term reliability of a component, however, can only be truly understood by subjecting the component to the given loading conditions over a long period of time. However, this can significantly increase the time to market for some components. In order to predict performance over time, components are subjected to stresses higher than their rated loading conditions, and the time it takes for them to fail is recorded. These accelerated stress tests (ASTs) help record the failure behavior of a component, without requiring a long-term stress test to be conducted. An example of this in the electronics industry is the time-dependent dielectric breakdown (TDDB) test conducted to analyze the lifetime of dielectric thin films in capacitive structures (like the gate of a CMOS transistor). In this case, a dielectric is subjected to variable stresses (more than the rated condition) and the average time for breakdown is recorded as a function of applied electrical stress. The breakdown time can then be extrapolated back to the rated conditions to determine the expected lifetime of a component at rated load. These tests include a time component for assessing the long-term reliability of a component and its susceptibility to fatigue even with loading conditions at or lower than the rated values.

19.2   RELIABILITY IN FLEXIBLE ELECTRONICS

The reliability of electronic components is one of the key reasons behind the success of the electronics industry. However, the reliability of flexible electronics has only recently gained attention. Indeed, for the commercial viability of flexible electronic systems and their eventual success, it is important to fabricate flexible systems that are as reliable as, if not more than, the current rigid electronic systems. In the case of flexible electronics, the notion of reliability is more complex compared to rigid systems because both electronic and mechanical reliability need to be considered. Further, flexible electronic systems are expected to operate seamlessly during bending, as well as after several thousand bending cycles. The problem is that the electronic properties of materials change under mechanical strain. This is clearly apparent in silicon electronics with the change in the band structure, bandgap, charge mobility, etc., with applied strain. Thus, if a flexible electronic system consisting of flexible monocrystalline silicon processors is flexed, the flexible silicon piece can malfunction because of the applied strain (Figure 19.2). Hence, the process window for the fabrication of silicon CMOS devices meant for flexible electronic applications is much tighter than that for rigid applications. This is also true of all other material systems and integration strategies discussed in this book. Thus, the final application and specifications of a flexible electronic system influence the material and design choices all the way up to the transistor and interconnect level. Thus, just as each generic rigid electronic component comes with a range of temperature, voltage, and other parameters for guaranteed performance, a generic flexible electronic component will consist of a minimum bending radius, maximum bending cycles, and other relevant mechanical flexing specifications as part of its datasheet.

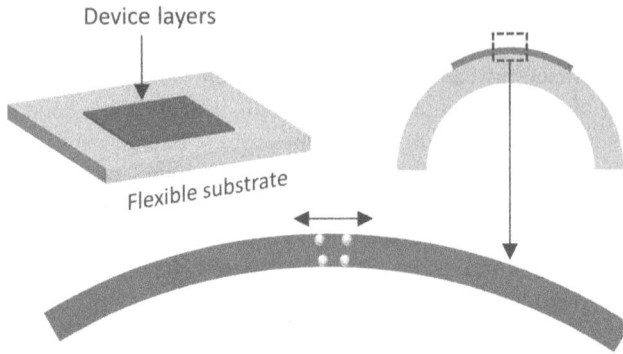

FIGURE 19.2   The strain caused because of bending of a substrate can change the electronic properties of the device layers deposited on the flexible substrates.

Flexible electronic components are designed to sustain electrical and mechanical stress simultaneously. The repeated bending cycles experienced by the system put mechanical stress on specific parts of the system, leading to a possibility of mechanical fatigue in the long term. Additionally, the use of the system for electronic applications puts electrical stress on the system, leading to potential electric failure. These are well-studied failure mechanisms in their own domains. However, flexible electronic systems also include the unique possibility of mechanical failure due to repeated electrical stress and electrical failure due to repeated mechanical stress. This is because the application of electrical stress causes localized hot spots and conductive filaments that can also contribute to the mechanical degradation of the system. On the other hand, the application of mechanical stress can cause localized lattice deformation, leading to a change in the electrical properties of the system. The mechanisms governing these phenomena are far less understood and require further investigation in the future. Literature reports on demonstrations of flexible electronic systems increasingly include the performance of the system after several (thousands of) bending cycles or during bending after a said number of cycles. This is an important step toward analyzing, predicting, and reporting the reliability of flexible electronic systems.

19.3   FUTURE OUTLOOK

We started this book with an optimistic view of the future of flexible electronic systems, the end should be no different. It is important to realize the inevitability of the ubiquitous permeance of flexible electronic systems in our lives in the future, simply because of the physically flexible and curvilinear structure of human and animal forms. If artificial electronic systems are to integrate seamlessly into the natural electronic systems (human brain), as is predicted by most science fiction writers, then the electronic systems will have to conform to the curvilinear structures. Further, because natural muscles and skin can undergo a significant amount of strain reversibly, the future of electronics may in fact be beyond flexible electronics – in stretchable and free-form electronic systems. Of course, before such a future is realized, there are several challenges that need to be solved:

- The performance of flexible electronic systems should be at par, or better than, the state-of-the-art conventional electronic systems. This will require research at the device level with different material systems and integration strategies.
- The integration of different material systems such as organic, metal oxide, inorganic semiconductors, along with nanostructured materials should be seamlessly carried out on a flexible substrate.
- The packaging of the complete system should be such that it can protect it from oxygen/water ingress, have a means of connecting with other circuits, and withstand repeated bending cycles.
- Finally, the reliability of the systems should be studied with respect to mechanical and electrical stress cycles.

References

1. A. P. Hanson, "Electric cable.," United States, US782391A, 1905.
2. A. M. Hussain and M. M. Hussain, "CMOS-Technology-Enabled Flexible and Stretchable Electronics for Internet of Everything Applications," *Advanced Materials*, vol. 28, no. 22, pp. 4219–4249, 2016.
3. C. Downs and T. E. Vandervelde, "Progress in Infrared Photodetectors Since 2000," *Sensors*, vol. 13, no. 4, pp. 5054–5098, 2013.
4. Z.-Q. Fang *et al.*, "Light Management in Flexible Glass by Wood Cellulose Coating," *Scientific Reports*, vol. 4, no. 1, p. 5842, 2014/07/28 2014.
5. Y. L. Kong *et al.*, "3D Printed Quantum Dot Light-Emitting Diodes," *Nano Letters*, vol. 14, no. 12, pp. 7017–7023, 2014/12/10 2014.
6. S. Li and D. Chu, "A Review of Thin-Film transistors/circuits Fabrication with 3D Self-Aligned Imprint Lithography,"*Flexible and Printed Electronics*, vol. 2, no. 1, p. 013002, 2017/03/01 2017.
7. M. M. Shulaker *et al.*, "Carbon Nanotube Computer," *Nature*, vol. 501, no. 7468, pp. 526–530, 2013/09/01 2013.
8. G. Hills *et al.*, "Modern Microprocessor Built from Complementary Carbon Nanotube Transistors," *Nature*, vol. 572, no. 7771, pp. 595–602, 2019/08/01 2019.
9. A. M. Hussain, G. A. T. Sevilla, K. R. Rader, and M. M. Hussain, "Chemical vapor deposition based tungsten disulfide (WS2) thin film transistor," in *2013 Saudi International Electronics, Communications and Photonics Conference*, 2013, pp. 1–5.
10. K. Fukuda *et al.*, "A 4 V Operation, Flexible Braille Display Using Organic Transistors, Carbon Nanotube Actuators, and Organic Static Random-Access Memory," *Advanced Functional Materials*, vol. 21, no. 21, pp. 4019–4027, 2011.
11. H. Ebihara, N. Karaki, S. Hirabayashi, S. Inoue, T. Kodaira, and T. Shimoda, "P-39: A Flexible 16kb SRAM Based on Low-Temperature Poly-Silicon (LTPS) TFT Technology," *SID Symposium Digest of Technical Papers*, vol. 37, no. 1, pp. 339–342, 2006.
12. F. D. Roose *et al.*, "A Thin-Film, a-IGZO, 128b SRAM and LPROM Matrix with Integrated Periphery on Flexible Foil," *IEEE Journal of Solid-State Circuits*, vol. 52, no. 11, pp. 3095–3103, 2017.
13. M. T. Ghoneim and M. M. Hussain, "Review on Physically Flexible Nonvolatile Memory for Internet of Everything Electronics," *Electronics*, vol. 4, no. 3, pp. 424–479, 2015.
14. Y. Ji *et al.*, "Stable Switching Characteristics of Organic Nonvolatile Memory on a Bent Flexible Substrate," *Advanced Materials*, vol. 22, no. 28, pp. 3071–3075, 2010.
15. H. G. Yoo, S. Kim, and K. J. Lee, "Flexible One diode–One Resistor Resistive Switching Memory Arrays on Plastic Substrates," *RSC Advances*, 10.1039/C4RA02536A vol. 4, no. 38, pp. 20017–20023, 2014.
16. M. T. Ghoneim *et al.*, "Thin PZT-based Ferroelectric Capacitors on Flexible Silicon for Nonvolatile Memory Applications," *Advanced Electronic Materials*, vol. 1, no. 6, pp. 1500045, 2015.
17. S.-M. Yoon, S. Yang, and S.-H. K. Park, "Flexible Nonvolatile Memory Thin-Film Transistor Using Ferroelectric Copolymer Gate Insulator and Oxide Semiconducting Channel," *Journal of The Electrochemical Society*, vol. 158, no. 9, p. H892, 2011.
18. J. M. Yoon *et al.*, "Fabrication of High-Density In3Sb1Te2phase Change Nanoarray on Glass-Fabric Reinforced Flexible Substrate," *Nanotechnology*, vol. 23, no. 25, p. 255301, 2012/05/31 2012.

19. Y. Jeon, M. Lee, T. Moon, and S. Kim, "Flexible Nano-Floating-Gate Memory With Channels of Enhancement-Mode Si Nanowires,"" *IEEE Transactions on Electron Devices*, vol. 59, no. 11, pp. 2939–2942, 2012.

20. C. C. Wu *et al.*, "Integration of Organic LEDs and Amorphous Si TFTs onto Flexible and Lightweight Metal Foil Substrates," *IEEE Electron Device Letters*, vol. 18, no. 12, pp. 609–612, 1997.

21. W. Shockley and H. J. Queisser, "Detailed Balance Limit of Efficiency of p-n Junction Solar Cells," *Journal of Applied Physics*, vol. 32, no. 3, pp. 510–519, 1961.

22. S. Xu *et al.*, "Stretchable Batteries with Self-Similar Serpentine Interconnects and Integrated Wireless Recharging Systems," *Nature Communications*, vol. 4, no. 1, p. 1543, 2013/02/26 2013.

23. S. Kang *et al.*, "Stretchable Lithium-Ion Battery Based on Re-entrant Micro-honeycomb Electrodes and Cross-Linked Gel Electrolyte," *ACS Nano*, vol. 14, no. 3, pp. 3660–3668, 2020/03/24 2020.

24. Y. M. Song *et al.*, "Digital Cameras with Designs Inspired by the Arthropod Eye," *Nature*, vol. 497, no. 7447, pp. 95–99, 2013/05/01 2013.

25. A. M. Hussain, E. B. Lizardo, G. A. Torres Sevilla, J. M. Nassar, and M. M. Hussain, "Ultrastretchable and Flexible Copper Interconnect-Based Smart Patch for Adaptive Thermotherapy," *Advanced Healthcare Materials*, vol. 4, no. 5, pp. 665–673, 2015.

26. Z. Yu, X. Niu, Z. Liu, and Q. Pei, "Intrinsically Stretchable Polymer Light-Emitting Devices Using Carbon Nanotube-Polymer Composite Electrodes," *Advanced Materials*, vol. 23, no. 34, pp. 3989–3994, 2011.

27. J.-H. So, J. Thelen, A. Qusba, G. J. Hayes, G. Lazzi, and M. D. Dickey, "Reversibly Deformable and Mechanically Tunable Fluidic Antennas," *Advanced Functional Materials*, vol. 19, no. 22, pp. 3632–3637, 2009.

Index

297

For Product Safety Concerns and Information please contact our EU
representative GPSR@taylorandfrancis.com
Taylor & Francis Verlag GmbH, Kaufingerstraße 24, 80331 München, Germany

www.ingramcontent.com/pod-product-compliance
Lightning Source LLC
Chambersburg PA
CBHW060334220326
41598CB00023B/2704